T0296636

Aerial Robotic Workers

Design, Modeling, Control, Vision, and Their Applications

Aerial Robotic Workers

Design, Modeling, Control, Vision, and Their Applications

Edited by

George Nikolakopoulos

Sina Sharif Mansouri

Christoforos Kanellakis

Butterworth-Heinemann is an imprint of Elsevier
The Boulevard, Langford Lane, Kidlington, Oxford OX5 1GB, United Kingdom
50 Hampshire Street, 5th Floor, Cambridge, MA 02139, United States

ISBN: 978-0-12-814909-6

For information on all Butterworth-Heinemann publications
visit our website at https://www.elsevier.com/books-and-journals

Publisher: Mara E. Conner
Acquisitions Editor: Sonnini R. Yura
Editorial Project Manager: Emily Thomson
Production Project Manager: Manju Paramasivam
Cover Designer: Mark Rogers

Typeset by VTeX

To my mother Aleka

Contents

10. Machine learning for ARWs

Anton Koval, Sina Sharif Mansouri, and Christoforos Kanellakis

11. Aerial infrastructures inspection

Vignesh Kottayam Viswanathan, Sina Sharif Mansouri, and Christoforos Kanellakis

12. ARW deployment for subterranean environments

Björn Lindqvist, Sina Sharif Mansouri, Christoforos Kanellakis, and Vignesh Kottayam Viswanathan

Contributors

Yifan Bai, Department of Computer, Electrical and Space Engineering, Luleå University of Technology, Luleå, Sweden

Avijit Banerjee, Department of Computer, Electrical and Space Engineering, Luleå University of Technology, Luleå, Sweden

Sumeet Gajanan Satpute, Department of Computer, Electrical and Space Engineering, Luleå University of Technology, Luleå, Sweden

Jakub Haluska, Department of Computer, Electrical and Space Engineering, Luleå University of Technology, Luleå, Sweden

Christoforos Kanellakis, Department of Computer, Electrical and Space Engineering, Luleå University of Technology, Luleå, Sweden

Samuel Karlsson, Department of Computer, Electrical and Space Engineering, Luleå University of Technology, Luleå, Sweden

Vignesh Kottayam Viswanathan, Department of Computer, Electrical and Space Engineering, Luleå University of Technology, Luleå, Sweden

Anton Koval, Department of Computer, Electrical and Space Engineering, Luleå University of Technology, Luleå, Sweden

Björn Lindqvist, Department of Computer, Electrical and Space Engineering, Luleå University of Technology, Luleå, Sweden

George Nikolakopoulos, Department of Computer, Electrical and Space Engineering, Luleå University of Technology, Luleå, Sweden

Andreas Papadimitriou, Department of Computer, Electrical and Space Engineering, Luleå University of Technology, Luleå, Sweden

Akash Patel, Department of Computer, Electrical and Space Engineering, Luleå University of Technology, Luleå, Sweden

Achilleas Santi Seisa, Department of Computer, Electrical and Space Engineering, Luleå University of Technology, Luleå, Sweden

Sina Sharif Mansouri, Autonomous Driving Lab in Scania Group, Stockholm, Sweden Department of Computer, Electrical and Space Engineering, Luleå University of Technology, Luleå, Sweden

Ilias Tevetzidis, Department of Computer, Electrical and Space Engineering, Luleå University of Technology, Luleå, Sweden

Chapter 1

Introduction

George Nikolakopoulos
Department of Computer, Electrical and Space Engineering, Luleå University of Technology, Luleå, Sweden

1.1 Introduction

This book aims to address the major challenges and knowledge regarding the creation of the next generation of Aerial Robotic Workers (ARWs). The term ARW stems from the enhancement of the classical Unmanned Aerial Vehicles (UAVs) by providing the ability to autonomously interact with the environment while enabling advanced levels of autonomy with respect to environmental perception, 3D reconstruction, active aerial manipulation, intelligent task planning, and multi-agent collaboration capabilities.

Such a team of ARWs will be capable of autonomously inspecting infrastructure facilities and acting executing maintenance or complex generic task by aerial manipulation and exploiting multi-robot collaboration. Furthermore, the book aims to investigate the emerging scientific challenges of multi-robot collaboration, path-planning, control for aerial manipulation, aerial manipulator design, autonomous localization, sensor fusion, and cooperative environmental perception and reconstruction.

Emphasizing transforming research excellence in specific and realistic technological innovation, this book also deals with the investigation of an approach with very promising returns in the areas of infrastructure inspection, repair & maintenance, leading to big savings in costs while maximizing personnel/asset safety. With such a potential impact, the technological concept of ARWs has been placed at the forefront of bringing Robotics to the basis necessary in real applications, where they can make a real-life difference and true impact. An illustration of the ARW concept for the case of wind turbine maintenance is depicted in Fig. 1.1.

The concept of ARWs advocates that important civil applications can be accomplished efficiently and effectively by a swarm of ARWs with unprecedented onboard capabilities and collaborating autonomously for the execution of complex common tasks and missions.

Towards this vision, this book integrates a collection of multi-disciplinary research in the field of ARWs that are integrated into robust, reliable, and ready-to-operate technological solutions that form the fundamental components of the

Aerial Robotic Workers. https://doi.org/10.1016/B978-0-12-814909-6.00007-X

FIGURE 1.1 Concept of Inspection and Maintenance of a wind turbine based on multiple ARWs.

ARW technology. An overview of the scientific and technological directions is provided below.

1. **Dexterous Aerial Manipulator Design and Development**
   Among the first main goals of the book is the realization of an aerial manipulation system that will be able to equip the aerial vehicles to perform physical interaction and mutual robot-robot interaction toward collaborative work-task execution. The design of the manipulator will be presented such that it limits the influence on the stability of the aerial platform through the proper selection of the mechanisms and an adequate configuration and morphology. At the same time, it allows for additional co-manipulation tasks of the same object from two or more ARWs. Towards this objective, a complete aerial manipulator will be presented, including the corresponding modeling and control scheme for enabling dexterous aerial manipulation.
2. **Collaborative Perception, Mapping and Vision for Manipulation**
   For automated inspection and manipulation to become feasible, the ARWs need to build up the right level of scene perception to encode the spatial awareness necessary before they can autonomously perform the tasks at hand. As a result, the first step in developing the ARWs' perception has been the utilization of the sensor suite onboard that each ARW has (comprised primarily of visual and inertial sensors) to perceive both their ego-motion and their workspace, essentially forming the backbone of each ARW's autonomy. Following promising leads from previous work, here there have been multiple sensor fusion approaches in order to account for the dynamic camera motions expected by agile robots, such as the rotorcraft ARWs and the challenging industrial environments of potentially GPS denied SubT and featureless areas, deviating from traditional scenarios of office-like, urban or natural sceneries.
3. **Aerial Robotic Workers Development and Control**
   The ARWs constitute a great research challenge, both with respect to its development, as well as its control, targeted for active aerial (co-) manipulation and tool-handling interaction for the execution of work-tasks. The zero-liability prerequisite for its intended deployment within critical infrastructure environments imposes careful investigation to determine those practices that add to its dependability (fail-safe redundancy, real-time diagnostics, vehicle

design aspects) that make it almost-inherently unable to cause asset damage or place human lives at risk. The control-related research efforts focused on addressing the problems of single task manipulation, interaction, and work task execution. The formulated control strategies encompassed each complete ARW's configuration while achieving flight stabilization, yielding increased levels of reactive safety and excellent robustness against collisions. The synthesized control schemes focused on achieving high manipulation performance and compliant motion during work-task interaction, mainly in complex and entirely unknown (exploration missions) environments.

4. **Collaborative Autonomous Structural Inspection and Maintenance**
New methods for collaboration of multiple heterogeneous aerial robots have been developed and addressed the problem of autonomous, complete, and efficient execution of infrastructure inspection and maintenance operations. The presented methodology aims to provide decentralized, local control laws and planners to the individual vehicles that can accommodate heterogeneity through a model-free approach. To facilitate the necessary primitive functionalities, an inspection path-planner that can guide a single ARW to efficiently and completely inspect a structural model will be presented. At the same time, these features allowed for attributes such as system adaptability to unexpected events, heterogeneity and reactivity to new online tasks, and further enhancing the overall system. The topics that will be presented will focus on the: a) single and multi-robot "anytime" and efficient collaboration for autonomous structural inspection, b) collaborative manipulation of tools and objects, c) complex work-task adaptation during exploration missions in SubT environments.
5. **Deployment of Autonomous Aerial Robotic Infrastructure Solutions**
The collaborative team of ARWs will be able to autonomously plan, execute, and adapt online the plan for complete infrastructure inspection missions. This essentially corresponds to a step change that will benefit the infrastructure services market by both increasing the safety levels (reducing personnel risks and asset hazards) and reducing the direct and indirect inspection costs, mostly by minimizing the inspection times, executing certain tasks during operating time of the facility, providing repeatable tools, and enabling formal analysis of the results.

1.2 Structure of the book

To reach the vision of the book in the future utilization of ARWs, the presented materials have been divided into two major parts. The first part focuses on the theoretical concepts around the design, modeling, and control of ARWs, while the second part deals with characteristic applications and demonstrations of the ARW utilization in realistic use cases.

As such, the first part will begin by presenting analytically the fundamental hardware modules that consist of the ARWs in Chapter 2, followed by establishing the most popular mathematical modeling and control approaches for

ARWs in Chapters 3 and 4, respectively. In these Chapters, we have selected to highlight only the most popular modeling and control approaches that can be directly utilized in experiments as an effort to increase the overall focus and impact of the book. In the sequel, Chapter 5 will focus on the perception aspects of the ARWs. In contrast, Chapter 6 will initiate the integration of the previous Chapters into more complete and integrated missions, and thus it will consider the navigation problem of ARWs. Chapter 7 will discuss the exploration task for ARWs in demanding complex and GPS-denied environments, e.g., the case of sub-terranean use cases. Chapter 8 will analyze methods for measuring and estimating the external forces acting on ARWs, while Chapter 9 will purely focus on aerial manipulation, including modeling, control, and visual servoing tasks. Chapter 10 will conclude the first part of the book by presenting the latest machine learning (data-driven) approaches and applications in the area of ARWs, as an alternative approach to the classical modeling and control tasks, for increasing the capabilities and the utilization of ARWs in more application oriented missions.

The second part of the book will focus on real-life applications of ARWs, and as such, Chapter 11 will discuss establishing a framework for the collaborative aerial inspection of wind turbines based on multiple ARWs. Chapter 12 will present and analyze a complete framework for reactive (online) exploration of fully unknown sub-terranean environments and in full autonomy, while Chapter 13 will conclude this book by presenting an edge-based networked architecture for offloading fundamental algorithmic modules that are legacy performed on board an aerial vehicle, to the edge and as such increasing the overall performance and the resiliency of the mission.

Chapter 2

The fundamental hardware modules of an ARW

Anton Koval, Ilias Tevetzidis, and Jakub Haluska
Department of Computer, Electrical and Space Engineering, Luleå University of Technology, Luleå, Sweden

2.1 Introduction

ARWs are an example of advanced integration and synchronization among numerous hardware and software components, all designed and integrated to operate onboard the ARW and thus enabling the proper autonomy levels for completing demanding missions. The latest developments in embedded computers, electric motors, sensorial systems, and batteries are capable of introducing radical new developments in the area of UAVs and, more specifically, on the topic of Aerial Robotic Workers. In this case, the ARWs are envisioned in the next years to find a huge interest from the robotics community, especially in the development of innovative field robotic technologies. As such, in the rest of this Chapter, the fundamental components and technologies for developing an ARW will be presented.

2.2 Design of the ARWs

This Section introduces the basics of the design of ARWs, focusing on ARWs in the form of multi-rotor aircrafts. Furthermore, the purpose of this Section is to give a basic overview of possible configurations and design approaches and provide very general guidance for their design based on the experiences from the user perspective and lessons learned.

2.2.1 Frame design

The core element of all multi-rotor aircrafts is the frame. Essentially, its structure needs to be rigid while being able to lower or cancel vibrations that are coming from the motors. Before we move any further in the design, let us define the coordinate frame of a generic ARW. We place the center of origin at the geometrical center of the ARW. The $X(+)$ axis is going toward the front

Aerial Robotic Workers. https://doi.org/10.1016/B978-0-12-814909-6.00008-1

of the drone, and the $Z(+)$ axis upwards, as shown in Fig. 2.1. At this point, it should also be noted that in this book, we will follow this convention when talking about coordinates and axis.

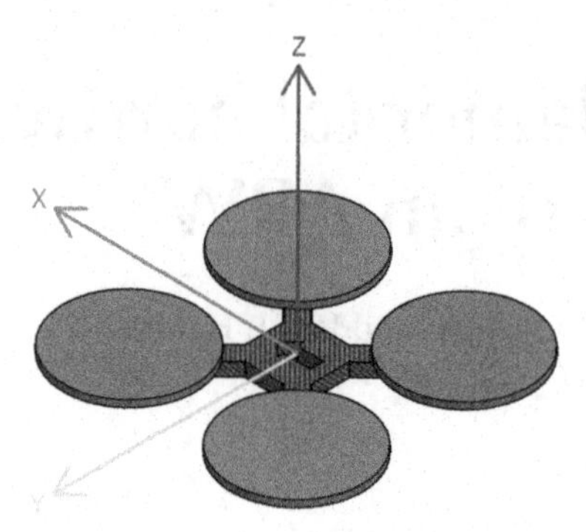

FIGURE 2.1 Definition of the coordinate frame for a generic ARW.

The flight of an ARW consists of three basic movements, namely the roll, pitch, and yaw, while these movements are depicted in Fig. 2.2.

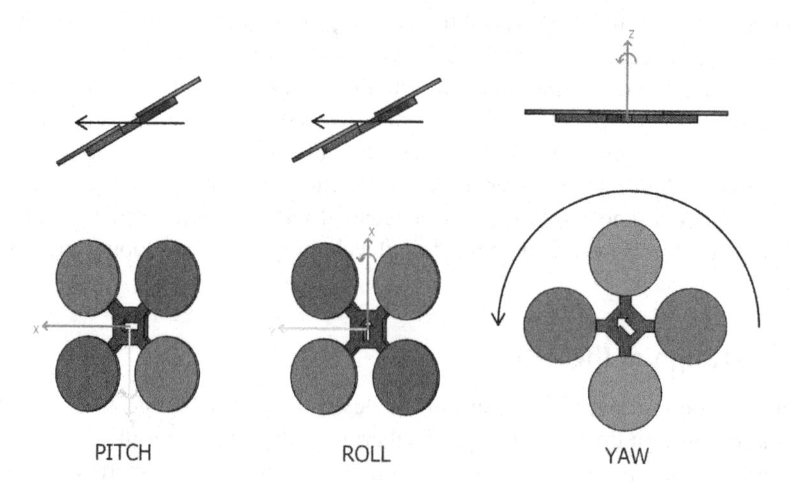

FIGURE 2.2 The basic movements of ARW. Roll, pitch, and yaw.

One of the main factors determining the shape of an ARW is the number of rotors used for the propulsion. The most commonly used configurations one can mention are the Tricopter, the Quadrotor, the Hexarotor, and the Octarotor. These configurations are depicted in Fig. 2.3.

The configurations above describe ARWs with a single motor and a propeller mounted on each arm. However, there are special cases with co-axial motors as depicted in Fig. 2.4, where each arm carries two motors, and the propellers are placed on top of each other while rotating in opposite directions. Using this ap-

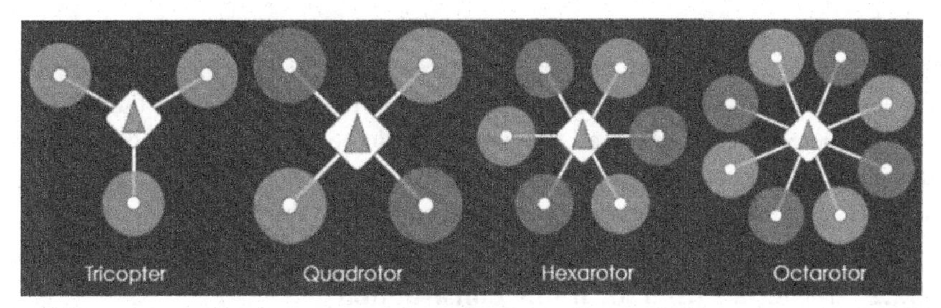

FIGURE 2.3 Selection of Rotor configurations of the multi-rotor ARWs supported by the PX4 flight controller.

FIGURE 2.4 Single rotor configuration – left. Coaxial rotor configuration – right.

proach, one can increase the thrust generated by the propulsion system without changing the footprint or the frame configuration of the ARW.

Another parameter defining the shape of the ARW's frame is the motor arrangement. For example, the quadrotor frame can have the following motor arrangements: Plus, H, True X, Modified X, and Dead cat X, as depicted in Fig. 2.5.

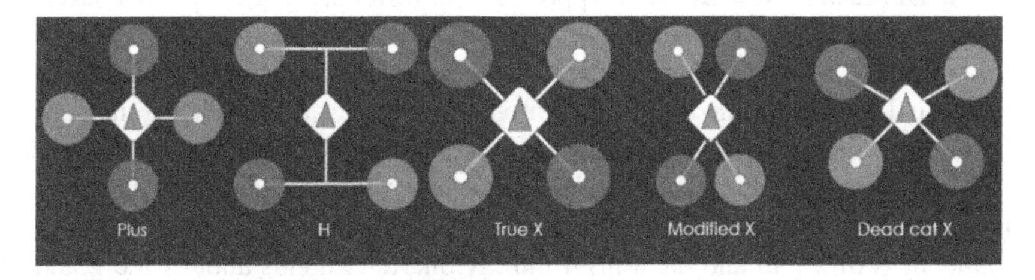

FIGURE 2.5 Examples of quadrotor configurations.

The optimal frame design should be geometrically symmetrical along the XZ and YZ planes. If the frame is fully symmetrical, the ARW will have the same flying behavior in both the pitch and roll maneuvers. In certain cases, modifications of the ARW frame can result in a positive impact on the behavior of the ARW during the flight. For example, a quadrotor with elongated X frame will be more stable during the Pitch maneuver (compared to the Roll maneuver). This can be helpful when the platform is designed to fly at high speeds with aggressive pitch.

Another reason for the utilization of a non-symmetrical frame can be due to the need for a specific sensor payload on board the vehicle. For example, the

Dead cat X configuration, Fig. 2.5 can be used to remove propellers from the field of view of the front-facing camera. In any case, the weight distribution of the sensor payload on the frame should be well balanced to allow the load to be distributed evenly across all the motors while hovering. Besides the weight distribution, it is necessary to keep in mind the placement of the flight controller, which is commonly placed in the center of rotation of the ARW.

2.2.2 Materials used for frame construction

Mass-produced, of-the-shelf multi-rotor aircrafts, like any other products, consist of various parts manufactured using various technologies. From plastic injection molding and CNC machining to components made out of composite materials. The purpose of this Section is to give a basic overview and inspiration of materials and technologies used for building custom "Do It Yourself" ARWs, while other drone builders might utilize different machines and technology for manufacturing.

In general, for the structural parts of the ARWs, carbon fiber materials (sheets and tubes) could be utilized. A computer numerical control (CNC) router can be utilized to manufacture the components out of the carbon fiber sheets. The rest of the components (sensor mounts, covers, landing gears) could be manufactured using 3D printers. Most 3D printed components are made from fused filament fabrication (FFF) 3D printing technology. In case some components require higher precision than FFF, 3D printing can be provided; one can also print them on Stereolithography (SLA) 3D printers.

As an illustration of different approaches of different types of ARWs developed, the first is a re-configurable drone that is called "R-Shafter". The research idea behind this drone was an exploration of narrow openings, where each individual arm is actuated by a servo motor, and the drone can change its configuration while flying. Fig. 2.6 shows the concept of the configuration change.

From the materials and construction point of view, the construction the drone is made out of carbon fiber cutouts and 3D printed parts. The body plates have a thickness of 2 mm and the arms 4 mm. R-Shafter weights about 1500 g and uses 5 inch propellers (diameter 80 A). In the "X" configuration as shown in the Fig. 2.7, the x, y dimensions (without the propellers) are about 250×250 mm.

The second example is an ARW called "Outsider", Fig. 2.8. The frame is constructed from a combination of carbon fiber cutouts and carbon fiber tubes. The thickness of the top and bottom plates is 2 mm. The tubes used for the arms have a diameter of 22 mm and a thickness of 1 mm. This drone uses 3D printed parts as covers and sensor mounts. Outsider weights about 5000 g and uses 14 inch propellers (diameter 355.6 mm). The maximal width of the platform (excluding the propellers) is 910 mm.

The arms are made out of carbon fiber tubes (or profiles), and it could be lighter than arms from carbon fiber sheets, while keeping the same stiffness. Another reason to use carbon fiber tubes in this case is that many CNC machines

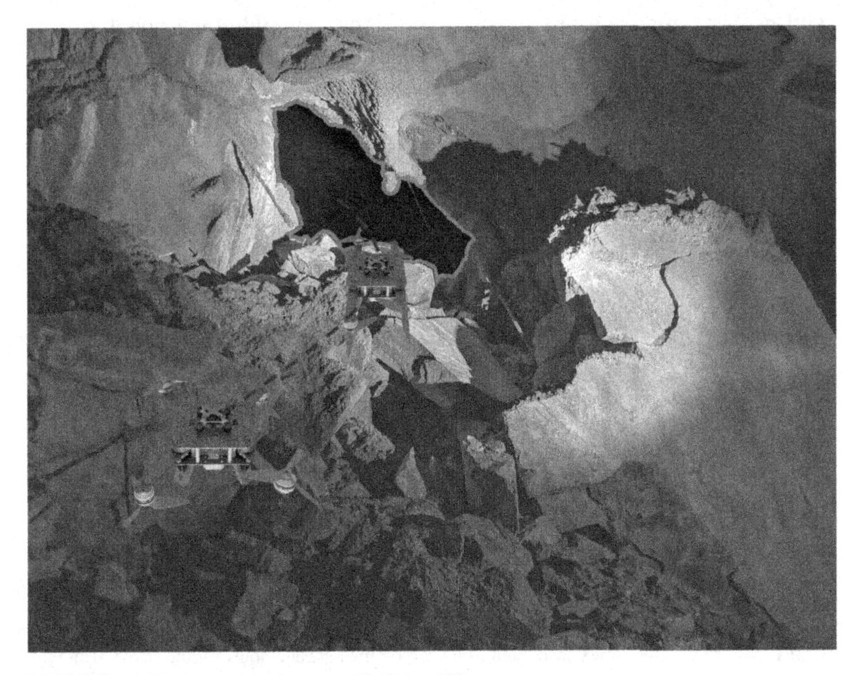

FIGURE 2.6 Configuration change while flying [1].

FIGURE 2.7 "R-Shafter" – re-configurable multi-rotor ARW that was constructed using carbon fiber sheet cutouts.

could not accommodate this in their work area. An additional bonus of this approach is the possibility to utilize the inner space of the tubes, for example, as a guide and protection for the motor cables.

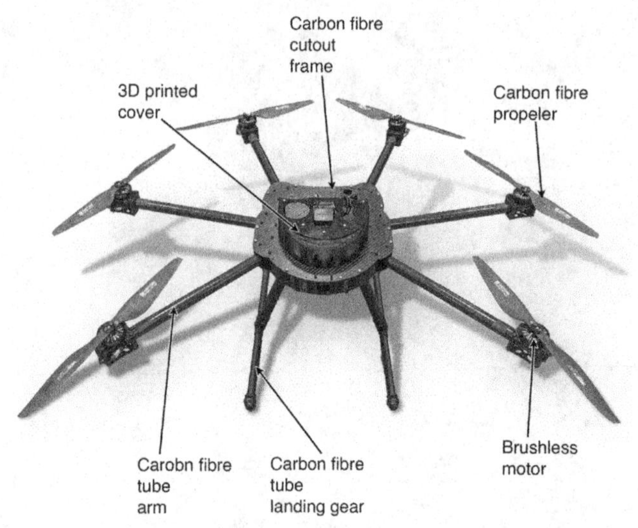

FIGURE 2.8 "Outsider" was constructed using carbon fiber sheet cutouts for the frame, while the arms were made of carbon fiber tubes.

The last example of a custom-made ARW is a drone called "Shafter", Fig. 2.9. Its design approach combines a custom-made frame out of carbon fiber cutouts. Shafter's arms are built from re-used of- the-shelf spare parts for commercial hexa-copter. The carbon fiber plates that are used for the frame have thickness of 3 mm. The platform uses 12 inch propellers (diameter 304.8 mm) and weights about 4500 g. Shafter's x, y dimensions are 385 × 515 mm. The idea of utilizing already engineered and functional components has a drawback. Once the original platform or parts are discontinued, it might be problematic to find the spare parts for the ARW.

In most cases, these ARW platforms have a payload that consists of sensors used for infrastructure inspection and autonomous navigation in open space and GPS-denied environments. The most expensive and fragile sensor mounted on ARW is the Velodyne 3D LiDAR. To protect this sensor, many drones, for example, the Shafter drone, integrate a protection cage, which is also constructed from carbon fiber cutouts of 3 mm thickness, as depicted in Fig. 2.10.

Generally, the design of a multi-rotor aircraft should be robust in case of collisions. It should also be repairable with easy replacement of the damaged or broken parts. The components like an onboard computer, which may require access to its ports for debugging and troubleshooting, should be accessible without tearing the whole platform apart. At the same time, the design should be "smart" enough to allow replacing the components without extensive disassembly of the whole platform.

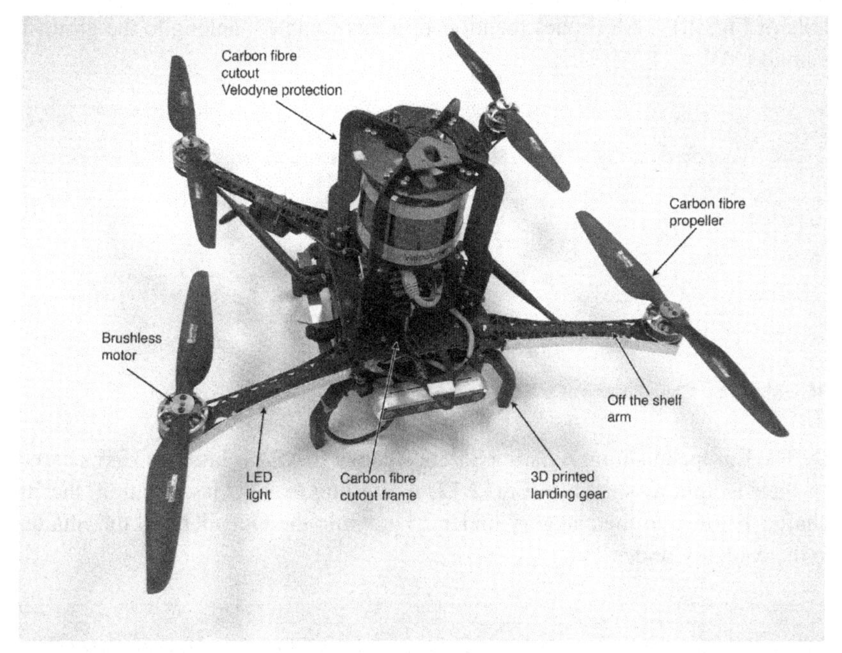

FIGURE 2.9 "Shafter" was constructed using a combination of carbon fiber sheet cutouts and commercially available arms.

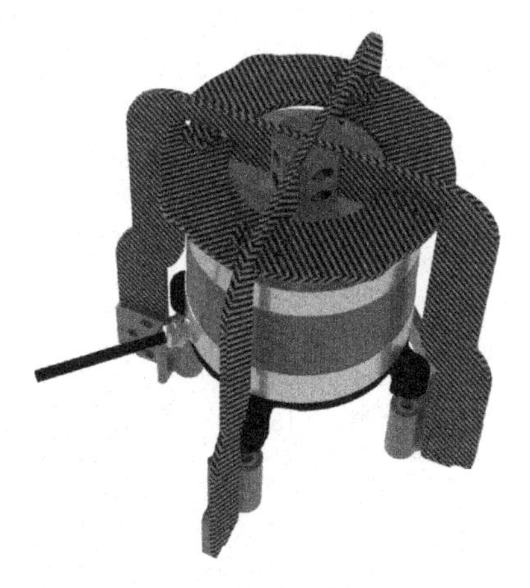

FIGURE 2.10 Render of the protective cage for the Velodyne and sensor itself.

2.2.3 Frame sizes and classification

There are many approaches to dividing and classifying ARWs into different categories. For example, based on the take-off weight, according to the US department of defense, the UAVs can be divided into 5 groups as shown in the table of Fig. 2.11. All drones mentioned in this Chapters belong to the group 1 – small UAV.

Category	Size	Maximum Gross Takeoff Weight (MGTW) (lbs)	Normal Operating Altitude (ft)	Airspeed (knots)
Group 1	Small	0-20	<1,200 AGL*	<100
Group 2	Medium	21-55	<3,500	<250
Group 3	Large	<1320	<18,000 MSL**	<250
Group 4	Larger	>1320	<18,000 MSL	Any airspeed
Group 5	Largest	>1320	>18,000	Any airspeed

FIGURE 2.11 Classification of UAVs [2].

The European Union Aviation Safety Agency (EASA) classifies UAVs based on their weight as shown in Fig. 2.12. According to this classification, the R-Shafter belongs to the category under 2 kg, while the Outsider and the Shafter to the category under 25 kg.

UAS	Operation		Drone operator/pilot		
Max weight	Subcategory	Operational restrictions	Drone operator registration	Remote pilot competence	Remote pilot minimum age
< 250 g	A1 (can also fly in subcategory A3)	– No flight expected over uninvolved people (if it happens, overflight should be minimised) – No flight over assemblies of people	No, unless camera / sensor on board **and** the drone is not a toy	– No training required	No minimum age
< 500 g			Yes	– Read carefully the user manual – Complete the training and pass the exam defined by your national competent authority or have a 'Proof of completion for online training' for A1/A3 'open' subcategory	16*
< 2 kg	A2 (can also fly in subcategory A3)	– No flying over uninvolved people – Keep a horizontal distance of 50 m from uninvolved people	Yes	– Read carefully the user manual – Complete the training and pass the exam defined by your national competent authority or have a 'Remote pilot certificate of competency' for A2 'open' subcategory	16*
< 25 kg	A3	– Do not fly near or over people – Fly at least 150 m away from residential, commercial or industrial areas	Yes	– Read carefully the user manual – Complete the training and pass the exam defined by your national competent authority or have a 'Proof of completion for online training' for A1/A3 'open' subcategory	16*

FIGURE 2.12 EASA classification of UAVs [3].

When it comes to the custom-build research platforms, there is no practical classification of the ARWs. Typically the ARWs used as research platforms are designed for a specific purpose; thus, they excel in one aspect and suffer in the rest, in comparison with universal, commercially built ARWs. For example, ARWs similar to the Shafter platform are generally small platforms and can fly in relatively narrow mine corridors and still carry all the sensors needed for autonomous navigation and exploration but for missions with a smaller flying duration.

The rule of thumb can be as follows: Heavier ARWs use larger propellers, and so the size of the frame is bigger. Depending on the requirements of the ARW, the design can be restricted based on the ARW size. The size of the ARW directly affects the propeller size selection and, consequently, the maximal take-off weight. In general, the take-off weight can be restricted (sensors, onboard computers, etc.), which will, in the sequel, determine the size of the ARW, while different applications pose different requirements and restrictions on the selection of the ARW type or the ARW-specific configuration and size.

2.3 Battery life

This Section introduces the process of battery selection according to different use cases that takes into consideration the sensor payload, the onboard computer, the actuators, and the propulsion. The selection is based on the desired battery chemistry Lithium-Polymer (Li-Po) or Lithium-Ion (Li-Ion), capacity, discharge rate, and voltage that is derived from the particular requirements of each individual configuration. As we will discuss in the next subsection, the platform design will also affect the selection process of the battery to power it.

2.3.1 Different use cases

This subsection will explain how some electrical components can affect the battery selection process more than other. In the case of an ARW, it will be the propulsion system of the vehicle that will dictate some of the battery characteristics. As, in most cases, the motors and the electronic speed controllers (ESC) used for propulsion are designed by the factory to work with specific voltages and draw a maximum amount of current that is significantly higher than any other component that might be included on the ARW. In these cases, the battery needs not only to satisfy the voltage but also to have a high enough discharge rate to be able to provide the maximum current of the motors safely. Discharge rates and other battery characteristics will be discussed in a later subsection.

In cases where propulsion systems are not included, the discharge rating of the battery is not as important as in the example above. Then, it is a matter of defining the voltage suited for the electrical components and calculating the battery capacity depending on the desired total running time of the platform.

2.3.2 Battery characteristics

In this subsection, we will discuss the different battery characteristics, among which are more and less important. Let us begin with the three most important characteristics that need to be tailored for each individual ARW: the capacity, the discharge rate, and the voltage.

Battery capacity describes how much current can be drawn from the battery in one hour; therefore, the units used are amp-hours. Calculating the battery capacity needed is the most complex of the three characteristics described. As there are more variables that need to be accounted for, for example, the total weight, often called all up weight (AUW) of the ARW, the average amp draw (AAD) and also the motor-propeller combination of the ARW.

Motors with different propellers and voltages will provide different thrust at different current draw, while the operating voltages will be described later in this subsection. Manufacturers usually provide data tables of the motors with different propeller combinations that show how much thrust one motor provides at different workloads, see Fig. 2.13.

Item No.	Volts (V)	Prop	Throttle	Amps (A)	Watts (W)	Thrust (G)	RPM	Efficiency (G/W)	Operating temperature(°C)
MT2216 KV900	11.1	T-MOTOR 10*3.3CF	50%	2.8	31	300	5200	9.68	38
			65%	3.7	42	360	5700	8.57	
			75%	4.7	53	420	6300	7.92	
			85%	6.2	69	520	6900	7.54	
			100%	7.4	81	600	7400	7.41	
		T-MOTOR 11*3.7CF	50%	3	35	350	4900	10.00	40
			65%	4.4	50	420	5400	8.40	
			75%	5.7	64	530	5900	8.28	
			85%	7.3	82	630	6500	7.68	
			100%	8.9	98	720	7000	7.35	
		T-MOTOR 12*4CF	50%	3.5	41	420	4200	10.24	43
			65%	5.8	65	560	5000	8.62	
			75%	7.8	86	680	5450	7.91	
			85%	10	110	820	5900	7.45	
			100%	12	129	920	6350	7.13	
	14.8	T-MOTOR 9*3CF	50%	3.4	52	370	7000	7.12	46
			65%	4.4	65	410	7800	6.31	
			75%	5.3	79	470	8500	5.95	
			85%	7.4	108	610	9300	5.65	
			100%	8.5	126	700	10000	5.56	
		T-MOTOR 10*3.3CF	50%	4.1	62	460	6500	7.42	50
			65%	5.6	84	570	7300	6.79	
			75%	7.1	106	690	7900	6.51	
			85%	9.5	139	830	8600	5.97	
			100%	11.2	165	940	9300	5.70	
Notes: The test condition of temperature is motor surface temperature in 100% throttle while the motor run 10 min.									

FIGURE 2.13 Thrust table for different voltages and propeller configurations.

A rule of thumb for choosing the correct motor-propeller configuration is that the workload on the motor should be between 50% and 60% to create the thrust equal to the total weight of the platform. These data tables can be used to calculate the AAD of the ARW, which will help with calculate the battery capacity needed for specific flight time, based on the equation below:

$$AUW \times I_1 = AAD, \tag{2.1}$$

where I_1 is the current that the motors require to lift one kilogram. With AAD known, the battery capacity can be calculated as follows:

$$\frac{T_h \times AAD}{D_{\%}} = capacity, \tag{2.2}$$

where T_h is the desired flight time in hour, and $D_{\%}$ is the discharge percentage in decimal, in other words, how much of the total capacity of the battery is going to be used. For Li-Po and Li-Ion batteries, that is between 70–80% without risking damaging the battery pack.

Depending on the use case of the drone, the calculated capacity will not be totally adequate for the real-life application. If the ARW spends most of its time hovering and does not move much, then the real-time flight will be approximately 75% of the calculated flight time. If the ARW faces strong winds and moves more, then the real time flight will be approximately 50% of the calculated flight time.

There is also a balance between battery capacity and weight that needs to be considered, because the higher the capacity of a battery pack gets, the heavier the battery pack will be, thus decreasing the total flight time.

The discharge rate describes how much current can be drawn from the battery without damaging it. The discharge rate is often referred to as C rating, and it is multiplied by the battery capacity to determine how much current the battery can provide. Battery manufacturers often have two C ratings on the specifications. The first one and the most important refers to how much current one can draw from the battery constantly without damaging it. The second C rating refers to the maximum instantaneous, often called "burst", current draw that the battery is capable of delivering for a specific amount of time, often specified by the manufacturer.

For example, if the designed ARW utilizes 4 motor propulsion systems and each one draws a maximum current of 20 A, the total maximum current draw of the propulsion system will be 80 A. Let us say that the chosen battery capacity needed is 5000 mAh, then the discharge rating is calculated by dividing the total maximum current draw of the propulsion system by the battery capacity. At this point, it should always be highlighted that special attention should be paid to converting mAh into Ah:

$$\frac{5000}{1000} = 5 \text{ Ah}. \tag{2.3}$$

Then calculate the discharge rate as:

$$\frac{80}{5} = 16C. \tag{2.4}$$

It is always good practice to get a battery with a bit higher discharge rate than the calculated one so that the battery is not discharging at its maximum capacity all the time.

Batteries consist of so-called cells. Depending on the voltage needed, the number of cells that make up each battery pack will be different. More than often, Li-Po and Li-Ion batteries have a nominal voltage of 3.7 V per cell. Connecting those cells in serial configuration will add voltage to the number of cells in the battery pack. The cell count is often indicated in the battery packs by the number of cells followed by the letter "S". For example, a 4-cell battery is called a 4S battery pack. Fig. 2.14 depicts a Li-Po battery. As shown in Fig. 2.14, all the information about the battery characteristics described in this subsection is often printed on the battery pack itself.

FIGURE 2.14 Example of Li-Po battery.

As mentioned above, the motors and electronic speed controller (ESC), when manufactured, are designed to work with specific voltages, which translates to specific cell counts in a battery pack.

2.3.3 Battery selection process

In this subsection, we will explain the battery selection process when considering all the information provided in subsections 2.3.1 and 2.3.2.

There are many online flight time calculators that can be used to determine the battery capacity needed for a specific flight time, but in the cases brought up in this book, the ARWs also carry different sensors and computational units. Each component individually may not consume much power, but adding them together will result in increased power consumption, which needs to be considered when choosing the proper battery pack.

The selection process begins with determining the total weight of the ARW and the specifications of the propulsion system. With this information, it is possible to calculate the battery capacity needed to drive only the propulsion system without considering the payload power consumption. Then, the payload power consumption is calculated to add the battery capacity needed to accommodate that. The essential information that is needed from the specifications are the input voltage (V_{in}) and the maximum current draw (A_{max}) of the different payload components. If the maximum current draw is not given, then it can be calculated using the input voltage and the related power consumption (W) and Ohm's Law.

To determine how much extra battery capacity is needed to accommodate the extra payload, the current draw of each component is added. When added, the total current draw of the payload is multiplied by the desired flight time converted to hours as presented below:

$$A_p \times T_h = Ah_p, \tag{2.5}$$

where A_p is the current draw of the payload, and Ah_p is the extra capacity needed to accommodate the extra consumption. It is also necessary to mention that the calculations are seldom one hundred percent accurate, and real-life tuning and measuring are always needed.

Let us take an example with the help of the thrust table in Fig. 2.13. A quadrotor ARW with the total weight of 2 kg, including the battery, has the requirement to fly for 5 minutes, a discharge of the battery of a maximum of 80% and payload current draw of 5 A, while the ARW will fly slow in an enclosed environment.

In the thrust table in Fig. 2.13, there are two different operating voltages and four different propellers. The first step is to decide the operating voltage, or the cell count of the battery pack, that meets the requirements on thrust. Dividing 2 kg by four gives 500 g, which means that each motor needs to produce 500 g of thrust, when the throttle is between 50–60%. According to the thrust table, the best option is a nominal voltage of 14.8 V, which is a 4-cell battery with 10 inch propellers (diameter 254 mm), and this solves the voltage requirement.

The next step is to calculate the total capacity of the battery pack to meet the flight time requirement. To do this, the AAD needs to be calculated first. From the thrust table in Fig. 2.13, the motor needs 11.2 A to lift one kilogram. From Eq. (2.1), we can calculate that:

$$2\text{ kg} \times 11.2\text{ A} = 22.4 \tag{2.6}$$

In the sequel, by converting minutes to hours, we get that:

$$\frac{5}{60} = 0.083 \tag{2.7}$$

Now that the AAD, the flight time and the maximum discharge are known, the calculation of battery capacity is possible. Inserting the results from Eqs. (2.6) and (2.7) to Eq. (2.2) and with a total discharge of 80%, we calculate that:

$$\frac{0.083 \times 22.40}{0.80} = 2.32\text{ Ah} \tag{2.8}$$

Compensating for the real-life application by increasing by 25%, it results in a battery capacity of approximately 2.91 Ah.

Now that the battery capacity for the propulsion system has been calculated, the next step is to accommodate for the payload power consumption. Thus, inserting the flight time and the payload current draw into Eq. (2.5) gives:

$$0.083 \times 5 \approx 0.42 \text{ Ah} \tag{2.9}$$

Thus, by adding the previously obtained results, we can obtain a total battery capacity of 3.33 Ah.

The final step is to calculate the discharge rate required for the battery pack to be able to safely supply the current to the motors. That is done by multiplying the maximum current draw of the motor by four, since in this case, it is a quadrotor ARW, and adding the current draw of the payload to that. It will result in a total maximum current draw of 49.8 A. From Eq. (2.4), dividing 44.8 A by the battery capacity previously calculated will give a discharge rating for the battery pack of approximately 15C. As mentioned before, it is always a good practice to get a higher discharge rate battery pack than the one theoretically calculated for safety reasons.

2.4 Propulsion system

The propulsion system, Fig. 2.15, commonly consists of the battery, the ESC, motor, and propeller. In this section, we are discussing it for a quadrotor ARW.

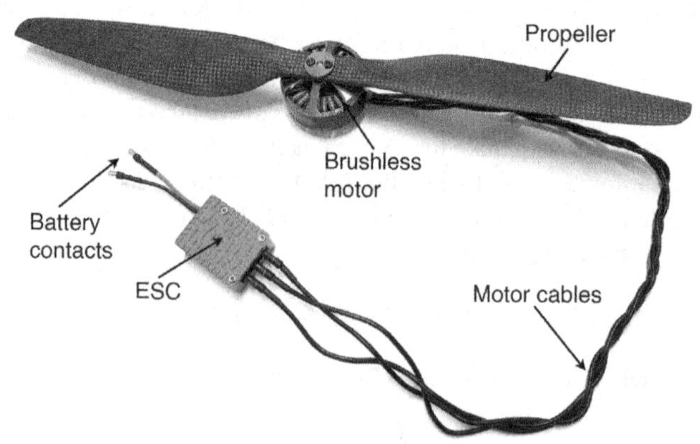

FIGURE 2.15 Example of the propulsion system of ARW consisting of a battery (represented as connectors, ESC, motor cables, and motor with propeller [4].

2.4.1 ARW flying and maneuvering principles

In this case, we will explain the principle of how the ARW is flying based on the example of a quadrotor ARW. The gist of it are four motors that are stiffly placed on the drone frame. The flight controller unit (FCU) is responsible for giving commands to the ESC with the so-called Pulse Width Modulated (PWM) signals to either increase or decrease the thrust of the motors that will consequently maneuver the ARW by producing more or less thrust force. Besides the thrust force (F1, F2, F3, F4), each motor generates a moment (M1, M2, M3, M4) in the direction opposite to the rotation of the motor. More discussions about forces and moments will be given in Chapter 3. The ARW uses the principle of canceling moments. By spinning the two diagonal opposite motors clockwise (CV) and the other two diagonally opposite motors counter clockwise (CCW), the moment of each pair of motors is canceled as can be seen in Fig. 2.16.

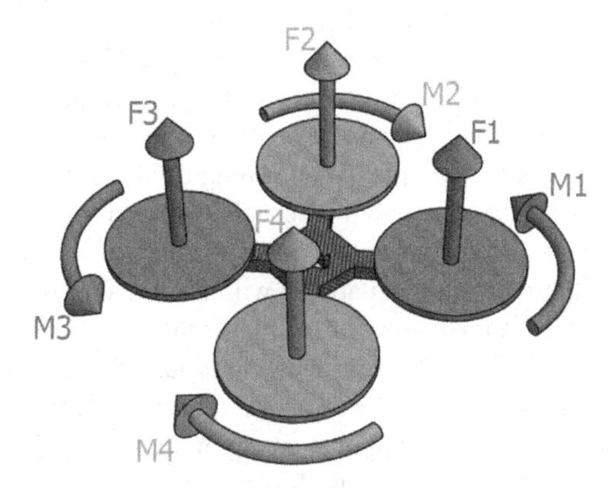

FIGURE 2.16 Thrust forces and moments generated by the ARWs motors [5].

The ARW performs various maneuvers by continuously changing the motor speeds, resulting in the change in the thrust and moments generated by each motor.

The roll, pitch, and yaw, Fig. 2.2, maneuvers are performed using specific combinations of thrust forces as described below, while the total sum of the thrust forces from the motors must be greater than or equal to the weight of the platform. Generally, there are six basic states of flight of the ARW: hover, climb, descent, pitch, roll, and yaw.

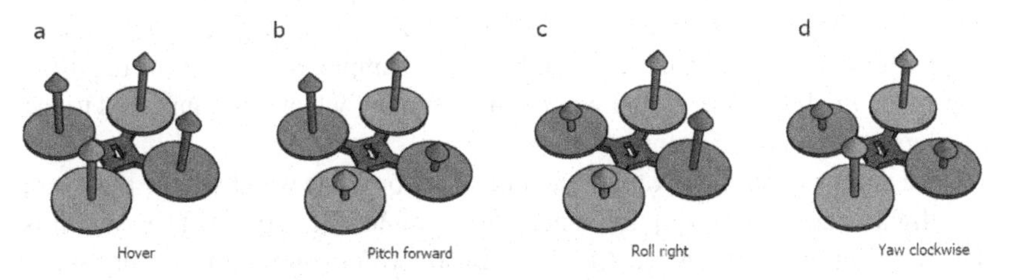

FIGURE 2.17 Different flight modes and their respective thrusts per rotors.

1. Hovering. In this state, when the platform hovers, all of the motors produce an equal thrust, which equals the weight of the platform. The platform is stable as shown in Fig. 2.17a.
2. Climbing. In this state, all the motors produce equal thrust, but the total sum is greater than the weight of the platform. The platform is climbing from the ground or from the default state of hovering.
3. Descent. This state is opposite to the Climbing. During the descent, all the motors produce equal thrust, but the total sum is lower than the weight of the platform. The platform is descending from the default state of hovering.
4. Pitching maneuver. For example, when pitching forward, the back two motors produce more thrust than the two motors in the front. It causes the platform to tilt and fly forward as shown in Fig. 2.17b. Respectively, if the front motors produce more thrust, the platform will move backward. The total thrust is slightly larger than the weight of the platform.
5. Rolling maneuver. For example, when rolling right, left two motors are producing more thrust than the two on the right. This causes the platform to tilt and fly to the right, as shown in Fig. 2.17c. Respectively, if the right motors are producing more thrust, the platform will move to the left. The total thrust is slightly larger than the weight of the platform.
6. Yawing maneuver. The platform is rotating around the line passing through the center of mass, parallel to the thrust vector. Yaw maneuver is done in the way that two motors on the diagonal are spinning faster, so producing more thrust than the opposite motors, as shown in Fig. 2.17d. At the same time, the torque of the motors is not canceled, and the platform yaws. The total sum of the thrust produced by all the motors is equal to the weight of the platform.

2.4.2 Propulsion system design

The goal of this subsection is not to give complete detailed instruction on how to design the propulsion system of an ARW, but rather give a basic overview and tips from field experience in deploying ARWs. The "propulsion system" of a propeller-based ARW consists of the following 4 sets of components: the battery, ESC, the rotor (motor+propeller). The simplified diagram of the propulsion system of the ARW, including the data and power chains, is shown in Fig. 2.18. The maximal performance of the whole system is given by the maximal performance of the weakest member. For example, a motor requiring 40 A to provide the full power will not perform optimally when it is connected to the ESC with the maximal output of 30 A.

The defining parameter of the actuator selection is the weight of the platform, and flight characteristic requirements (for example, flight time). The system is designed optimally when the ARW needs 50–60% of throttle to hover with a fully charged battery.

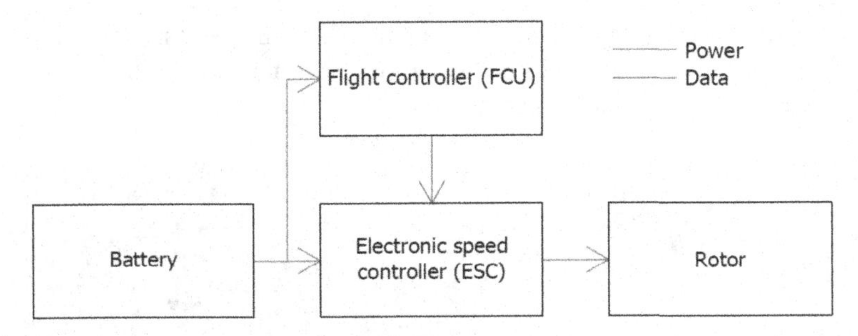

FIGURE 2.18 Block diagram of the propulsion system with power and data chains.

1. The battery must be able to provide enough current to the ESC in any phase of the flight. At the same time, the battery voltage can not drop under a certain level, as the brushless motors are voltage driven.
2. The ESC must provide enough current to the motors in any phases of the flight.
3. The rotor typically consists of a brushless motor and a propeller. The motor and the propeller must be selected together for an optimal operation. The general rule is that smaller propellers need faster motors (higher RPMs). The RMPs of a brushless motor are defined by parameter kv. The simplified equation to calculate the RPM of the motor is $RPM = KV \times \textit{Voltage}$.

The selection of the components is typically based on the component specifications provided from the data sheets by the manufacturers, while it is well known that the manufacturers tend to present the best-case scenario results. When selecting the components, the parameters then must be taken with some reserve; thus, it is wise to verify the parameters with in-house testing.

A good guide for the components selection are the provided calculations from the manufacturers. For example, KDE direct [6] can recommend optimal selection of ESC and Rotor from given parameters (Flight time requirement, weight of the platform, etc.)

If the power chain is not optimal, then it is pushed to and behind the limit of its capabilities. The flight performance of the ARW is compromising over the time of flight. For example, the ARW can perform as it should in all maneuvers after take-off, but at some point of the flight, the motors start to saturate. The saturation means that the motor reached the maximum RMPs for the given voltage of the battery. The motor saturation is especially dangerous during the YAW maneuver. So the user of the ARW must be aware of this limitation and avoid aggressive maneuvers and plan the flight accordingly.

2.5 Sensor hardware

In this Section, we will introduce sensors for ARW, which incorporate 3D LiDAR, 2D LiDAR, IMU, UWB node, RGB, RGB-D, LiDAR depth, event, thermal, 360 and action cameras [21]. Table 2.2 summarizes specifications of all sensors, while Fig. 2.19 represents the output from some of these sensors' recorded at Luleå underground tunnel.

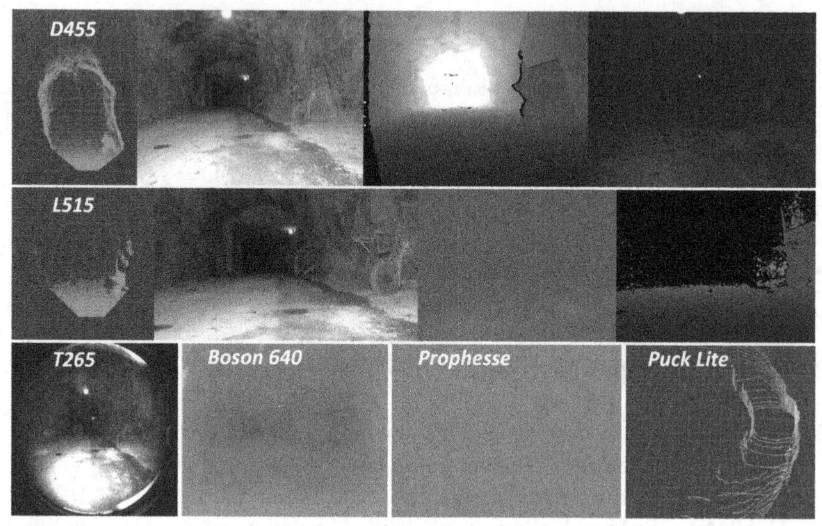

FIGURE 2.19 Sample data output and from the Luleå underground tunnel collected with the Intel RealSense cameras D455 – first row, L515 – second row, T265, thermal camera Flir Boson 640, Prophesse event camera, and 3D LiDAR Velodyne Puck Lite – third row.

2.5.1 Vision sensors

In this subsection, we introduce sensors that can be equipped on ARW or ground robots for autonomous navigation, obstacle avoidance, infrastructure inspection, 3D scanning and mapping, etc.

Typically, ARW obtains information about the surrounding environment through the perception sensors, either visual or ranging. This information can be used for state estimation, obstacle avoidance, Simultaneous Localization and Mapping (SLAM), which can produce a local or a global map that is later used for path planning and path tracking. Depending on the ARW requirements and its operating environment, it can be equipped with visual or ranging sensors for state estimation. For better performance, multiple sensors can be fused to improve its robustness and accuracy.

For visual navigation, inspection, and surveillance missions, the typical sensors utilized are visual sensors like RGB cameras. Compared to ranging sensors, they can acquire a lot of environmental information, including color, shape, illumination, etc. Additionally, these sensors are cheaper, lighter, and can be easily deployed. However, they also could have limitations like reduced performance under insufficient illumination and noise. Depending on the dimensions of the indoor environments, RGB cameras with fisheye lenses might be used. These cameras have a wider field of view that might be required in narrow areas. Additionally for obtaining the depth information of the environment, one RGB camera is not enough. For that, it should be used a stereo camera, which is nothing more than a pair of RGB cameras. An example of such a camera is Intel RealSense T265 [7],[1] depicted in Fig. 2.20. T265 can be used for collecting RGB images and evaluating the visual odometry methods. This is a lightweight camera sensor with low power consumption that includes two fisheye lens sensors with resolution 848 × 800 pixels, an IMU with a sample rate of 0.005 sec and vision processing unit (VPU). Another option can be to utilize RGB-D camera, which can acquire depth information together with visual data. Example of such a camera is the Intel RealSense D455 [8][2] depicted in Fig. 2.20. This depth camera has an IMU with sample rate of 0.005 sec and an RGB camera with 1280 × 800 at 30 fps and depth camera with IR projector to collect the depth data up to 1280 × 720 pixels. On the other hand, there is also LiDAR camera. Intel RealSense L515 [8],[3] depicted in Fig. 2.20, is a high-resolution LiDAR depth camera with power consumption of about 3.5 Watt, which provides depth data with a resolution of 1024 × 768 pixels per depth frame at 30 fps and a maximum range of 9 m.

FIGURE 2.20 Intel RealSense cameras. Left: T265, middle: D455, and right: L515.

Recently, event cameras [9] gained significant interest in robotic applications such as collision avoidance, especially due to their low latency and high dynamic range. Prophesse EVALUATION KIT – Gen3M VGA-CD 1.1, Fig. 2.21, is an event trigger camera, which provides up to 55 kHz asynchronous pixel vise events, high dynamic range, and enables seeing events with low illumination of 0.08 lx and has a resolution of 640 × 480 pixels. The intrinsic camera calibrations were performed with Prophesses calibration tool.[4] In Table 2.1, the event camera biases used is shown.

The ARW vision system can also be composed of a thermal camera [10]. The modality of such a camera allows perceiving the environment in poor illumination conditions and pitch black areas. This capability can be useful for navigation in dark subterranean environments, human search and rescue missions, infrastructure inspection, etc. The FLIR Boson 640, Fig. 2.21, is a longwave infrared (LWIR) professional-grade thermographic camera that senses infrared

[1] https://www.intelrealsense.com/tracking-camera-t265/.
[2] https://www.intelrealsense.com/depth-camera-d455/.
[3] https://www.intelrealsense.com/lidar-camera-l515/.
[4] https://docs.prophesee.ai/stable/metavision_sdk/.

TABLE 2.1 Event camera bias used during data collection.

bias_refr	1800
bias_hpf	2600
bias_diff_off	220
bias_diff_on	460
bias_diff	300
bias_fo	1670
bias_pr	1500

radiation, which makes it suitable for low-light environments. It is equipped with a 4.9 mm lens and a resolution of 640 × 512 pixels, the frame rate of 60 Hz, and thermal sensitivity < 50 mK.

The 360-degree cameras [11] is a new technology gaining a lot of attention due to its capability of full coverage of complex environments, like construction sites. Insta 360 EVO camera, Fig. 2.21, is capable providing omnidirectional videos with a resolution up to 5760 × 2880 at 30 fps and images with resolution of 6080 × 3040.

The last class of visual sensors we will mention in this subsection is action cameras [12], which can also be frequently used in robotic applications and machine learning. Action camera GoPro Hero7 Black camera is capable providing videos of 3840 × 2160 pixels with 60 fps.

FIGURE 2.21 Left: Event camera Gen3 VGA, middle: Flir Boson 640, and right: Insta 360 EVO.

2.5.2 Infrared ranging sensors

LiDAR sensors allow for estimating of the distance to the object by measuring the return time of a reflected signal. Their ability to provide high-definition scans of the environment made LiDARs a baseline sensor in mapping, reconstruction, and inspection applications [13]. This subsection introduces 2D LiDAR and 3D LiDAR state-of-the-art sensors mostly targeting applications in indoor environments with poor illumination conditions, which application areas are primarily focused on collision avoidance, SLAM, 3D reconstruction, etc.

Among the 2D omnidirectional rotating LiDARs, one can highlight Slamtec RpLiDARs that are ROS compatible and gained significant community interest for their usage in SLAM [14] and collision avoidance applications [15]. The major reasons for that are their solid performance, weight comparable with vision sensors, compact size and low power consumption. The RpLiDAR A2 and S1 sensors commonly used for autonomous navigation in subterranean environments are depicted in Fig. 2.22. Both sensors have a data publishing rate of 10 Hz, while A2, providing 8000 samples per second, has a measuring range up to 16 meters with an angular resolution of 0.9 degrees, and S1, providing 9200 samples per second, has a measuring range of 40 meters with a resolution of 0.4 degrees in average.

FIGURE 2.22 Left: RpLiDAR A2, middle: RpLiDAR S1, right: Velodyne Puck Lite.

The more advanced ranging sensors are the 3D LiDARs, which add vertical dimension to the scanning area. Among them, one can mention the Velodyne Puck Lite 3D LiDAR, depicted in Fig. 2.22, which demonstrates solid performance for robot autonomy development, mapping, reconstruction and infrastructure inspection applications that demand 3D scanning [16]. This LiDAR has 100 meters range with 360 degrees × 30 degrees horizontal and vertical field of view.

2.5.3 Inertial measurement unit

VectorNav vn-100 IMU is the main core element of the localization system for robotic applications in variety of applications, from underwater, to ground, aerial, and space robotics. Commonly it can consist of a 3-axis accelerometer, gyroscope, and magnetometer. By sampling these data from the IMU, the orientation of a rigid body can be estimated, which is fundamental for most data fusion methods. In our studies, we commonly utilize VectorNav vn-100, a small size unit with a weight of 15 g, that provides a 3-axis accelerometer, gyroscope, and magnetometer. The vn-100 provides sampled data of up to 400 Hz rate filtered data, and RMS of 0.5° and 2.0° for heading (magnetic) and pitch/role measurements. For the installation of VectorNav, damping vibration dampers should be used to reduce the frame vibration effect on IMU measurements.

2.5.4 UWB ranging sensor

UltraWideBand (UWB) is a relatively new technology for indoor short-range localization, which is capable providing positioning accuracy in the order of tens of centimeters [17]. In our studies, we utilize Decawave MDEK1001 UWB modules, which can provide a ranging accuracy within 10 cm, with a 3-axis accelerometer and size of 19.1 mm × 26.2 mm × 2.6 mm; finally, it uses 6.8 Mbps data rate IEEE802.15.4-2011 UWB compliant protocol. Localization based on the UWB ranging technology can be utilized for both single and multi-robot navigation [17]. In the latter case, it can also enable a relative localization or utilization for the robots' coordination and formation control. To improve position estimation accuracy, it can also be fused with IMU.

TABLE 2.2 Specification of all the sensors in the data collection platform.

Sensor	Field of view [degrees]	Dimensions, W x H x D: [mm]	Weight [gram]
VectorNav vn-100	–	36 × 33 × 9	15
RpLiDAR A2	360°, horizontal plane		190
RpLiDAR S1	360°, horizontal plane	55.5 × 55.5 × 51	105
Velodyne Puck Lite	360° × 30°, horizontal x vertical FOV		590
Flir Boson 640	95° (HFOV) 4.9 mm	21 × 21 × 11 (without lens)	140
Intel RealSense T265	Two Fisheye lenses with combined 163±5° FOV	108 × 24.5 × 12.5	55
Intel RealSense D455	Depth Field of View (FOV): 86° × 57° (±3°) RGB sensor FOV (H × V): 86° × 57° (±3)	124 × 26 × 29	103
Intel RealSense L515	Depth Field of View (FOV): 70° × 55° (±2°) RGB Sensor FOV (H × V): 70°±3 × 43°±2	D x H: 61 × 26	100
Prophesse event trigger camera	Diagonal FOV: 70	60 × 50 × 78	140
Decawave MDEK1001	–	19.1 × 26.2 × 2.6	41
Insta 360 EVO	Spherical	50.3 × 49 × 52.54	113
GoPro Hero 7	Horizontal FOV 94.4, Vertical FOV 122.6	62 × 28 × 4.5	92.4

2.6 ROS node interfacing sensors and actuators

Robot Operating System (ROS) [18] is a software framework that provides tools to interface sensors, actuators using nodes and services. Moreover, it has the capability to store sensor data in a ROS bag format for further analysis. To access and analyze sensor data in the bag files, ROS has a built-in set of tools called rosbag,[5] which provides a set of commands for easy data access. Additionally, for the cases when it is required to have processing time synchronized with sensor data, timestamps, as they were recorded into a clock server topic `/clock` should be used. In order to enable time synchronization, it is required to set parameter `/use_sim_time = true` and to play a bag file using the `--clock` option. The data can be accessed both from individual bag files as `rosbag play sensor_set_id_timestamp.bag` and from multiple bag files as `rosbag play sensor_set_id_timestamp*`.

ROS can be installed on a variety of board-size computers like Up Board or Intel NUC. The selection of one or another computer relies on the sensor payload, computational requirements, and ARW size. For example, for autonomous navigation utilizing 2D LiDAR would be sufficient to use Up Board [19], while for scenarios where 3D LiDAR is required, NUC can provide the required computational power [20].

2.7 Compact manipulator for ARW

One of the fundamental modules for ARWs is a compact aerial manipulator, further referred to as CARMA [22] depicted in Fig. 2.23. The manipulator is designed to provide for a multi-rotor ARW capabilities to physically interact with the environment, more specifically for pick-and-place operations, load transportation, contact inspections, or repairs using dedicated tools.

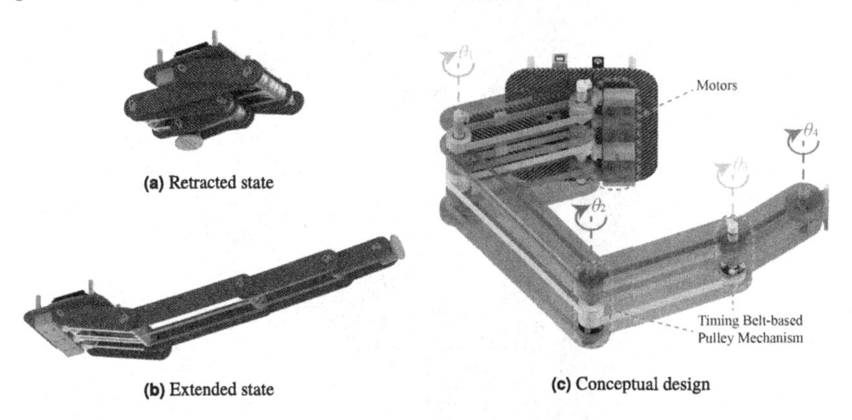

FIGURE 2.23 Visual representation of CARMA manipulator.

Manipulator's conceptual design is depicted in Fig. 2.23. In it, mechanical power from the motors that are located in the base is transmitted to the joints via bevel gears, in series with a timing belt and pulleys mechanism. These belts were deliberately not enclosed in order to allow the airflow generated by the propellers to circulate through the frame, thus minimizing the disturbances due to the resulting drag.

5 http://wiki.ros.org/rosbag.

The presented design has the following features:

- Low mass (350 g) and low inertia with respect to the ARW's center of gravity (CoG), thus minimizing the impact on the UAV dynamics.
- Wide workspace, overstepping the propeller blades perimeter while remaining compact in the retracted state as depicted in Fig. 2.23a, and thus allowing not to interfere during take-off and landing phases by fitting a 200 mm × 130 mm × 130 mm envelope.
- Low manufacturing complexity, assuming carbon fiber plate milling and PLA (Polylactic Acid) 3D printing to be available.
- The number of links was set to four in order to achieve long extensions while keeping a compact retracted state as depicted in Fig. 2.23b. Lengths of the links are 173 mm, 122 mm, 71 mm, and 20 mm, respectively.

The designed prototype of the manipulator is presented in Fig. 2.24. In it, the links consist of superimposed carbon fiber plates, 6 mm width MXL rubber timing belts that are tensioned using spiral springs tensioners, Bourns PTA4543-2015DPA103 linear slide potentiometers, injection molded acetal co-polymer bevel gears, and Pololu 298:1 Micro Metal Gearmotor HP. The micro-controller board is an Arduino Mega 2560 endowed with an Adafruit Motor Shield V2.3 as a motor control board.

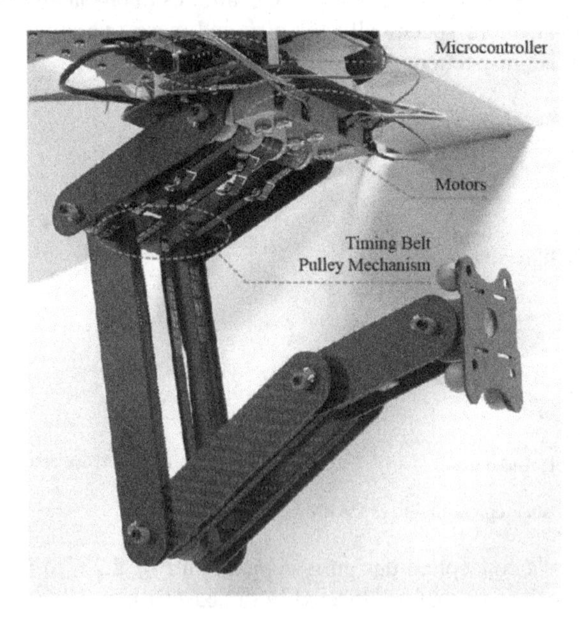

FIGURE 2.24 CARMA prototype.

References

[1] A. Papadimitriou, S. Mansouri, C. Kanellakis, G. Nikolakopoulos, Geometry aware NMPC scheme for morphing quadrotor navigation in restricted entrances, in: 2021 European Control

Conference (ECC), IEEE, 2021, pp. 1597–1603.

[2] Q.A. Abdullah, Classification of the unmanned aerial systems [Online]. Available: https://www.e-education.psu.edu/geog892/node/5, 2020. (Accessed January 2022).

[3] EASA, Classification of the unmanned aerial systems [Online]. Available: https://www.easa.europa.eu/domains/civil-drones-rpas/open-category-civil-drones, 2021. (Accessed January 2022).

[4] Zubax, UAV propulsion kits [Online]. Available: https://zubax.com/products/uav_propulsion_kits, 2021. (Accessed January 2022).

[5] P. Jaiswal, Demystifying drone dynamics [Online]. Available: https://towardsdatascience.com/demystifying-drone-dynamics-ee98b1ba882f, 2018. (Accessed January 2022).

[6] K. Direct, Build your system application [Online]. Available: https://www.kdedirect.com/blogs/news/kde-direct-build-your-system-calculator-for-uavs, 2021. (Accessed January 2022).

[7] P. Hausamann, C.B. Sinnott, M. Daumer, P.R. MacNeilage, Evaluation of the Intel RealSense T265 for tracking natural human head motion, Scientific Reports 11 (1) (2021) 1–12.

[8] M. Servi, E. Mussi, A. Profili, R. Furferi, Y. Volpe, L. Governi, F. Buonamici, Metrological characterization and comparison of D415, D455, L515 RealSense devices in the close range, Sensors 21 (22) (2021) 7770.

[9] G. Gallego, M. Gehrig, D. Scaramuzza, Focus is all you need: Loss functions for event-based vision, in: Proceedings of the IEEE/CVF Conference on Computer Vision and Pattern Recognition, 2019, pp. 12280–12289.

[10] S. Daftry, M. Das, J. Delaune, C. Sorice, R. Hewitt, S. Reddy, D. Lytle, E. Gu, L. Matthies, Robust vision-based autonomous navigation, mapping and landing for MAVS at night, in: International Symposium on Experimental Robotics, Springer, 2018, pp. 232–242.

[11] D. Holdener, S. Nebiker, S. Blaser, Design and implementation of a novel portable 360 stereo camera system with low-cost action cameras, The International Archives of the Photogrammetry, Remote Sensing and Spatial Information Sciences 42 (2017) 105.

[12] C.M.D. Junior, S.P. da Silva, R.V. da Nobrega, A.C. Barros, A.K. Sangaiah, P.P. Reboucas Filho, V.H.C. de Albuquerque, A new approach for mobile robot localization based on an online IOT system, Future Generations Computer Systems 100 (2019) 859–881.

[13] I. Puente, H. González-Jorge, J. Martínez-Sánchez, P. Arias, Review of mobile mapping and surveying technologies, Measurement 46 (7) (2013) 2127–2145.

[14] M.S. Aslam, M.I. Aziz, K. Naveed, U.K. uz Zaman, An RpLiDAR based slam equipped with IMU for autonomous navigation of wheeled mobile robot, in: 2020 IEEE 23rd International Multitopic Conference (INMIC), IEEE, 2020, pp. 1–5.

[15] T. Madhavan, M. Adharsh, Obstacle detection and obstacle avoidance algorithm based on 2-D RpLiDAR, in: 2019 International Conference on Computer Communication and Informatics (ICCCI), IEEE, 2019, pp. 1–4.

[16] S.S. Mansouri, C. Kanellakis, F. Pourkamali-Anaraki, G. Nikolakopoulos, Towards robust and efficient plane detection from 3D point cloud, in: 2021 International Conference on Unmanned Aircraft Systems (ICUAS), IEEE, 2021, pp. 560–566.

[17] J.P. Queralta, C.M. Almansa, F. Schiano, D. Floreano, T. Westerlund, UWB-based system for UAV localization in GNSS-denied environments: Characterization and dataset, in: 2020 IEEE/RSJ International Conference on Intelligent Robots and Systems (IROS), IEEE, 2020, pp. 4521–4528.

[18] M. Quigley, K. Conley, B. Gerkey, J. Faust, T. Foote, J. Leibs, R. Wheeler, A.Y. Ng, et al., ROS: an open-source robot operating system, in: ICRA Workshop on Open Source Software, vol. 3 (3.2), Kobe, Japan, 2009, p. 5.

[19] S. Sharif Mansouri, C. Kanellakis, E. Fresk, B. Lindqvist, D. Kominiak, A. Koval, P. Sopasakis, G. Nikolakopoulos, Subterranean MAV navigation based on nonlinear MPC with collision avoidance constraints, arXiv e-prints, 2020.

[20] B. Lindqvist, C. Kanellakis, S.S. Mansouri, A.-A. Agha-Mohammadi, G. Nikolakopoulos, Compra: A compact reactive autonomy framework for subterranean MAV based search-and-rescue operations, arXiv preprint, arXiv:2108.13105, 2021.

[21] A. Koval, S. Karlsson, S. Sharif Mansouri, C. Kanellakis, I. Tevetzidis, J. Haluska, A. Agha-mohammadi, G. Nikolakopoulos, Dataset collection from a SubT environment, in: Robotics and Autonomous Systems, Elsevier, 2022, pp. 1–8.

[22] D. Wuthier, D. Kominiak, C. Kanellakis, G. Andrikopoulos, M. Fumagalli, G. Schipper, G. Nikolakopoulos, On the design, modeling and control of a novel compact aerial manipulator, in: 2016 24th Mediterranean Conference on Control and Automation (MED), IEEE, 2016, pp. 665–670.

Chapter 3

Modeling for ARWs

Akash Patel, Andreas Papadimitriou, and Avijit Banerjee
Department of Computer, Electrical and Space Engineering, Luleå University of Technology, Luleå, Sweden

3.1 Notation

The axes follow the *ENU* coordinate system. The local body frame is called $\{\mathcal{B}\}$ with x-forward, y-left, and z-up. The global frame is denoted by $\{\mathcal{G}\}$ with X-east Y-north Z-up.

Furthermore, $\boldsymbol{p} = [p_x, p_y, p_z]$ is the position of the aerial platform; $\boldsymbol{v} = [v_x, v_y, v_z]$ is the linear velocity of the aerial platform; $\boldsymbol{\eta} = [\phi, \theta, \psi]$ is the roll, pitch, and yaw rotation angles of the aerial platform, while the angular velocities are defined as $\boldsymbol{\omega} = [\omega_x, \omega_y, \omega_z]$. Rotation is defined in quaternion as well with $\boldsymbol{q} = [q_x, q_y, q_z, q_w]$, and the rotor speeds are denoted as Ω.

The time is denoted by t, while sample-steps are denoted by k. The sampling time is d_t, and the frequency $f_s = 1/d_t$. Mass of a body is m, and the gravitational acceleration is g.

In general, scalars are defined as non-bold characters, lower-case or uppercase p_x, T. Vectors are defined as bold lower-case $\boldsymbol{p}$. Matrices are bold capital upper-case letters like $\boldsymbol{A}$. The $\mathbb{R}$ denotes the set of all real numbers, while $\mathbb{Z}$ represent integers. A set accompanied by the superscripts $\mathbb{R}^{+,-}$ defines non-zero positive or negative values and, if there is a numerical superscript, like $\mathbb{Z}^{3\times1}$, represents the dimensions of a vector or matrix.

3.2 Introduction

The mathematical modeling of complicated systems, like the ARWs, is essential to maximize the chances of a successful mission while minimizing the risks of collision and collateral damages. ARW models are mathematical descriptions of the dynamics, kinematics, and physical properties that govern an ARW [1]. There are numerous reasons behind the necessity of such mathematical descriptions. These models can describe complex physic phenomena with mathematical equations that can be used to simulate and analyze the system's capabilities. The mathematical description of an ARW can be used to visualize the states of the system, analyze the behavior of an ARW, estimate unknown variables of the

Aerial Robotic Workers. https://doi.org/10.1016/B978-0-12-814909-6.00009-3

vehicle or the environment, and predict the future state evolution. Furthermore, Modeling is the essential base for developing various control algorithms used for the navigation of the ARWs, while models can also be utilized in tuning the design parameters of prototypes through iterative simulations, without developing multiple physical prototypes and potentially resulting in lower expenditures. The actual phenomena that govern an ARW system are highly complicated, and it is challenging to model them precisely [2]. Luckily, not all components and coefficients of a model have the same influence on the dynamics of an ARW. Thus, we can use simplifications and assumptions, resulting in more straightforward yet effective models. A less complex model results in a lower computation burden, and it can be used by an onboard computation unit, which has decreased power requirements, for example, to facilitate software responsible for the control and state estimation of the ARW. Finally, the mathematical models of ARWs can be used to access the performance of the platform and the results of a mission through simulation. The mathematical description of ARWs can be implemented in simulation environments (e.g., Gazebo, MATLAB® simulink), which can be used to simulate different scenarios. In this approach, the development of novel control algorithms, navigation techniques, obstacle avoidance frameworks, etc. can be put to the test. Suppose a simulation fails and the ARW collides with an obstacle. In that case, we could easily update the algorithm and execute the next simulation, while in reality, we would have to replace it with a new platform.

To define the equations of motion of an ARW quadrotor, it is important to define the frame of reference first. For a quadrotor, two frames of reference are considered. The first global frame of reference $\mathcal{G}$ that is always stationary, and the second coordinate frame that is the quadrotor's body frame of reference $\mathcal{B}$. The structure and motion of the vehicle are described in the body frame of reference, where the position commands are provided in the world frame of reference.

As denoted in Section 3.1, the global and body frame directions are presented in Fig. 3.1. The control of the quadrotor is achieved by adjusting the motor speeds. Due to the difference in motor speeds, torques are generated because of the coupled forces from opposite motors. There are two standard configurations with respect to which a quadrotor motion is defined. These are the $\times$ and $+$ configurations. In this Chapter, for the modeling and control of a quadrotor, a $+$ configuration is considered. As shown in Fig. 3.1, the opposite rotors spin in the same direction, and the adjacent rotors spin in the opposite direction. The six degrees of freedom motion of the quadrotor is defined in terms of forward and backward directions, lateral motion, vertical movement, roll, pitch, and yaw movements. These motions can be narrowed down to thrust and torques around x, y, and z directions to achieve the control of a quadrotor as will be presented in the following chapter.

When the four speeds of the rotor are the same, the reactive torque is balanced from all the four motors and the quadrotor stays in place. When the four

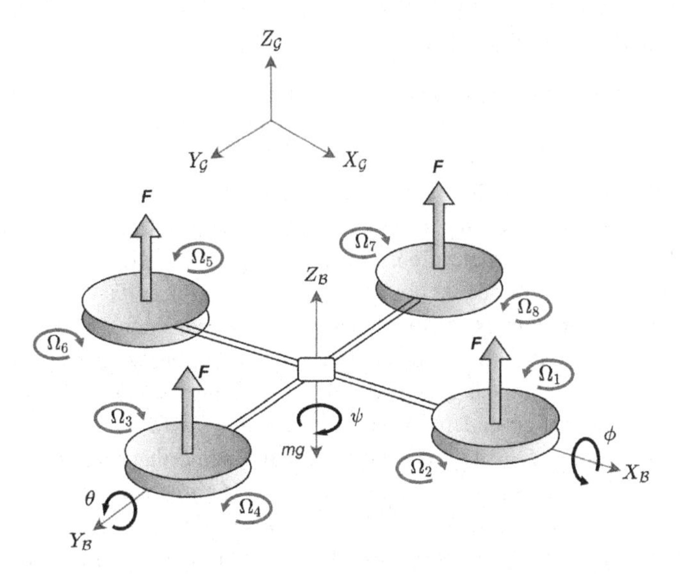

FIGURE 3.1 Global and Body frame of reference for a quadrotor.

input speeds are not exactly the same, the reactive torque from motors is different and depends upon the difference in the speeds, the quadrotor rolls, pitches, or yaws. When the speeds of all four rotors increase or decrease together by the same amount, the quadrotor will either go up or down. Essentially, there are four control inputs (thrust, roll, pitch, and yaw torques), and there are twelve states, as it is a 6 DOF system. Thus the quadrotor is considered an underactuated system [3]. To formulate the mathematical model of the quadrotor, some assumptions are made:

1. The quadrotor is a rigid body
2. The quadrotor structure is symmetric
3. The ground effect is ignored

The quadrotor mathematical model is provided by 12 output states that are the p_x, p_y, and p_z positions; v_x, v_y, and v_z are the translational velocities, ϕ, θ, and ψ are the roll, pitch, and yaw angles; ω_x, ω_y, and ω_z are the angular velocities for roll, pitch, and yaw moments.

3.3 Modeling of ARWs

This section deals with the modeling of an ARW, while presenting the mathematical models that describe the translational and rotational dynamics. Initially, the attitude mathematical representation will be shown, followed by the full nonlinear model of ARW. A simplified position model will be presented, and finishing with the modeling of a reconfigurable ARW.

3.3.1 Attitude modeling

This section will describe the mathematical representation of rigid body orientation called attitude representation. The mathematical description of attitude is a set of parameters and matrices that describe the orientation of a reference frame with respect to another frame. The three commonly used ways to express and store an attitude representation [4] are a) Euler Angles, b) Direct Cosine Matrix (DCM), and c) Quaternion. Fundamentally, the attitude of a rigid body describes the orientation of its body axis with respect to a fixed/ground frame of reference. In the context of the quadrotor, let us consider that a body-fixed frame $\{\mathcal{B}\}$ is attached with it, indicated by three orthogonal unit vectors as $X_{\mathcal{B}}$, $Y_{\mathcal{B}}$, $Z_{\mathcal{B}}$. The orientation of the quadrotor is to be represented with respect to a three-dimensional ground frame of reference $\{\mathcal{G}\}$, with its axis, denoted as unit vector $X_{\mathcal{G}}$, $Y_{\mathcal{G}}$, $Z_{\mathcal{G}}$. At this point, we will assume that the origin of both the ground frame and the body frame coincides. The mathematical description of attitude representations is characterized by a set of parameters. A minimum number of three parameters are required to uniquely define the attitude. However, there a various approaches to attitude representation. Standard practices typically include the Euler angle, Direction cosine matrix, and quaternion. Note that the attitude representations in these forms are interchangeable, and one representation form can be converted to any other.

3.3.1.1 *Euler angles*

Leonhard Euler introduced Euler angles that are used to define the orientation of a rigid body. According to the Euler definition, any orientation can be achieved by a maximum set of three successive rotations. The Euler angles are also used to describe the orientation of the reference frame relative to another reference frame, and they are also implemented in transforming the coordinates of a point from one frame of reference to another. The elemental rotation is defined as the rotation about the axis of the reference frame. The definition of the Euler angle represents three elemental rotations. The combination of rotation metricize, used to define Euler angles, is shown in Eqs. (3.1), (3.2), and (3.3).

$$\mathbf{R}_x(\phi) = \begin{bmatrix} 1 & 0 & 0 \\ 0 & \cos(\phi) & -\sin(\phi) \\ 0 & \sin(\phi) & \cos(\phi) \end{bmatrix} \tag{3.1}$$

$$\mathbf{R}_y(\theta) = \begin{bmatrix} \cos(\theta) & 0 & \sin(\theta) \\ 0 & 1 & 0 \\ -\sin(\theta) & 0 & \cos(\theta) \end{bmatrix} \tag{3.2}$$

$$\mathbf{R}_z(\psi) = \begin{bmatrix} \cos(\psi) & -\sin(\psi) & 0 \\ \sin(\psi) & \cos(\psi) & 0 \\ 0 & 0 & 1 \end{bmatrix} \tag{3.3}$$

In this Chapter, the set of Euler angles used is the ZYX Euler angles. The world frame coordinates and the body frame are correlated by the rotation $R_{zyx}(\phi, \theta, \psi)$.

$$\mathbf{R}_{zyx}(\phi, \theta, \psi) = \mathbf{R}_z(\psi) \cdot \mathbf{R}_y(\theta) \cdot \mathbf{R}_x(\phi)$$

3.3.1.2 Directional cosine matrix

Directional Cosine Matrix (DCM) is the fundamental and perhaps the simplest way of attitude representation of a rigid body in three-dimensional space. As shown in Fig. 3.2, it is intuitively obvious that any arbitrary vector V in $\{\mathcal{G}\}$ frame makes three angles with each coordinate axis of it. Direction cosines are cosines of angles between a vector and a base coordinate frame. In other words, it is the projection of a unit vector along the reference frame axis. With a similar approach, now, let us consider the body-fixed frame $\{\mathcal{B}\}$, attached with an arbitrary oriented rigid body quadrotor, the orientation of which is intended to be represented with respect to the $\{\mathcal{G}\}$ frame. Hence, each of the body i.e. $X_\mathcal{B}$, $Y_\mathcal{B}$, $Z_\mathcal{B}$ potentially makes three angles with the axis of the ground frame $\{\mathcal{G}\}$. For example, the components of the unit vector $X_\mathcal{B}$ in the base frame are represented as

$$[< X_\mathcal{G} \cdot X_\mathcal{B} >, < Y_\mathcal{G} \cdot X_\mathcal{B} >, < Z_\mathcal{G} \cdot X_\mathcal{B} >]^T$$

The other components of the body axis $Y_\mathcal{B}$ and $Z_\mathcal{B}$ can also be expressed similarly. The cosine of these angles is collectively represented as the direction cosine matrix, which represents the orientation of the quadrotor with respect to the $\mathcal{G}$ frame. To obtain the orientation of the rigid body with respect to the base frame of reference, it is required to express the direction cosine of each body frame axis. As mentioned above, combining these together, essentially, DCM is formed as a 3×3 rotation matrix T that can transform the body coordinate system into a base reference system.

$$T = \begin{bmatrix} \cos(\mathbf{X}_\mathcal{G}, \mathbf{X}_\mathcal{B}) & \cos(\mathbf{Y}_\mathcal{G}, \mathbf{X}_\mathcal{B}) & \cos(\mathbf{Z}_\mathcal{G}, \mathbf{X}_\mathcal{B}) \\ \cos(\mathbf{X}_\mathcal{G}, \mathbf{Y}_\mathcal{B}) & \cos(\mathbf{Y}_\mathcal{G}, \mathbf{Y}_\mathcal{B}) & \cos(\mathbf{Z}_\mathcal{G}, \mathbf{Y}_\mathcal{B}) \\ \cos(\mathbf{X}_\mathcal{G}, \mathbf{Z}_\mathcal{B}), & \cos(\mathbf{Y}_\mathcal{G}, \mathbf{Z}_\mathcal{B}) & \cos(\mathbf{Z}_\mathcal{G}, \mathbf{Z}_\mathcal{B}) \end{bmatrix} \tag{3.4}$$

It is apparent from the above discussion that the DCM representation uniquely determines the attitude and essentially requires a total of nine parameters. The DCM matrix (also referred to as the rotation matrix) is an important concept in orientation kinematics. If we know the coordinates of an arbitrary vector in the body frame, we can use it to obtain its global coordinates and vice versa.

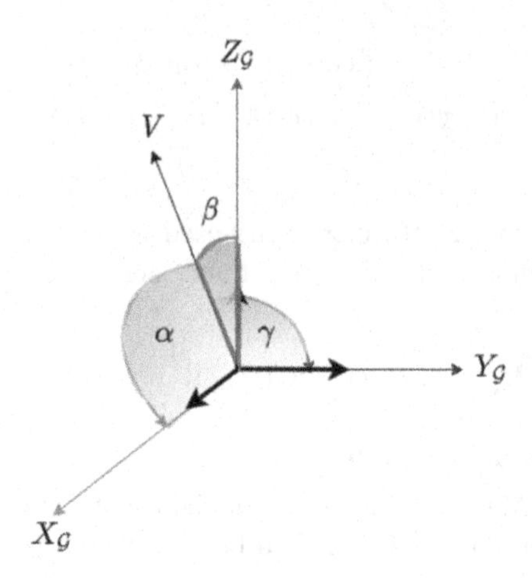

FIGURE 3.2 Schematic representation of vector orientation with respect to DCM angles.

3.3.1.3 Quaternion

A quaternion is an extension of the complex numbers of rank four and belongs to the hyper-complex numbers. Quaternions can be used to represent rotations to avoid the inherent geometrical singularity when representing rigid body dynamics with Euler angles or the complexity of having coupled differential equations with the DCM [5]. There are more than one conventions to represent a quaternion. The two most commonly used are the Hamilton and JPL, where the first one is used in robotics and the latter in the aerospace domain. For the ARW, the Hamilton convention is often used as many libraries, functions, and simulation environments use the same system. A quaternion consists of two parts, the vector or imaginary part q_x to q_z, and the scalar or real part q_w, which are represented by a vector $\boldsymbol{q}$ [6].

$$\boldsymbol{q} = [q_w \; q_x \; q_y \; q_z]^\top \tag{3.5}$$

As it will be required for the definition of the quaternion attitude modeling, initially, some quaternion properties and notations will be presented. For two quaternions $\boldsymbol{q}$ and $\boldsymbol{p}$, while the summation of the two quaternions is commutative, i.e., $\boldsymbol{q}+\boldsymbol{p}=\boldsymbol{p}+\boldsymbol{q}$, the quaternion product, defined as a Kronecker product and denoted with the symbol $\otimes$ is presented in Eq. (3.6), in not commutative $\boldsymbol{q}\otimes\boldsymbol{p}\neq\boldsymbol{p}\otimes\boldsymbol{q}$. However, both the summation and the product are associative, i.e., considering the third quaternion $(\boldsymbol{p}+\boldsymbol{q})+\boldsymbol{r}=\boldsymbol{p}+(\boldsymbol{q}+\boldsymbol{r})$ and $\boldsymbol{r}$, $(\boldsymbol{p}\otimes\boldsymbol{q})\otimes\boldsymbol{r}=\boldsymbol{p}\otimes(\boldsymbol{q}\otimes\boldsymbol{r})$.

$$\boldsymbol{p}\otimes\boldsymbol{q} = \begin{bmatrix} p_0q_w - p_1q_x - p_2q_y - p_3q_z \\ p_0q_x + p_1q_w + p_2q_z - p_3q_y \\ p_0q_y - p_1q_z + p_2q_w + p_3q_x \\ p_0q_z + p_1q_y - p_2q_x + p_3q_w \end{bmatrix} \tag{3.6}$$

Regarding additional quaternion algebra related properties, the norm and conjugate of a quaternion are similar to other complex numbers. More specifically, the conjugate of a quaternion is defined by Eq. (3.7):

$$\boldsymbol{q}^* = \begin{bmatrix} q_w & -q_x & -q_y & -q_z \end{bmatrix}^T \tag{3.7}$$

while the norm of a quaternion utilizing the quaternion product and conjugate properties is defined as shown in Eq. (3.8):

$$\|\boldsymbol{q}\| \triangleq \sqrt{\boldsymbol{q} \otimes \boldsymbol{q}^*} = \sqrt{\boldsymbol{q}^* \otimes \boldsymbol{q}} = \sqrt{q_w^2 + q_x^2 + q_y^2 + q_z^2} \in \mathbb{R}. \tag{3.8}$$

The quaternions used in robotics and, more specifically, in the presented approach have their norm equal to 1 and thus are called unit quaternions. As a result, the inverse of a quaternion normally is represented by Eq. (3.9):

$$\boldsymbol{q}^{-1} = \frac{\boldsymbol{q}^*}{\|\boldsymbol{q}\|}, \tag{3.9}$$

for *unit quaternions* $\boldsymbol{q}^{-1} = \boldsymbol{q}^*$, which is a degenerate form of the inverse quaternion when then quaternion has length one.

To model the attitude dynamics of the ARW, we follow the same notation as in Fig. 3.1. The ARW is considered a symmetrical rigid body. Thus its center of mass coincides with the body frame. In this case, we will utilize the Euler-Newton equations for translational and rotational dynamics defined by Eq. (3.10), where $\boldsymbol{\omega}$ is the angular rate $\boldsymbol{\omega} = [\omega_x \; \omega_y \; \omega_z]^\top$. This approach leaves open the connection between the control signal and the corresponding torques, which can be described by the dynamics of the ARW's rigid body rotations [7].

$$\begin{bmatrix} F \\ \tau \end{bmatrix} = \begin{bmatrix} m & 0 \\ 0 & \boldsymbol{I}_{cm} \end{bmatrix} \begin{bmatrix} a_{cm} \\ \dot{\boldsymbol{\omega}} \end{bmatrix} + \begin{bmatrix} 0 \\ \boldsymbol{\omega} \times (\boldsymbol{I}_{cm} \cdot \boldsymbol{\omega}) \end{bmatrix} \tag{3.10}$$

The time derivative of the unit quaternion $\boldsymbol{q}$ is the vector of quaternion rates $\dot{\boldsymbol{q}}$ that is related to the angular velocity with their relation mapped by [4]:

$$\dot{\boldsymbol{q}}_{\omega}(\boldsymbol{q}, \boldsymbol{\omega}) = \frac{1}{2} \boldsymbol{q} \otimes \begin{bmatrix} 0 \\ \boldsymbol{\omega} \end{bmatrix} \tag{3.11}$$

$$\dot{\boldsymbol{q}}_{\omega'}(\boldsymbol{q}, \boldsymbol{\omega}') = \frac{1}{2} \begin{bmatrix} 0 \\ \boldsymbol{\omega}' \end{bmatrix} \otimes \boldsymbol{q} \tag{3.12}$$

Considering that the angular rates $\boldsymbol{\omega}$ are in the body frame of reference and combining (3.10) and (3.11), the rotation dynamics of the ARW utilizing the quaternion can be described by Eq. (3.13):

$$\begin{cases} \dot{q} = -\frac{1}{2}\begin{bmatrix} 0 \\ \omega \end{bmatrix} \otimes q \\ \dot{\omega} = I_{cm}^{-1} \cdot \tau - I_{cm}^{-1}[\omega \times (I_{cm} \cdot \omega)] \end{cases} \tag{3.13}$$

Lastly, it remains the actual mapping of the control signal to the body frame torques, which can be tackled in various ways. One way would be to take into account the physics that govern the motor-rotor-propeller combination, while a simpler approach, but yet effective, would be to consider a linear relation between torques and angular velocity of the motors, like in (3.24) with the use of a control allocation $\boldsymbol{A}$.

3.3.2 Complete nonlinear model of a quadrotor ARW

To describe the three-dimensional motion of the quadrotor, a mathematical model of a quadrotor is formulated using the Newton and Euler equations [8,9]. The aim of formulating a mathematical model is to provide a reliable dynamic model for simulating and controlling the quadrotor behavior for 6 DOF motion. The linear and angular position vector is given as $[p_x \ p_y \ p_z \ \phi \ \theta \ \psi]^T$, and this position vector is with respect to the global frame. The linear and angular velocity vector is given as $[v_x \ v_y \ v_z \ \omega_x \ \omega_y \ \omega_z]^T$ and the velocity vector is with respect to the body frame of reference. For control purposes, the following transformation is considered to transform the velocities from the body frame to the inertial frame of reference. $V_{\mathcal{B}}$ is the velocity vector in the body frame and $V_{\mathcal{G}}$ is the velocity vector in the global frame of reference. Therefore,

$$V_{\mathcal{G}} = \mathbf{R}_{zyx} \cdot V_{\mathcal{B}} \tag{3.14}$$

$V_{\mathcal{B}} = [v_x \ v_y \ v_z]^T$ is transformed into the velocity in the global frame and is given by $V_{\mathcal{G}} = [\dot{x} \ \dot{y} \ \dot{z}]^T$. In the same way, the angular velocities can be transformed from the body frame to the global frame of reference or:

$$\omega = \mathbf{T} \cdot \omega_{\mathcal{B}} \tag{3.15}$$

where $\omega = [\dot{\phi} \ \dot{\theta} \ \dot{\psi}]^T$ is the vector of the angular velocity in the inertial frame of reference. The Transformation vector T is defined as in Eq. (3.16) [10]:

$$\mathbf{T} = \begin{bmatrix} 1 & s(\phi)t(\theta) & c(\phi)t(\theta) \\ 0 & c(\phi) & -s(\phi) \\ 0 & \frac{s(\phi)}{c(\theta)} & \frac{c(\phi)}{c(\theta)} \end{bmatrix} \tag{3.16}$$

From Newton's second law of motion, the total force acting on the quadrotor's body is represented as:

$$F_B = m(\omega_{\mathcal{B}} \times V_{\mathcal{B}} + \dot{V}_{\mathcal{B}})$$

where $F_{\mathcal{B}} = [F_x \ F_y \ F_z]^T$ is the force vector. The total torque applied to the quadrotor body is also given as,

$$\tau_{\mathcal{B}} = I \cdot \dot{\omega_{\mathcal{B}}} + \omega_{\mathcal{B}} \times (I \cdot \omega_{\mathcal{B}})$$

where $\tau_{\mathcal{B}} = [\tau_x \ \tau_y \ \tau_z]^T$ is the torque vector. **I** is represented as the moment of inertia matrix as:

$$\mathbf{I} = \begin{bmatrix} I_{xx} & 0 & 0 \\ 0 & I_{yy} & 0 \\ 0 & 0 & I_{zz} \end{bmatrix} \tag{3.17}$$

$$T_i = K_T \Omega_i^2, \ i = 1, 2, ..7, 8$$

where $g = 9.81$ m/s^2 is the gravitational acceleration on the Earth, B is the aerodynamic friction coefficient, and $T = \sum T_i$ is the total thrust in the body frame. Considering the Newton's second law of motion in the global coordinate frame, we get:

$$m\dot{V} = \begin{pmatrix} 0 \\ 0 \\ mg \end{pmatrix} - \mathbf{R} \begin{pmatrix} 0 \\ 0 \\ T \end{pmatrix} - BV \tag{3.18}$$

Furthermore, the distance from the rotor hub to the center of gravity of the vehicle is d, and therefore the roll moment of the vehicle about the X axis is given as:

$$\tau_x = dK_T(\Omega_4^2 - \Omega_2^2) \tag{3.19}$$

The pitch moment of the vehicle about Y axis is given as:

$$\tau_y = dK_T(\Omega_3^2 - \Omega_1^2) \tag{3.20}$$

If the coefficient of drag is K_D, then the yaw moment of the vehicle about Z axis is given as:

$$\tau_z = K_D(\Omega_2^2 + \Omega_4^2 - \Omega_1^2 - \Omega_3^2) \tag{3.21}$$

The total thrust T is represented as throttle control. The overall rotor speed at any time instance is represented as:

$$\Omega_r = \Omega_2 + \Omega_4 - \Omega_1 - \Omega_3 \tag{3.22}$$

The total rotational acceleration of the vehicle is described as:

$$\mathbf{I}\dot{\omega} = -\omega \times I\omega + \tau \tag{3.23}$$

Here, ω is the angular velocity vector in the inertial frame of reference. If only the thrust force is considered, in the inertial frame of reference, then the total forces and torque matrix are given as:

$$\begin{pmatrix} T \\ \tau_x \\ \tau_y \\ \tau_z \end{pmatrix} = \begin{pmatrix} -K_T & -K_T & -K_T & -K_T \\ 0 & -dK_T & 0 & dK_T \\ -dK_T & 0 & dK_T & 0 \\ -K_D & K_D & -K_D & K_D \end{pmatrix} \begin{pmatrix} \Omega_1^2 \\ \Omega_2^2 \\ \Omega_3^2 \\ \Omega_4^2 \end{pmatrix} = \mathbf{A} \begin{pmatrix} \Omega_1^2 \\ \Omega_2^2 \\ \Omega_3^2 \\ \Omega_4^2 \end{pmatrix}$$

It is assumed that the Thrust coefficient, K_T, the Drag coefficient, K_D, and the distance from the center of gravity, d, are constants and positive values. The rotor speeds for the input can be computed from the thrust and the torque required values. Therefore, the rotor speeds are formulated as shown in Eq. (3.24):

$$\begin{pmatrix} \Omega_1^2 \\ \Omega_2^2 \\ \Omega_3^2 \\ \Omega_4^2 \end{pmatrix} = \mathbf{A}^{-1} \begin{pmatrix} T \\ \tau_x \\ \tau_y \\ \tau_z \end{pmatrix} \tag{3.24}$$

where **A** is the allocation matrix for the low-level motor mixer. If we want to do the same formulations for a coaxial quadrotor, which has eight rotors mounted as pair of two rotors on each end, then the allocation matrix can be defined in the following manner:

$$\tau_x = dK_T(\Omega_7^2 + \Omega_8^2 - \Omega_3^2 - \Omega_4^2) \tag{3.25}$$

$$\tau_y = dK_T(\Omega_5^2 + \Omega_6^2 - \Omega_1^2 - \Omega_2^2) \tag{3.26}$$

$$\tau_z = K_D(\Omega_2^2 + \Omega_4^2 + \Omega_6^2 + \Omega_8^2 - \Omega_1^2 - \Omega_3^2 - \Omega_5^2 - \Omega_7^2) \tag{3.27}$$

$$\Omega_r = -\Omega_1 + \Omega_2 - \Omega_3 + \Omega_4 - \Omega_5 + \Omega_6 - \Omega_7 + \Omega_8 \tag{3.28}$$

$$\begin{pmatrix} T \\ \tau_x \\ \tau_y \\ \tau_z \end{pmatrix} = \begin{pmatrix} -K_T & -K_T & -K_T & -K_T & -K_T & -K_T & -K_T & -K_T \\ 0 & 0 & -dK_T & -dK_T & 0 & 0 & dK_T & dK_T \\ -dK_T & -dK_T & 0 & 0 & dK_T & dK_T & 0 & 0 \\ -K_D & K_D & -K_D & K_D & -K_D & K_D & -K_D & K_D \end{pmatrix}$$

$$\times \begin{pmatrix} \Omega_1^2 \\ \Omega_2^2 \\ \Omega_3^2 \\ \Omega_4^2 \\ \Omega_5^2 \\ \Omega_6^2 \\ \Omega_7^2 \\ \Omega_8^2 \end{pmatrix}$$

$$= \mathbf{A}\begin{pmatrix} \Omega_1^2 & \Omega_2^2 & \Omega_3^2 & \Omega_4^2 & \Omega_5^2 & \Omega_6^2 & \Omega_7^2 & \Omega_8^2 \end{pmatrix}'$$

$$\begin{pmatrix} \Omega_1^2 & \Omega_2^2 & \Omega_3^2 & \Omega_4^2 & \Omega_5^2 & \Omega_6^2 & \Omega_7^2 & \Omega_8^2 \end{pmatrix}' = \mathbf{A}^{-1}\begin{pmatrix} T & \tau_x & \tau_y & \tau_z \end{pmatrix}' \tag{3.29}$$

Considering the assumptions made for the quadrotor, in this case, the equations of motion can also be formulated using the Newton-Euler concept, where we assume that the origin of the body frame of reference coincides with the center of gravity of the vehicle. The linear acceleration in the X direction is provided as:

$$\dot{v_x} = [\sin\psi \sin\phi + \cos\psi \sin\theta \cos\phi]\frac{T}{m} \tag{3.30}$$

The linear acceleration in Y direction is given as:

$$\dot{v_y} = [-\cos\psi \sin\phi + \sin\psi \sin\theta \cos\phi]\frac{T}{m} \tag{3.31}$$

and the linear acceleration in Z direction is given as:

$$\dot{v_y} = [\sin\theta \cos\phi]\frac{T}{m} - g \tag{3.32}$$

Similarly, the rotational dynamics can be described by the angular acceleration in the roll, pitch, and yaw directions, while the angular accelerations are given as:

$$\dot{\omega_x} = [\omega_y\omega_z(I_{yy} - I_{zz}) - J_r\omega_y\Omega_r + \tau_x]\frac{1}{I_{xx}} \tag{3.33}$$

$$\dot{\omega_y} = [\omega_x\omega_z(I_{zz} - I_{xx}) - J_r\omega_x\Omega_r + \tau_y]\frac{1}{I_{yy}} \tag{3.34}$$

$$\dot{\omega_z} = [\omega_x\omega_y(I_{xx} - I_{yy}) + \tau_z]\frac{1}{I_{zz}} \tag{3.35}$$

In Eqs. (3.33) and (3.34), J_r is the total rotational inertia of the vehicle.

3.3.3 Simplified position modeling

A common practice in ARWs is to utilize one mathematic model to describe the attitude of the rigid body, i.e., roll, pitch, and yaw rotations with respect to the inertia frame, and another to describe its position modeling. While this method is a significant simplification yet results in a pretty accurate model. For the simplified position modeling, the first assumption comes from considering the response of the roll and pitch angles of the ARW that follow the response of first-order systems with a time constant τ and a gain K [11]. The next simplification comes from considering the yaw rotation independent of the system states, which has zero effect on the position and velocity states of the ARW. Under the above assumptions, the states of the nonlinear system of the ARW, considering the body and inertia frames in Fig. 3.1 and for the state vector $\boldsymbol{x} = [p_x, p_y, p_z, v_x, v_y, v_z, \phi, \theta]^\top$, are modeled by (3.36) as:

$$\dot{\boldsymbol{p}}(t) = \boldsymbol{v}(t) \tag{3.36a}$$

$$\dot{\boldsymbol{v}}(t) = \boldsymbol{R}_{x,y}(\theta, \phi) \begin{bmatrix} 0 \\ 0 \\ T_d \end{bmatrix} + \begin{bmatrix} 0 \\ 0 \\ -g \end{bmatrix} - \begin{bmatrix} A_x & 0 & 0 \\ 0 & A_y & 0 \\ 0 & 0 & A_z \end{bmatrix} \boldsymbol{v}(t), \tag{3.36b}$$

$$\dot{\phi}(t) = 1/\tau_\phi (K_\phi \phi_d(t) - \phi(t)), \tag{3.36c}$$

$$\dot{\theta}(t) = 1/\tau_\theta (K_\theta \theta_d(t) - \theta(t)), \tag{3.36d}$$

where $\boldsymbol{p} = [p_x, p_y, p_z]^\top \in \mathbb{R}^3$ is the position vector, and $\boldsymbol{v} = [v_x, v_y, v_z]^\top \in \mathbb{R}^3$ is the vector of the linear velocities, $\phi, \theta \in \mathbb{R} \cap [-\pi, \pi]$ are the roll and pitch angles, respectively, $\boldsymbol{R}_{x,y}$ is the rotation matrix about the x and y axes, $T_d \in [0, 1] \cap \mathbb{R}$ is the mass-normalized thrust, g is the gravitational acceleration, A_x, A_y, and $A_z \in \mathbb{R}$ are the normalized mass drag coefficients. The low-level control system is approximated by the first-order dynamics driven by the reference pitch and roll angles ϕ_d and θ_d with gains of $K_\phi, K_\theta \in \mathbb{R}^+$ and time constants of $\tau_\phi \in \mathbb{R}^+$ and $\tau_\theta \in \mathbb{R}^+$. To retrieve the discrete-time equivalent dynamical system $\boldsymbol{x}_{k+1} = f(\boldsymbol{x}_k, \boldsymbol{u}_k)$ the forward Euler discretization method for a sampling time d_t can be followed.

The system inputs are the thrust T_d, the roll ϕ_d, and pitch θ_d angles. The actual mapping of the inputs can be described by a second mathematical model, which expresses the attitude of the ARW while the mapping of the inputs to specific motor commands that can drive the robot will be undertaken by a secondary low-level controller. The simplified position modeling of the ARW is ideal for nonlinear model-based control schemes and linear ones after the linearization of the state equations around the hovering equilibrium. Examples of such control schemes will be presented in the next Chapter.

3.3.4 ARW foldable quadrotor modeling

Until now, we have presented modeling of the more traditional ARW designs. It is important to note that less traditional and novel ARWs designs can be utilized for accomplishing tasks that would not be possible by the traditional ARWs [12,13]. This section focuses on modeling an ARW designed as a morphing quadrotor [14]. It has to be noted that different morphing aerial platforms require slightly different handling for the modeling based on the structural reformations they follow. Fig. 3.3 presents a conceptual design of such morphing ARW depicted in an isometric view for different formations. The quadrotor's arms are connected to the structural frame of the ARW via 2DoF joints controlled by servo motors. The servo motors are installed so that the arms rotate around the body z-axis. It is essential to highlight that the formation changes of the quadrotor (Fig. 3.4) come directly from the arms rotation around the z-axis of the ARW. Thus, the geometry varies only for the x, y axes resulting in a planar-varying geometry depicted in Fig. 3.5.

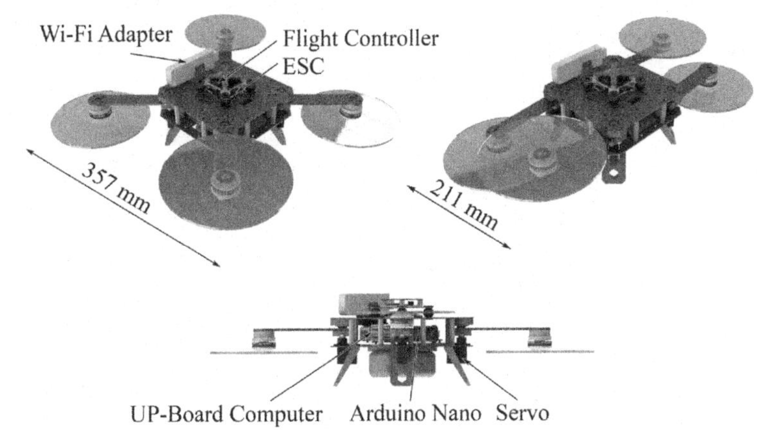

FIGURE 3.3 Isometric view of the morphing ARW design in X and H morphology and a side view displaying the various components.

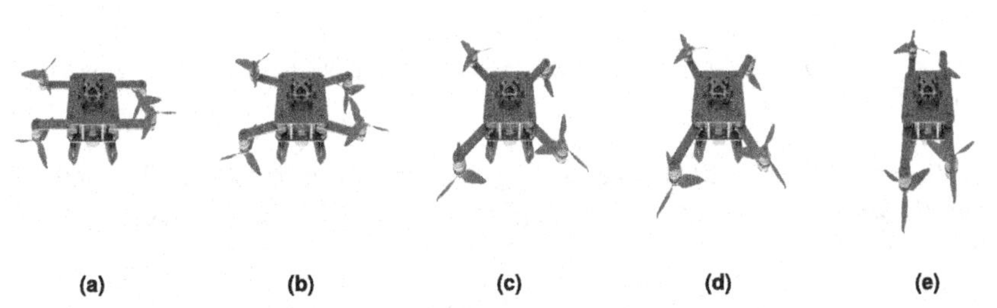

FIGURE 3.4 A reconfigurable ARW, depicted for different configurations, which utilizes four servo motors to fold each arm individually around its main structural frame.

Actuation of the servo motors angle results in the arms updated position with respect to the body frame. This position change affects mainly the Center of Gravity (CoG) and the Moments of Inertia (MoI) of the rigid body. Let the CoG of the rigid body locate at $\boldsymbol{r}_{\mathrm{CoG}} \in \mathbb{R}^3$ distance from the Geometric Center (GC). The offset vector $\boldsymbol{r}_{\mathrm{CoG}}$ is calculated by taking into account every component of the quadrotor CoG position vector $\boldsymbol{r}_{(.)} \in \mathbb{R}^3$. To simplify the calculations and develop a slightly more generic model, we consider the following dominant components that characterize the geometry of the ARW. These are the main body of the quadrotor with mass m_b and dimensions $(2l \times 2w \times 2h)$ denoting the length, width, and height, respectively. The arms located at the four corners of the ARW with mass $m_{a,i}$ and offset $\boldsymbol{r}_{a,i}$, where $i \in \mathbb{Z}[1,4]$ is the identification number of the individual components. Finally, the combination of the motor, rotor, and propeller is considered one component with mass m_c and offset $\boldsymbol{r}_{c,i}$ from the GC and its own CoG.

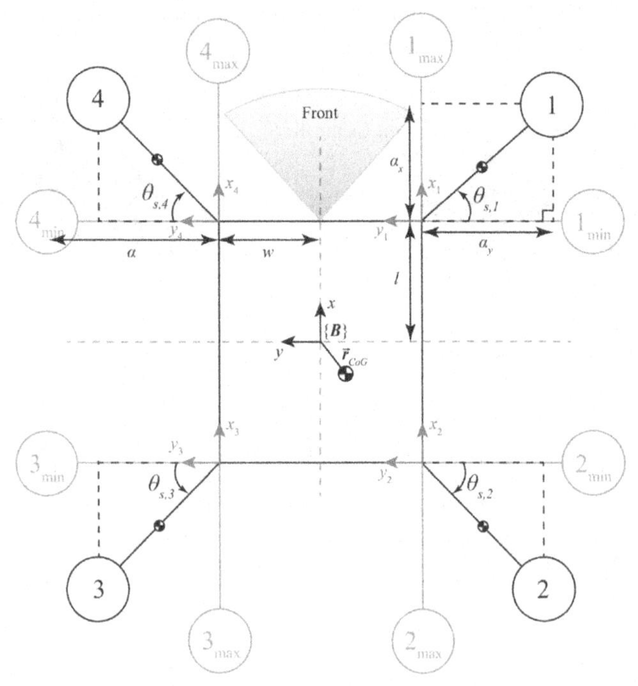

FIGURE 3.5 2D representation of the foldable quadrotor with highlighted the main geometrical properties.

In Fig. 3.5, $\theta_{s,i} \in \mathbb{S}^1$ denote the angles of the servo motors used to actuate the ARW's arms, while the offset $\boldsymbol{r}_{CoG}$ between the GC of the vehicle is defined by (3.37):

$$\boldsymbol{r}_{CoG} = \frac{m_b \boldsymbol{r}_b + \sum_{i=1}^{4}(m_a \boldsymbol{r}_{a,i} + m_c \boldsymbol{r}_{c,i})}{m_b + \sum_{i=1}^{4}(m_a + m_c)} \tag{3.37}$$

The total mass of the ARW remains independent of the arms rotation, and it is defined only by the considered dominant components in (3.38):

$$m = m_b + \sum_{i=1}^{4}(m_a + m_c), \tag{3.38}$$

since the distance of the components' CoG is a function of the servo angles $\theta_{s,i}$, (3.37) can be written as:

$$\boldsymbol{r}_{CoG} = \frac{1}{m}\left(m_b\boldsymbol{r}_b + \sum_{i=1}^{4}(m_a\boldsymbol{r}_{a,i}(\theta_{s,i}) + m_{c,i}\boldsymbol{r}(\theta_{s,i}))\right). \tag{3.39}$$

The distance from the GC to the servo is constant and equal to the dimensions of the body $[w,\ l]^\top$. The offset vectors $\boldsymbol{r}_{(.)}$ can be calculated by the angle $\theta_{s,i}$. The offset on the z-axis of every component to the GC is constant since it does not change when they rotate around z-axis. However, the offsets $\boldsymbol{r}_{.,x}$ and $\boldsymbol{r}_{.,y}$ need to be re-calculated. The angle θ_s of the servo is assumed to be known (Fig. 3.5). The position $(.)_{\min}$ in Fig. 3.5 denotes the minimum angle that the arm can rotate 0°, while the $(.)_{\max}$ denotes the maximum angle 90°. The minimum and maximum rotation angles are defined only due to the physical limitations considering the specific conceptual design of morphing ARW; thus, in a different design we could have different limit angles.

Assuming that the CoG of the motor, rotor, and propeller combination is located in their center, the Euclidean distance of the CoG from the servo is α. Based on the geometrical properties, the new position, where the thrust is generated for every arm, can be calculated by:

$$r_{c,1} = \begin{bmatrix} -w - \alpha\cos\theta_{s,1} - r_{\text{CoG},y}, & l + \alpha\sin\theta_{s,1} - r_{\text{CoG},x}, & z \end{bmatrix}^\top \tag{3.40a}$$

$$r_{c,2} = \begin{bmatrix} -w - \alpha\cos\theta_{s,2} - r_{\text{CoG},y}, & -l - \alpha\sin\theta_{s,2} - r_{\text{CoG},x}, & z \end{bmatrix}^\top \tag{3.40b}$$

$$r_{c,3} = \begin{bmatrix} w + \alpha\cos\theta_{s,3} - r_{\text{CoG},y}, & -l - \alpha\sin\theta_{s,3} - r_{\text{CoG},x}, & z \end{bmatrix}^\top \tag{3.40c}$$

$$r_{c,4} = \begin{bmatrix} w + \alpha\cos\theta_{s,4} - r_{\text{CoG},y}, & l + \alpha\sin\theta_{s,4} - r_{\text{CoG},x}, & z \end{bmatrix}^\top \tag{3.40d}$$

The varying positions of the arms result in a varying control allocation since it affects the position of the motors and directly the torques around the x, y-axis.

$$\begin{bmatrix} T \\ \tau_x \\ \tau_y \\ \tau_z \end{bmatrix} = \begin{bmatrix} K_T & K_T & K_T & K_T \\ r_{c,1_x} & r_{c,2_x} & r_{c,3_x} & r_{c,4_x} \\ r_{c,1_y} & r_{c,2_y} & r_{c,3_y} & r_{c,4_y} \\ -K_D & K_D & -K_D & K_D \end{bmatrix} \begin{bmatrix} f_1 \\ f_2 \\ f_3 \\ f_4 \end{bmatrix} \tag{3.41}$$

where r_{c,i_x} and r_{c,i_y} denote the first and second element of the $\boldsymbol{r}_{c,i}$, respectively. Finally, K_T and K_D are coefficients related to the thrust and torque, respectively.

In addition, the MoI of the platform varies as the arms change their position. Each component is characterized by its own MoI matrix. The total MoI of the platform can be calculated from the individual MoI with the use of the parallel axis theorem at the body frame $\mathcal{B}$.

$$(\boldsymbol{I}_{ARW})_{\mathcal{B}} = (\boldsymbol{I}_b)_{\mathcal{B}} + \sum (\boldsymbol{I}_{a_i})_{\mathcal{B}} + \sum (\boldsymbol{I}_{m_c})_{\mathcal{B}} \tag{3.42}$$

At this point, we again need to decide on the derivation of the MoI numerical values. One approach would be to consider generalized shapes, like a cylinder, cube, rectangular prism, etc., dominant components, which then could be integrated in (3.42). On the other hand, if the developed design comes out, CAD software would be to utilize the values computed through the detailed design of the ARW.

The attitude model of the morphing quadrotor can be defined by the Newton-Euler law using the angular acceleration, rate, and torques in (3.43).

$$\dot{\boldsymbol{\omega}} = \boldsymbol{I}^{-1}(-\boldsymbol{\omega} \times \boldsymbol{I}\boldsymbol{\omega} + \boldsymbol{\tau}), \tag{3.43}$$

where we consider the inertia matrix with zero off-diagonal elements in (3.44) and updated on the basis of the arms position and as defined by Eq. (3.42).

$$(\boldsymbol{I}_{ARW})_B = \boldsymbol{I} = \begin{bmatrix} I_{xx} & I_{yy} & I_{zz} \end{bmatrix} \boldsymbol{I}_3 \tag{3.44}$$

For the torque dynamics, we can consider a linear approximation and assume that they follow the same dynamics as the first-order system.

$$\dot{\tau} = \frac{1}{\tau_\alpha}(\tau_d - \tau) \tag{3.45}$$

where τ_α is the time constant, and τ_d is the desired torque. Thus, the system of equations in (3.46) describes the attitude model of the morphing ARW with variable inertia matrix and varying control allocation matrix.

$$\begin{cases} \dot{\boldsymbol{\omega}} = \boldsymbol{I}^{-1} \cdot \boldsymbol{\tau} - \boldsymbol{I}^{-1} [\boldsymbol{\omega} \times (\boldsymbol{I} \cdot \boldsymbol{\omega})] \\ \dot{\tau} = \frac{1}{\tau_\alpha}(\tau_d - \tau) \end{cases} \tag{3.46}$$

This model is related only to the attitude of the ARW. Thus there is a need to use an additional model in conjunction with this one which would describe the position of the drone. Such a model can be the one presented in Section 3.3.3.

3.4 Conclusions

This Chapter presented the necessary elements for constructing various mathematical models of ARWs to describe their dynamical motion in the perspective of combined translational and rotational maneuvers. Different approaches to attitude representations have been presented, including the Euler angle, DCM, and quaternion. Firstly, a general nonlinear model of ARW was presented describing the rigid body dynamics with its variable components, followed by a simplified ARW modeling. The various simplifications and approximations, used in these modeling derivations, were discussed from the accuracy and convenience point of view. Finally, modeling of a special class of ARW, i.e., morphing quadrotors, has also been presented.

References

[1] R.W. Beard, T.W. McLain, Small Unmanned Aircraft: Theory and Practice, 2012.
[2] R. Featherstone, Rigid Body Dynamics Algorithms, 2008.
[3] F. Sabatino, Quadrotor Control: Modeling, Nonlinear Control Design, and Simulation, 2015.
[4] J. Diebel, Representing attitude: Euler angles, unit quaternions, and rotation vectors, Matrix 58 (2006).
[5] J. Solà, Quaternion kinematics for the error-state Kalman filter, CoRR, arXiv:1711.02508, 2017 [Online].
[6] J.B. Kuipers, Quaternions and Rotation Sequences, 2020.
[7] E. Fresk, G. Nikolakopoulos, Full quaternion based attitude control for a quadrotor, in: 2013 European Control Conference (ECC), 2013, pp. 3864–3869.
[8] P. Corke, Robotics, Vision and Control – Fundamental Algorithms in MATLAB®, second, completely revised, extended and updated edition, vol. 75, 2017.
[9] R. Mahony, V. Kumar, P. Corke, Multirotor aerial vehicles: Modeling, estimation, and control of quadrotor, IEEE Robotics & Automation Magazine 19 (2012).
[10] D. Lee, T.C. Burg, D.M. Dawson, D. Shu, B. Xian, E. Tatlicioglu, Robust tracking control of an underactuated quadrotor aerial-robot based on a parametric uncertain model, in: 2009 IEEE International Conference on Systems, Man and Cybernetics, IEEE, 2009, pp. 3187–3192.
[11] M. Kamel, T. Stastny, K. Alexis, R. Siegwart, Model predictive control for trajectory tracking of unmanned aerial vehicles using robot operating system, in: Robot Operating System (ROS), Springer, 2017, pp. 3–39.
[12] D. Falanga, K. Kleber, S. Mintchev, D. Floreano, D. Scaramuzza, The foldable drone: A morphing quadrotor that can squeeze and fly, IEEE Robotics and Automation Letters 4 (2) (2019) 209–216.
[13] M. Zhao, T. Anzai, F. Shi, X. Chen, K. Okada, M. Inaba, Design, modeling, and control of an aerial robot dragon: A dual-rotor-embedded multilink robot with the ability of multi-degree-of-freedom aerial transformation, IEEE Robotics and Automation Letters 3 (2) (2018) 1176–1183.
[14] A. Papadimitriou, G. Nikolakopoulos, Switching model predictive control for online structural reformations of a foldable quadrotor, in: IECON 2020 The 46th Annual Conference of the IEEE Industrial Electronics Society, 2020, pp. 682–687.

Chapter 4

Control of ARWs

Akash Patel, Avijit Banerjee, Andreas Papadimitriou, and Björn Lindqvist
Department of Computer, Electrical and Space Engineering, Luleå University of Technology, Luleå, Sweden

4.1 Introduction

The control of an Aerial Robotic Worker ARW refers to the set of tools required to track the desired states, defined as a reference trajectory or individual waypoints. As defined in the modeling Chapter 3, the states can be the position, velocity, and attitude of the platform. In general, there is a possibility of using different types of controllers, but also the control strategies can vary from ARW to ARW or for different application scenarios. For example, as depicted in Fig. 4.1, one case would be to use a single controller with the reference state $\boldsymbol{x}_{\text{ref}}^{\top}$ set by a path planner or an operator, and the controller would provide the necessary commands for the motors $\boldsymbol{n}^{\top}$ to drive the aerial platform to the desired states. Another method would be to split the controllers as depicted in Fig. 4.2. In that case, there are two separate control loops. The position controller, which provides the control input $\boldsymbol{u}$ that is responsible for achieving a specific position or velocity. At the same time, the attitude controller objective is to follow the desired angles and thrust provided by the position controller and give the motor commands $\boldsymbol{n}^{\top}$.

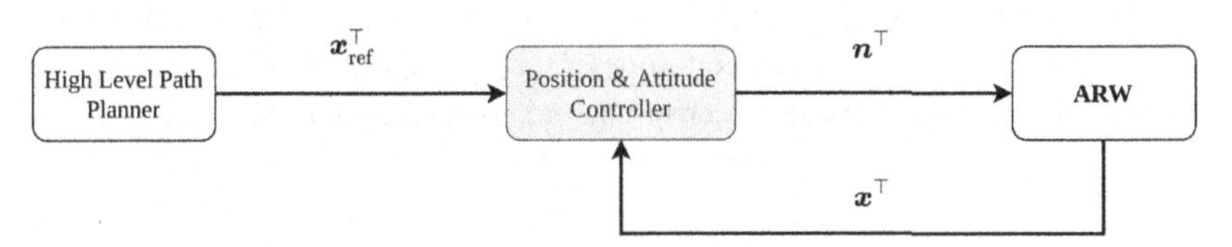

FIGURE 4.1 ARW platform control block diagram with a single controller.

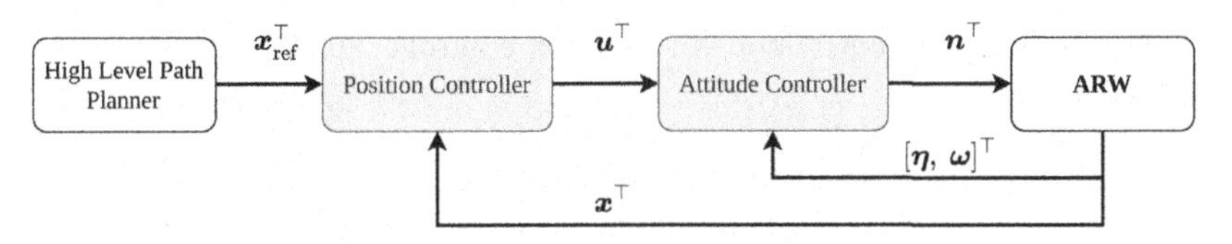

FIGURE 4.2 ARW platform control block with a position and attitude controller.

Aerial Robotic Workers. https://doi.org/10.1016/B978-0-12-814909-6.00010-X

4.2 PID control

Proportional integral derivative (PID) is the most common and simplified controller due to its general applicability for a wide range of applications. In practice, the PID control design does not explicitly require a mathematical system model. Therefore, PID control design proved to be most useful; even for a few critical systems, analytical control design methods are challenging due to a complicated or inaccurate mathematical model (obtaining a precise mathematical model is difficult). The PID controller essentially constructed based on the function of the error feedback which is mathematically expressed as:

$$u(t) = K_p e(t) + K_i \int e(t)dt + K_d \frac{de}{dt} \tag{4.1}$$

The PID controller uses the *error* value between the desired state and the measured state and applies a correction based on the three terms. In (4.1) K_p, K_d, and K_i are the proportional, integral, and derivative gains, which are typically design tuning parameters for the control design.

Assuming the model defined in the previous chapter in modeling for ARWs to control the position of the platform, there are going to be two control loops. The outer loop will be the position controller while the inner loop will be attitude controller. The control law of the outer loop minimizes the error between the current and the desired position of the platform. Starting from the inner loop, the goal is to regulate the roll/pitch/yaw angles and the altitude of the platform. It is important to highlight that the aerial platform attitude controller is often driven by an IMU sensor. The IMU due to various reasons introduces drifts to the measurements thus to prevent instability issues, the integral component of the PID is not used, and, instead, only a PD controller is used in practice. As shown in the modeling section, the rotational speeds of the motors are related with the thrust and torques through a static map. Due to that, the control output of the PID controller are the three torques τ_x, τ_y, τ_z and the thrust T.

Starting with the altitude, the error and the error rate are defined as:

$$\begin{aligned} p_{ze} &= p_{zd} - p_z \\ p_{\dot{z}e} &= p_{\dot{z}d} - \dot{p}_z \end{aligned} \tag{4.2}$$

Similarly for the outer loop XY position controller, the error rate can be defined as:

$$\begin{aligned} p_{xe} &= p_{xd} - p_x \\ p_{\dot{x}}e &= p_{\dot{x}d} - \dot{p}_x \\ p_{ye} &= p_{yd} - p_y \\ p_{\dot{y}}e &= p_{\dot{y}d} - \dot{p}_y \end{aligned} \tag{4.3}$$

And, therefore, a simple XY position PID controller can be formulated as:

$$\begin{aligned} p_x &= K_{p,x} p_{xe} + K_{d,x} \dot{p_{xe}} \\ p_y &= K_{p,y} p_{ye} + K_{d,y} \dot{p_{ye}} \end{aligned} \tag{4.4}$$

Thus, the Thrust,

$$T = K_{p,z} p_{ze} + K_{d,z} \dot{p_{ze}} + T_0 \tag{4.5}$$

where T_0 is a feed-forward term depending on the mass m of the platform and the gravitational coefficient g where $T_0 = mg$. Similarly, the errors of the angles ϕ_e, θ_e, and ψ_e are calculated.

$$\begin{aligned} \tau_x &= K_{p,\tau_x} \phi_e + K_{d,\tau_x} \dot{\phi}_e \\ \tau_y &= K_{p,\tau_y} \theta_e + K_{d,\tau_y} \dot{\theta}_e \\ \tau_z &= K_{p,\tau_x} \psi_e + K_{d,\tau_z} \dot{\psi}_e \end{aligned} \tag{4.6}$$

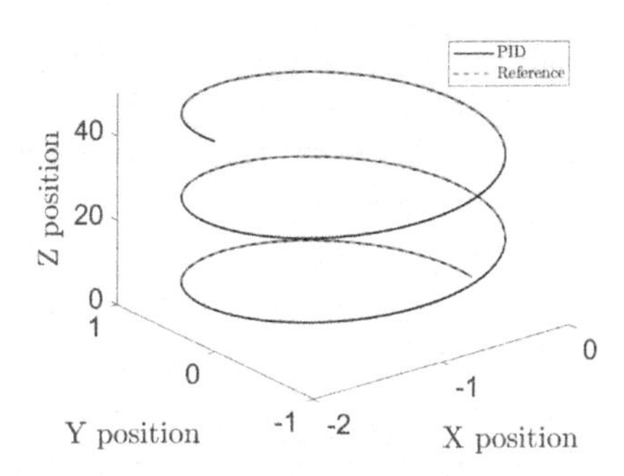

FIGURE 4.3 PID response on following a spiral trajectory.

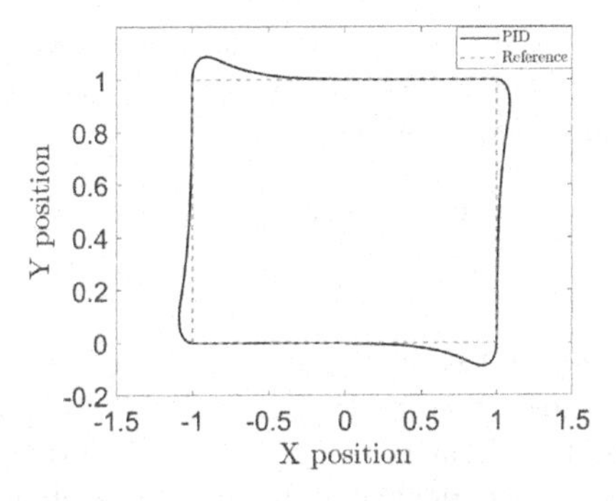

FIGURE 4.4 PID response on following a square trajectory.

In Fig. 4.3 and Fig. 4.4, two trajectory following responses are presented using a PID controller as formulated above. As from the definition of using a

PD controller, it is evident that on corners while following square trajectory, the quadrotor overshoots from the reference trajectory and slowly starts following the trajectory again as the error in position becomes smaller. Such behavior will be different in a Model Predictive Controller based control scheme, as it will be presented in the next section.

4.3 Linearized quadrotor model

The quadrotor dynamic model discussed in previous chapter is highly nonlinear. In order to design a linear Model Predictive Controller for the system, the model needs to be linearized around a reference point. In this case, the reference point considered is at near hover condition in equilibrium. Therefore, the following small angle approximations are valid for the near hover situation.

$$\omega_x \cong \omega_y \cong \omega_z \cong 0$$

$$\sin\phi \cong \phi; \ \sin\theta \cong \theta; \ \sin\psi \cong 0$$

In the near hover condition, the Yaw angle ψ and the Yaw rate $\dot{\psi}$ are considered to be nearly zero. Therefore, the positions in p_x and p_y are solely dependent on θ and ϕ, respectively. Considering the near hover condition, the nonlinear model of the Quadrotor can be simplified as:

$$\dot{v_x} = g\theta; \ \dot{v_y} = -g\phi; \ \dot{v_z} = \frac{T}{m}$$

$$\dot{\omega_x} = \frac{\tau_x}{I_{xx}}; \ \dot{\omega_y} = \frac{\tau_y}{I_{yy}}; \ \dot{\omega_z} = \frac{\tau_z}{I_{zz}}$$

The State Space representation of the Quadrotor system can be interpreted as a set of input, output, and state variables that are related by first-order differential equations. In the state space representation, the state space is the space that consists of all the state variables at its axis. The generic state space representation of the linear system is given as:

$$\dot{x}(t) = \mathbf{A}x(t) + \mathbf{B}u(t) \tag{4.7}$$

$$Y(t) = \mathbf{C}x(t) + \mathbf{D}u(t) \tag{4.8}$$

In Eqs. (4.7) and (4.8), $x(t)$ is the 'State Vector', $y(t)$ is the 'Output Vector', $u(t)$ is the 'Input Vector', $\mathbf{A}$ is the 'System Matrix', $\mathbf{B}$ is the 'Input Matrix', $\mathbf{C}$ is the 'Output Matrix', and $\mathbf{D}$ is the 'Feed forward Matrix'. The vector $[p_x \ p_y \ p_z \ \phi \ \theta \ \psi]$ represents the linear and angular positions of the Quadrotor in the World frame of reference. Whereas, the vector $[v_x \ v_x \ v_x \ \omega_x \ \omega_y \ \omega_z]$ represents the linear and angular velocities in the body-fixed frame of the reference. The linearized state space model can be derived at the equilibrium condition, where $x = x_0$ and $u = u_0$. x_0 and u_0 represent the reference values at the near hover condition. The mathematical model of the Quadrotor used for this chapter is an actuator-based model [1]. In order to make the system more realistic, the control inputs for the proposed mathematical model are the rotor speeds, instead of the corresponding force or torque. In order to linearize the proposed model around the near hover point, the A and B matrices for the state space representation are derived below:

$$\mathbf{A} = \left(\frac{\partial F(x,U)}{\partial x}\right)_{x_0,u_0} \quad ; \quad \mathbf{B} = \left(\frac{\partial F(x,U)}{\partial u}\right)_{x_0,u_0}$$

Thus, the state space equations for the Quadrotor system are mentioned below:

$$\overbrace{\begin{bmatrix} v_x \\ \dot{v_x} \\ v_y \\ \dot{v_y} \\ v_z \\ \dot{v_z} \\ \omega_x \\ \dot{\omega_x} \\ \omega_y \\ \dot{\omega_y} \\ \omega_z \\ \dot{\omega_z} \end{bmatrix}}^{\dot{x}(t)} = \overbrace{\begin{bmatrix} 0&1&0&0&0&0&0&0&0&0&0&0 \\ 0&0&0&0&0&0&0&0&g&0&0&0 \\ 0&0&0&1&0&0&0&0&0&0&0&0 \\ 0&0&0&0&0&0&-g&0&0&0&0&0 \\ 0&0&0&0&0&1&0&0&0&0&0&0 \\ 0&0&0&0&0&0&0&0&0&0&0&0 \\ 0&0&0&0&0&0&0&1&0&0&0&0 \\ 0&0&0&0&0&0&0&0&0&0&0&0 \\ 0&0&0&0&0&0&0&0&0&1&0&0 \\ 0&0&0&0&0&0&0&0&0&0&0&0 \\ 0&0&0&0&0&0&0&0&0&0&0&1 \\ 0&0&0&0&0&0&0&0&0&0&0&0 \end{bmatrix}}^{\mathbf{A}} \overbrace{\begin{bmatrix} p_x \\ v_x \\ p_y \\ v_y \\ p_z \\ v_z \\ \phi \\ \omega_x \\ \theta \\ \omega_y \\ \psi \\ \omega_z \end{bmatrix}}^{x(t)} + \overbrace{\begin{bmatrix} 0&0&0&0 \\ 0&0&0&0 \\ 0&0&0&0 \\ 0&0&0&0 \\ 0&0&0&0 \\ \frac{K_T}{m}&\frac{K_T}{m}&\frac{K_T}{m}&\frac{K_T}{m} \\ 0&0&0&0 \\ 0&-\frac{dK_T}{I_{xx}}&0&\frac{dK_T}{I_{xx}} \\ 0&0&0&0 \\ -\frac{dK_T}{I_{yy}}&0&\frac{dK_T}{I_{yy}}&0 \\ 0&0&0&0 \\ -\frac{K_D}{I_{zz}}&\frac{K_D}{I_{zz}}&-\frac{K_D}{I_{zz}}&\frac{K_D}{I_{zz}} \end{bmatrix}}^{\mathbf{B}} \overbrace{\begin{bmatrix} \Omega_1^2 \\ \Omega_2^2 \\ \Omega_3^2 \\ \Omega_4^2 \end{bmatrix}}^{u(t)}$$

Since the Quadrotor control system is designed for a position and heading controller, the output variables are $[x \ y \ z \ \psi]$, and Eq. (4.8) can be written as:

$$\overbrace{\begin{bmatrix} x \\ y \\ z \\ \psi \end{bmatrix}}^{y(t)} = \overbrace{\begin{bmatrix} 1&0&0&0&0&0&0&0&0&0&0&0 \\ 0&0&1&0&0&0&0&0&0&0&0&0 \\ 0&0&0&0&1&0&0&0&0&0&0&0 \\ 0&0&0&0&0&0&0&0&0&0&1&0 \end{bmatrix}}^{\mathbf{C}} \overbrace{\begin{bmatrix} p_x \\ v_x \\ p_y \\ v_y \\ p_z \\ v_z \\ \phi \\ \omega_x \\ \theta \\ \omega_y \\ \psi \\ \omega_z \end{bmatrix}}^{x(t)} + \overbrace{\begin{bmatrix} 0&0&0&0 \\ 0&0&0&0 \\ 0&0&0&0 \\ 0&0&0&0 \end{bmatrix}}^{\mathbf{D}} \overbrace{\begin{bmatrix} \Omega_1^2 \\ \Omega_2^2 \\ \Omega_3^2 \\ \Omega_4^2 \end{bmatrix}}^{u(t)}$$

4.4 LQR control

The Linear Quadratic Regulator (LQR) is a special class of optimal control approach, which considers linear system dynamics with quadratic cost function. In practice, it is quite often observed that a meaningful performance indices can be formulated in the quadratic form. Hence, the LQR plays a crucial role in understanding and approaching control related problems arising for ARWs systems.

As described in Section 4.3, the linearized dynamics for the aerial platform can be written in state space form as:

$$\dot{\boldsymbol{x}}(t) = \boldsymbol{A}\boldsymbol{x}(t) + \boldsymbol{B}\boldsymbol{u}(t) \tag{4.9}$$

$$\boldsymbol{y}(t) = \boldsymbol{C}\boldsymbol{x}(t) \tag{4.10}$$

The system dynamics considers 12 states construed of $3-D$ position, velocity, attitude, and angular velocities. Descriptions of the system matrices can be found in Eqs. (4.7)–(4.8). In principle, LQR essentially provides a state feedback control mechanism $\boldsymbol{u}(t) = -\boldsymbol{K}\boldsymbol{x}(t)$, where the feedback gain $\boldsymbol{K}$ is optimally designed based on minimization of an performance index presented as:

$$J = \int_0^\infty \left(x^T Q x + u^T R u\right) dt \tag{4.11}$$

The interesting fact about the LQR formulation is that it provides a closed form solution of the optimal state feedback gain presented as:

$$K = \left(R^{-1} B^T P\right) \tag{4.12}$$

where the P is the solution to the following algebraic Riccati Equation [2]:

$$PA + A^T P - PBR^{-1}B^T P + Q = 0 \tag{4.13}$$

Initially, in a MATLAB® system implementation, the optimal state feedback gain can be obtained using the command *lqr(A,B,Q,R)*. In this chapter, since we are interested in establishing a track following control mechanism for ARW system, we have introduced an error minimization factor in the cost function as follows:

$$J = \int_0^\infty \left(e^T Q e + u^T R u\right) dt \tag{4.14}$$

where $e(t) = z(t) - Cx(t)$ represents the error between the commanded signal $z(t)$ and the system output. A more detailed derivation for the command tracking LQR can be found in [3]. In the present context of track following controller design, the performance of the LQR controller to follow different trajectories are presented in Figs. 4.5(a) and 4.5(b). However, it should be noted that even though the LQR provides an optimal state feedback control, it is unable to address necessary constraints associated with the control magnitude and other constraints related to the system state. These types of additional constraints can better be addressed by introducing a more advanced optimal control design such as a Model predictive controller, which will be discussed in the following section.

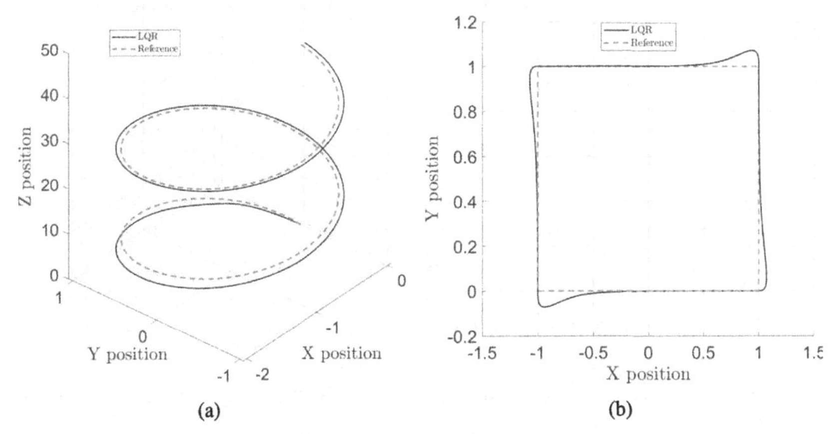

FIGURE 4.5 (a) Tracking performance of the LQR controller following a reference spiral trajectory; (b) the performance of LQR controller on tracking a square trajectory.

4.5 Linear model predictive control

Model Predictive Control, also known as Receding Horizon Control (RHC), uses the mathematical model of the system in order to solve a finite, moving horizon, and closed loop optimal control problem [4]. Thus, the MPC scheme is able to utilize the information about the current state of the system in order to predict future states and control inputs for the system [5]. The main advantage of using the MPC scheme is the ability to take into account the physical limitations of the system, as well as the external noise. Due to these considerations in the design process, the controller is able to predict the future output of the system and formulate an optimal control effort that brings the system to the desired state for a predefined reference trajectory [6]. In the controller design, the ability to introduce constrains allows the control inputs and output to be in a predefined bound for the desired performance. In general, the state and input constrains are set in relation to the application profile of the control scheme [7]. In the MPC design, the control signals need to be computed online for every sampling time, while considering the constrains at the same time. Forward shift operators are used in the conventional MPC design to optimize future control signals [8]. For a complex dynamic system, a large prediction and control horizon is required, which results in increased online computations. There have been studies that propose methods to decrease the factors affecting online computation as discussed in [9].

The optimal control problem is solved at each sampling time interval for the predefined prediction horizon. The first step of the solution of the optimization problem is applied to the system until the next sampling time [10]. This process can be repeated over and over in order to correctly compute the control inputs for the system to reach the desired state. The earlier discussed nonlinear mathematical model of the Quadrotor is linearized in order to design a linear MPC. Therefore, the equivalent discretized linear system for a sampling time T_s can be described as:

$$\Delta x_{k+1} = \mathbf{A}\Delta x_k + \mathbf{B}\Delta u_k$$

$$\Delta y_k = \mathbf{C}\Delta x_k$$

where, $\Delta x_k = x_k - x_T$ and $\Delta u_k = u_k - u_T$, and k is the current sample. In the quadrotor model, the state matrix $\mathbf{A} \in \mathbb{R}^{12\times 12}$, the input matrix $\mathbf{B} \in \mathbb{R}^{12\times 4}$, and the $\mathbf{C} \in \mathbb{R}^{4\times 12}$ output matrix are as described in the state space representation of the linearized system. The reference point is considered at near hover condition, while the nominal control input will be $u = \frac{mg}{4}[1, 1, \ldots, 1]_{4\times 1}$. Based on the linear state space model, the designed controller predicts the future states as a function of current state and future control inputs. Therefore, for $i = 1, 2, 3, ..., N$,

$$\Delta x_{k+i+1} = \mathbf{A}\Delta x_{k+i} + \mathbf{B}\Delta u_{k+i}$$

$$\Delta y_{k+i} = \mathbf{C}\Delta x_{k+i} + \mathbf{D}\Delta u_{k+i}$$

$$\Delta X_k = \mathbf{G}\Delta x_k + \mathbf{H}\Delta u_k$$

$$\Delta Y_k = \bar{\mathbf{C}}\Delta X_k + \bar{\mathbf{D}}\Delta U_k \tag{4.15}$$

Eq. (4.15) becomes the prediction equation in which

$$\Delta X_k = \begin{bmatrix} \Delta x_k \\ \Delta x_{k+1} \\ \Delta x_{k+2} \\ \vdots \\ \Delta x_{k+N-1} \end{bmatrix}; \Delta U_k = \begin{bmatrix} \Delta u_k \\ \Delta u_{k+1} \\ \Delta u_{k+2} \\ \vdots \\ \Delta u_{k+N-1} \end{bmatrix}; \Delta Y_k = \begin{bmatrix} \Delta y_k \\ \Delta y_{k+1} \\ \Delta y_{k+2} \\ \vdots \\ \Delta y_{k+N-1} \end{bmatrix}$$

$$\mathbf{G} = \begin{bmatrix} I \\ A \\ A^2 \\ \vdots \\ A^{N-1} \end{bmatrix}; \mathbf{H} = \begin{bmatrix} 0 & & & & \\ B & 0 & & & \\ AB & B & 0 & & \\ \vdots & \vdots & \ddots & \ddots & \\ A^{N-2}B & A^{N-3}B & \ldots & B & 0 \end{bmatrix}$$

$$\bar{\mathbf{C}} = \begin{bmatrix} C & & & \\ & C & & \\ & & C & \\ & & & \ddots & \\ & & & & C \end{bmatrix}; \bar{\mathbf{D}} = \begin{bmatrix} D & & & \\ & D & & \\ & & D & \\ & & & \ddots & \\ & & & & D \end{bmatrix}$$

In order to choose the optimal control input at each sampling time interval, a cost function formulation is considered for the MPC design. The aim of the cost function formulation is to minimize the control effort, while driving the predicted outputs to a reference trajectory. In the proposed controller design, the cost function penalizes the error in the reference trajectory and the current output state, as well as the control inputs.

$$\begin{aligned} Q(\Delta x_k, \Delta U_k) = & (\Delta x_k^{ref} - \Delta x_k)^T \bar{M}_x (\Delta x_k^{ref} - \Delta x_k) \\ & + (\Delta u_k^{ref} - \Delta u_k)^T \bar{M}_u (\Delta u_k^{ref} - \Delta u_k) \\ & + (\Delta u_k - \Delta u_{k-1})^T \bar{M}_{\Delta u} (\Delta uk - \Delta u_{k-1}) \end{aligned}$$

where $\bar{M}_x$, $\bar{M}_u$, and $\bar{M}_{\Delta u}$ are the weighing matrices for the states, the control inputs, and the rate of change in the control inputs. The linear MPC is validated using the linearized plant model of the Quadrotor mathematical model. The linearized model uses the state space form of linear dynamics, where a near hover condition is considered the reference point. Once the MPC design is validated, the same controller is used with the non-linear model of Quadrotor. The closed-loop control architecture is shown in Fig. 4.6. The term $\zeta_{ref} = [x \; y \; z \; \psi]^T$ represents reference trajectory to follow.

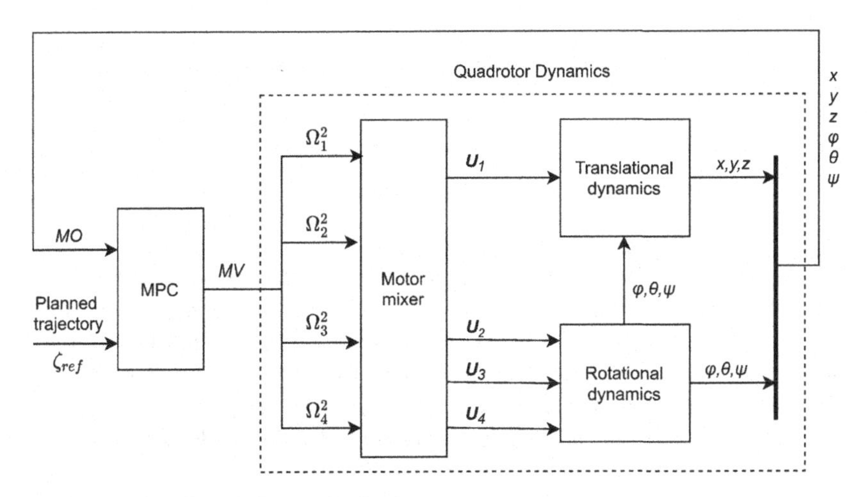

FIGURE 4.6 MPC block diagram for Quadrotor.

$$\min_{u_k,\ldots,u_{k+N-1}} \left\{ \sum_{j=0}^{N-1} \left\| \boldsymbol{x}_{k+j} - \boldsymbol{x}_r(k) \right\|_{Q_x}^2 + \left\| u_{k+j} - u_r(k) \right\|_{Q_u}^2 \right) \right\} \tag{4.16}$$

$$\text{s.t.: } x_{k+j+1} = f\left(x_{k+j}, u_{k+j}\right)$$
$$u_{\min} \leq u_{k+j} \leq u_{\max}$$

In Fig. 4.6, MO represents Measured Outputs, MV represents manipulated variables, U_1, U_2, U_3, and U_4 are Thrust, Roll, Pitch, and Yaw torques.

In Fig. 4.7 and Fig. 4.8, the MPC responses to the following different trajectories are shown. As in the definition of Model Predictive Control, the formu-

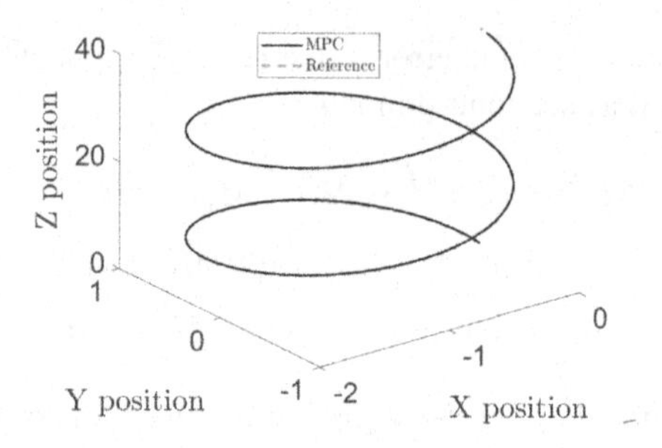

FIGURE 4.7 MPC response on following a spiral trajectory.

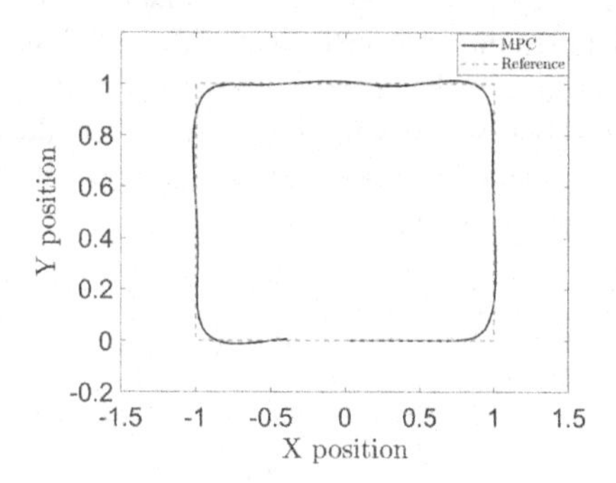

FIGURE 4.8 MPC response on following a square trajectory.

lated controller predicts the future states of the system and updates the control inputs, respectively. This behavior of MPC is clearly evident in Fig. 4.8 while comparing the same square trajectory scenario with a PID-based controller response as shown in Fig. 4.4. The difference between the two plots is that the MPC-based controller is able to predict the sharp turn at corners while following s square trajectory, and corresponding control inputs make the quadrotor turn before overshooting in such scenarios. Where, as in PID-based controller as shown in Fig. 4.4, the controller overshoots on corners on the following a square trajectory.

4.6 Nonlinear MPC

The nonlinear MPC is a model-based control framework that calculates control actions at each control interval using a combination of model-based prediction and constrained optimization. While it is very similar to the linear MPC version, it has several additional features. The NMPC model can use a nonlinear prediction model, while the various equality and inequality constraints can be either linear or nonlinear, and it can accept cost functions that are nonquadratic. While the advantages are evident over the linear MPC, a significant trade-off is the computation burden and the limited resources of the ARW onboard computation unit, which luckily will be mitigated as the computers are continuously improving, and they will be able to handle more complex problems.

To reduce the complexity of the nonlinear problem for an ARW control, instead of using the full nonlinear model and solving directly for torques or motor speed, it is possible to utilize a simplified position modeling like the one presented in Section 3.3.3 of Chapter 3. Subsequently, we need to assume that the ARW is equipped with an attitude controller that accepts thrust, roll, pitch, and yaw-rate references (note: the yaw angle can be controlled by a simple decoupled PD controller). The control scheme will have a very intuitive cascaded structure like the one presented in Fig. 4.2, and most commercial drones are equipped with such an attitude controller, as those control inputs are the commonly accepted commands for remote operation with an RC (roll and pitch references to move left/right, thrust for up/down, and yaw-rate to rotate).

Using the model presented in Section 3.3.3 of Chapter 3 in (3.36), let the state vector be denoted by $\boldsymbol{x} = [\boldsymbol{p}, \boldsymbol{v}, \phi, \theta]^\top$, and the control action by $\boldsymbol{u} = [T_d, \phi_d, \theta_d]^\top$. The system dynamics of the ARW is discretized with a sampling time d_t using the forward Euler method to obtain

$$x_{k+1} = f(x_k, u_k). \tag{4.17}$$

This discrete model is used as the *prediction model* of the NMPC. This prediction is made with a receding horizon, e.g., the prediction considers a set number of steps into the future. We denote this as the *prediction horizon*, $N \in \mathbb{N}$, of the NMPC. In some applications, the *control horizon* is distinct from N, but in this chapter, we will only consider the case where they are the same, meaning both control inputs and predicted states are computed in the same horizon without loss of generality. By associating a cost to a configuration of states and inputs at the current time and in the prediction, a nonlinear optimizer can be tasked with finding an optimal set of control actions, defined by the cost minimum of this *cost function*.

Let $x_{k+j|k}$ denote the predicted state at time step $k+j$, produced at the time step k. Also denote the control action as $u_{k+j|k}$. Let the full vectors of predicted states and inputs along N be denoted as $\boldsymbol{x}_k = (x_{k+j|k})_j$ and $\boldsymbol{u}_k = (u_{k+j|k})_j$. The controller aims to make the states reach the prescribed set points while delivering smooth control inputs. To that end, we formulate the following cost function as:

$$\begin{aligned} J(\boldsymbol{x}_k, \boldsymbol{u}_k; u_{k-1|k}) = \sum_{j=0}^{N} & \underbrace{\|x_{\text{ref}} - x_{k+j|k}\|^2_{Q_x}}_{\text{State cost}} \\ & + \underbrace{\|u_{\text{ref}} - u_{k+j|k}\|^2_{Q_u}}_{\text{Input cost}} + \underbrace{\|u_{k+j|k} - u_{k+j-1|k}\|^2_{Q_{\Delta u}}}_{\text{Input change cost}}, \end{aligned} \tag{4.18}$$

where $\boldsymbol{Q}_x \in \mathbb{R}^{8\times 8}$, $\boldsymbol{Q}_u$, $\boldsymbol{Q}_{\Delta u} \in \mathbb{R}^{3\times 3}$ are symmetric positive definite weight matrices for the states, inputs, and input rates, respectively. In (4.18), the first term denotes the *state cost*, which penalizes deviating from a certain state reference $\boldsymbol{x}_{\text{ref}}$. The second term denotes the *input cost* that penalizes a deviation from the steady-state input $u_{\text{ref}} = [g, 0, 0]$, i.e., the inputs that describe hovering in place. Finally, to enforce smooth control actions, the third term is added that penalizes changes in successive inputs, the *input change cost*. It should be noted that for the first time step in the prediction, this cost depends on the previous control action $u_{k-1|k} = u_{k-1}$.

We also directly apply constraints on the control inputs. As we are designing the NMPC to be utilized on a real ARW, hard bounds must be posed on the roll and pitch control references as the low-level attitude controller will only be able to operate with the desired flight behavior within a certain range of angles. Additionally, as the thrust of the ARW is limited, a similar hard bound must be posed on the thrust values. As such, constraints on the control inputs are posed as:

$$u_{\min} \le u_{k+j|k} \le u_{\max}. \tag{4.19}$$

Following the NMPC formulation, with model and cost function as previously described, the resulting NMPC problem becomes:

$$\underset{\boldsymbol{u}_k, \boldsymbol{x}_k}{\text{Minimize}}\; J(\boldsymbol{x}_k, \boldsymbol{u}_k, u_{k-1|k}) \tag{4.20a}$$

$$\begin{aligned} \text{s.t.: } & x_{k+j+1|k} = f(x_{k+j|k}, u_{k+j|k}), \\ & j = 0, \ldots, N-1, \end{aligned} \tag{4.20b}$$

$$u_{\min} \le u_{k+j|k} \le u_{\max},\ j = 0, \ldots, N, \tag{4.20c}$$

$$x_{k|k} = \hat{x}_k. \tag{4.20d}$$

This problem reads as: minimize the cost function $J(\boldsymbol{x}_k, \boldsymbol{u}_k, u_{k-1|k})$, subject to the discrete state-update function $f(x_{k+j|k}, u_{k+j|k})$ (simply meaning that the states evolve based on a computed control input in accordance with the model), and subject to the input constraints. The problem also implies that the state at the current time step ($j = 0$) is the measured state as $x_{k|k} = \hat{x}_k$. Such a nonlinear optimization problem can be solved by the optimization Engine [11] or OpEn for short, utilizing the PANOC (Proximal Averaged Newton-type method for Optimal Control) algorithm [12], which is an open-source and easy-to-use code generation software that generates a rust-based custom solver from a specified model, cost function, and set of constraints. It is also very fast and can be called directly in C++ and Python.

To demonstrate the performance of this controller, we will use the cascaded control method as shown in Fig. 4.2 with the NMPC as the position controller, and a PID controller as the attitude controller while for the ARW we will utilize the full nonlinear dynamics. The PID controller will track the T_d, ϕ_d, θ_d and $\psi = 0$ inputs from the NMPC and provide the necessary torques

τ to regulate the position of the ARW. The NMPC prediction horizon N has been set to 40 and the sampling time at $T_d = 0.1$ sec. The roll and pitch angles are constrained within $[-\pi/12 \leq \phi_d, \theta_d \leq \pi/12]$, rad. The weights for various matrices are selected empirically based on iterative trials until the tracking performance satisfies desired maximum error. The state deviation is penalized by $\boldsymbol{Q}_x = \text{diag}(10, 10, 10, 20, 20, 20, 1, 1)$, while the control action weights are $\boldsymbol{Q}_u = \text{diag}(20, 10, 10)$, and the weights for the input rate of change are $\boldsymbol{Q}_{\Delta u} = \text{diag}(10, 20, 20)$. The tracking performance of the position controller is shown in Fig. 4.9 for a spiral and a square trajectory.

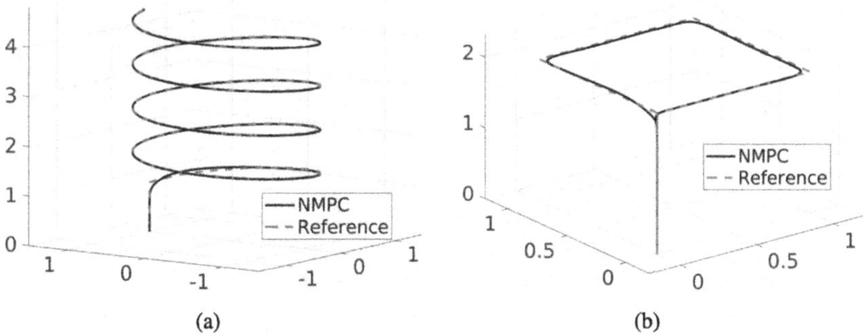

FIGURE 4.9 (a) Tracking performance of the NMPC position controller following a reference spiral trajectory while in (b) the same NMPC controller tracks a square trajectory.

4.6.1 Switching model based attitude controller of an ARW foldable quadrotor

For a system whose dynamics change due to structural reconfiguration during flight, like the one presented in Chapter 3, it will be essential to follow the cascaded control approach in Fig. 4.2. This approach will assist in dividing the tracking and attitude problem into two parts. Firstly, the position controller will give body frame control commands $\phi_{ref}, \theta_{ref}, T_{ref}$ essential for the ARW to track the position reference points. On the other hand, the attitude controller, whose job is now to track the given inputs from the position controller and consider the variations in the structural configuration of the ARW, thus, providing the necessary motor commands to maintain the ARW's stability. For position controller, one can utilize the same controller as in Section 4.6. But any controller that could provide either *Thrust*, *roll*, *pitch*, or rate commands considering independence from the heading yaw reference will be sufficient for that purpose.

After linearizing Eq. (3.46) from Chapter 3, at $\omega = 0$ and $\tau = 0$, it results in the following linear system:

$$\begin{bmatrix} \omega \\ \dot{\omega} \\ \dot{\tau} \end{bmatrix} = \begin{bmatrix} \mathbf{0} & \boldsymbol{I}_3 & \mathbf{0} \\ \mathbf{0} & \mathbf{0} & \boldsymbol{I}^{-1} \\ \mathbf{0} & \mathbf{0} & -\frac{1}{\tau_\alpha}\boldsymbol{I}_3 \end{bmatrix} \begin{bmatrix} \eta \\ \omega \\ \tau \end{bmatrix} + \begin{bmatrix} \mathbf{0} \\ \mathbf{0} \\ \frac{1}{\tau_\alpha}\boldsymbol{I}_3 \end{bmatrix} \tag{4.21}$$

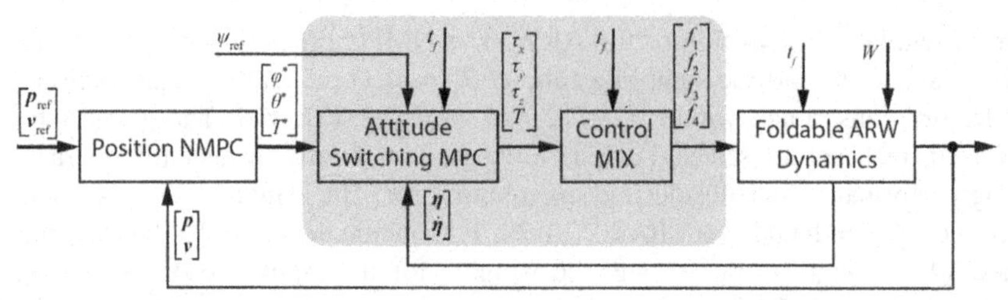

FIGURE 4.10 Block diagram displaying the overall ARW foldable quadrotor control scheme. Highlighted area indicates the low-level attitude control.

The attitude model of the ARW after the linearization stage depends on the variable inertia matrix. The inertia matrix depends on the position of the ARW arms based on the rotation of the four servo motors; thus, the ARW system is subject to parametric changes as the ARW updates its formation. For this example, we consider four different formations named after the final position of the arms based on the notation in Fig. 3.5. X for $\boldsymbol{\theta}_{s,[1,\ldots,4]} = [\pi/4]$, H for $\boldsymbol{\theta}_{s,[1,\ldots,4]} = [\pi/2]$,$Y$ for $\boldsymbol{\theta}_{s,[1,4]} = [\pi/4]$ and $\boldsymbol{\theta}_{s,[2,3]} = [\pi/2]$, lastly T for $\boldsymbol{\theta}_{s,[1,4]} = [0]$ and $\boldsymbol{\theta}_{s,[2,3]} = [\pi/2]$. The linear system can be represented as a Linear Parametric Varying (LPV) system based on the servo angles $\boldsymbol{\theta}_s$,

$$\boldsymbol{x}_{k+1} = \boldsymbol{A}(\theta_{S,k})\boldsymbol{x}(k) + \boldsymbol{B}(\theta_{S,k})\boldsymbol{u}(k) \tag{4.22}$$

where $\boldsymbol{x}(k) \in \mathbb{R}^{n\times 1}$ denotes the system states, and $\boldsymbol{u}(k) \in \mathbb{R}^{m\times 1}$ is the input vector. The linear switching MPC optimization problem is described by (4.23):

$$\begin{aligned}
&\text{minimize} && \sum_{k=0}^{N-1} \left(\Delta\boldsymbol{x}_k^\top \boldsymbol{Q}_x \Delta\boldsymbol{x}_k + \boldsymbol{u}_k^\top \boldsymbol{R}_u \boldsymbol{u}_k\right) && \text{(4.23a)}\\
&\text{subject to} && \Delta\boldsymbol{u}_{\min} \leq \Delta\boldsymbol{u}_k \leq \Delta\boldsymbol{u}_{\max},\ k = 1, \ldots, N_p, && \text{(4.23b)}\\
& && \boldsymbol{x}_{k+1|k} = f(\boldsymbol{x}_{k|k}, \boldsymbol{u}_{k|k}, \boldsymbol{\theta}_{s,k|k}),\ k \geq 0, && \text{(4.23c)}\\
& && \boldsymbol{x}_0 = \boldsymbol{x}(t_0) && \text{(4.23d)}
\end{aligned}$$

where $\Delta\boldsymbol{x}_k = \boldsymbol{x}_{k,\text{ref}} - \boldsymbol{x}_k$ and $\Delta\boldsymbol{u}_k = \boldsymbol{u}_{k|k} - \boldsymbol{u}_{k|k-1}$, while N_p denote the prediction horizon, respectively. $\boldsymbol{Q}_x \succeq 0$ and $\boldsymbol{R}_u \succ 0$ are the penalty on the state error and on the control input, respectively, while the bounds of the constraints are denoted as $(.)_{\text{min,max}}$. The state update of the optimization problem is a function of the current states, inputs, and the angle of the arms, $\boldsymbol{\theta}_s$ and $\boldsymbol{x}_0$ are the initial state conditions.

The complete control scheme is displayed in Fig. 4.10. The highlighted area indicates the attitude switching MPC that is updated each time to the appropriate model based on the switching variable t_f, which indicates the formation of the platform (X-H-Y and T). The computed torques and thrusts from the switching MPC are given to the parametric varying control mix as defined in the previous chapter, which results in the necessary forces for the motors.

Similar to the previous control design, the combined controller is for a square and spiral trajectory. Starting with the spiral trajectory, the ARW starts from X formation while it updates the position of the arms every 10 [sec] to H, Y, T and back to the initial X formation. The trajectory tracking of the linear MPC has a sampling time of 0.1 sec. The input weights are set to $\boldsymbol{R}_u = \text{diag}(25, 25, 8)$ for the roll pitch and thrust, while the yaw reference is kept at $\psi_{\text{ref}} = 0$. The states weights are set to $\boldsymbol{Q}_u = \text{diag}(40, 40, 60, 80, 80, 80, 0.1, 0.1)$. Fig. 4.11 shows the tracking performance of the ARW that successfully tracks the spiral trajectory despite the continuous formation changes.

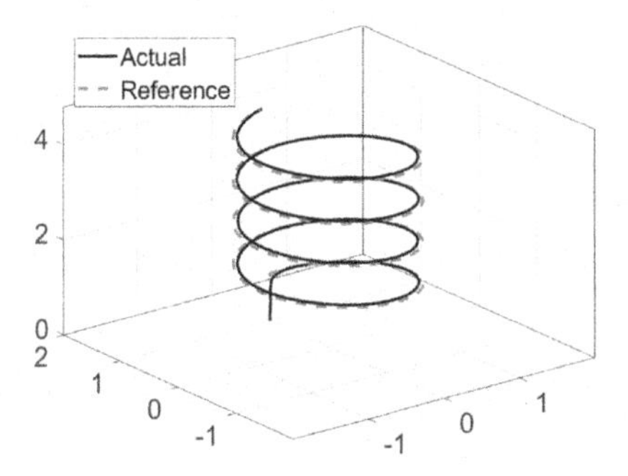

FIGURE 4.11 Position response of the foldable aerial worker following a spiral trajectory.

Fig. 4.12(a) shows the attitude control action response of the switching MPC. Fig. 4.12(b) illustrates the effect of the platform morphology impact on the required force by each motor. Note that while the required roll pitch and yaw and response of the attitude controller is not aggressive, to balance the ARW requires quite large variation that are handled successfully by the motor mixer.

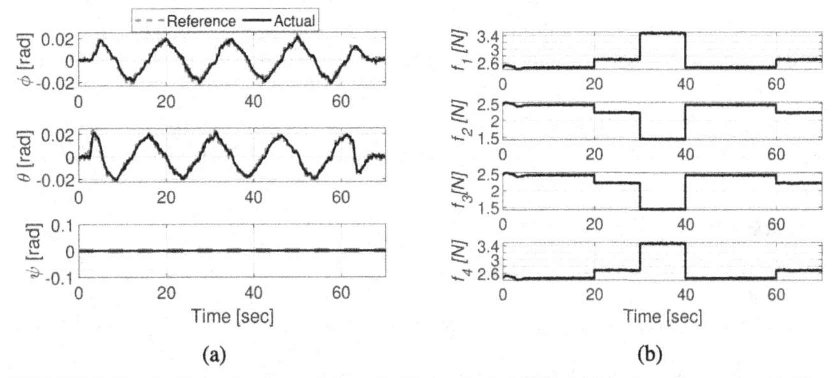

FIGURE 4.12 (a) Control actions of the switching attitude MPC and (b) motor forces provided by the motor mixer for a spiral reference trajectory.

Furthermore, in Fig. 4.13(a), the reconfigurable ARW tracking performance is evaluated for a square trajectory for same tuning of the controller. The ARW successfully tracks the given trajectory despite the formation updates even when the reformation occurs close to a turning of the square trajectory. The reference angles ϕ_{ref} and θ_{ref} provided by the position controller are shown in Fig. 4.13(b) (red dashed line), while $\psi_{\text{ref}} = 0$ is externally given and successfully tracked by the switching MPC (black line).

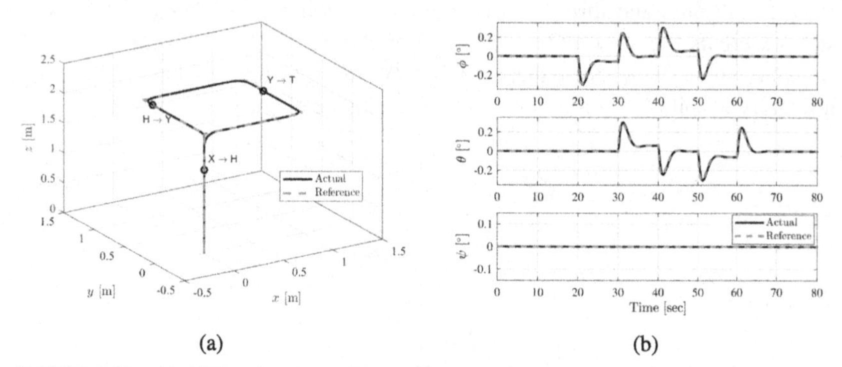

FIGURE 4.13 (a) ARW trajectory tracking performance for a given square path, and (b) control action response during a square trajectory tracking simulation of the switching attitude MPC.

4.7 Conclusions

This chapter focused on the control of ARWs that are utilized to track waypoints or trajectories and navigate to the desired destination, or track a desired path. Among the various control methods, both model-less controllers, as well as model-based controllers, are presented based on the models described in Chapter 3. *PID*, *LQR*, and *MPC* controllers have been tested under the tracking problem of a spiral and a square trajectory. Furthermore, based on a simplified position modeling, an NMPC is presented for the navigation of ARWs. The NMPC has a primary advantage in that it can be used directly with the nonlinear model instead of requiring an intermediate linearization step as well as accepting both linear and nonlinear constraints. Finally, a cascaded position NMPC and an attitude switching MPC are presented to cope with the model variations of an ARW re-configurable quadrotor, which updates its morphology in-flight.

References

[1] A. Patel, A. Banerjee, B. Lindqvist, C. Kanellakis, G. Nikolakopoulos, Design and model predictive control of mars coaxial quadrotor, arXiv preprint, arXiv:2109.06810, 2021.

[2] V. Kučera, Algebraic Riccati equation: Hermitian and definite solutions, in: The Riccati Equation, Springer, 1991, pp. 53–88.

[3] D.S. Naidu, Optimal Control Systems, CRC Press, 2002.

[4] R.V. Lopes, P. Santana, G. Borges, J. Ishihara, Model predictive control applied to tracking and attitude stabilization of a VTOL quadrotor aircraft, in: 21st International Congress of Mechanical Engineering, 2011, pp. 176–185.

[5] D.Q. Mayne, J.B. Rawlings, C.V. Rao, P.O. Scokaert, Constrained model predictive control: Stability and optimality, Automatica 36 (6) (2000) 789–814.

[6] K. Alexis, G. Nikolakopoulos, A. Tzes, Model predictive control scheme for the autonomous flight of an unmanned quadrotor, in: 2011 IEEE International Symposium on Industrial Electronics, IEEE, 2011, pp. 2243–2248.
[7] E. Small, P. Sopasakis, E. Fresk, P. Patrinos, G. Nikolakopoulos, Aerial navigation in obstructed environments with embedded nonlinear model predictive control, in: 2019 18th European Control Conference (ECC), 2019, pp. 3556–3563.
[8] X. Chen, L. Wang, Cascaded model predictive control of a quadrotor UAV, in: 2013 Australian Control Conference, IEEE, 2013, pp. 354–359.
[9] L. Wang, Model Predictive Control System Design and Implementation Using MATLAB®, Springer Science & Business Media, 2009.
[10] P. Ru, K. Subbarao, Nonlinear model predictive control for unmanned aerial vehicles, Aerospace 4 (2) (2017) 31.
[11] P. Sopasakis, E. Fresk, P. Patrinos, Open: Code generation for embedded nonconvex optimization, IFAC-PapersOnLine 53 (2) (2020) 6548–6554.
[12] L. Stella, A. Themelis, P. Sopasakis, P. Patrinos, A simple and efficient algorithm for nonlinear model predictive control, in: 2017 IEEE 56th Annual Conference on Decision and Control (CDC), 2017, pp. 1939–1944.

Chapter 5

Perception capabilities for ARWs

The art of perceiving the environment

Samuel Karlsson and Yifan Bai
Department of Computer, Electrical and Space Engineering, Luleå University of Technology, Luleå, Sweden

5.1 Introduction

There is a multitude of ways to perceive the surroundings using different types of sensors, which can sense different aspects of the surroundings being perceived. If the distance to the surrounding environment is desired, a lidar is useful. Lidar is a laser range measurement device that measures the distance in multiple directions. The exact amount of measurements and the coverage area depends on the specific model being used. For example, Velodyne VLP-16 gathers up to 300,000 measurements per second, spread over a field of view of 360° horizontally and 30° vertically [1]. The data collected by lidar is commonly referred to as a point cloud (An example of which can be seen in Fig. 5.1).

An RGB camera is a vision sensor that captures the colors in the area that is in the field of view of the camera. As the saying goes, "A picture is worth a thousand words" is as true for a robot as anywhere else. The challenge in robotics is to interpret the information and neural networks are commonly used for processing the images from the camera. A popular technique for the detection of objects in an image is the convolution neural network YOLO [2]. YOLO is capable of identifying the position of an object in an image and identifying the object (amongst a set of objects that it is trained to recognize). An image is a 2D representation of the environment, and therefore, only the position of an object in the 2D image frame can be directly discerned. For ARW's this means that only the direction to an object is known and not the distance of the object from the camera. There exist ways of deriving the distance to the detected objects from the 2D image. It is possible to use a neural network to estimate the depth in the image or combine the information in the image with more sensors that provide distance information. Another option is to use a stereo camera (two cameras with known relative position) and identify an object in the images of both cameras and use trigonometry to calculate the distance of the object from

Aerial Robotic Workers. https://doi.org/10.1016/B978-0-12-814909-6.00011-1

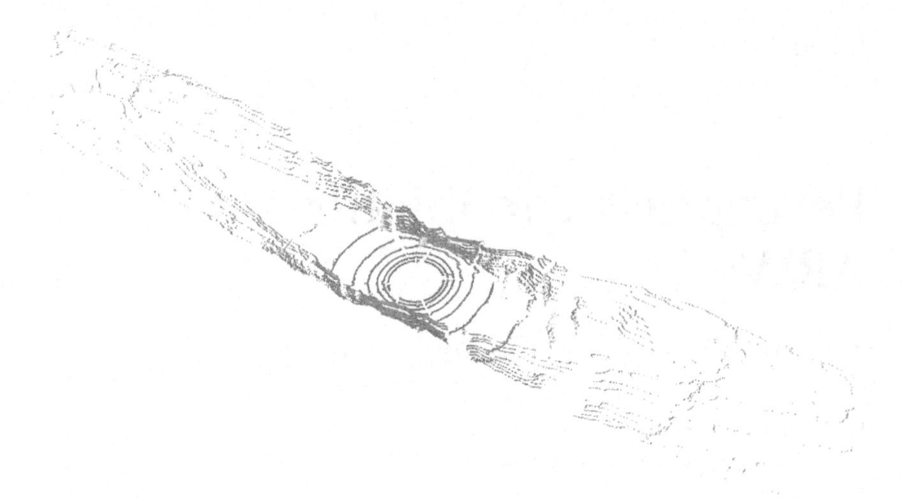

FIGURE 5.1 An example of a point cloud captured by a Velodyne VLP-16 lidar.

the camera. There exist cameras that have this behavior built in, and they are commonly referred to as RGB-D (Red Green Blue Depth) cameras. Even with a single camera is it possible to derive depth information, if reliable state estimation is possible during the movement of the robot, so that multiple images from different positions can be used to derive depth information as in a stereo camera [3].

Sensors like thermal cameras and sonar are the other option for ARW's and often a specific application dictates which of these needs to be added to the ARW's sensory repertoire. Sonars are useful in low-visibility scenarios (to see through the fog, for instance), and thermal cameras are useful for surveillance/inspection applications. When it comes to ARWs, there are limitations on the weight that can be carried, and it is crucial to use sensors that are appropriate for the task at hand.

5.2 Mapping

Building a map of the environment that the ARW has to pass through is crucial for multiple reasons. The map can be used for navigation or localization of the ARW. Maps of known environments can be loaded into the ARW's memory or built online (during operation) using the sensors on board the ARW. In both cases, the operation of the ARW is heavily reliant on accurate knowledge of its current position. There exist methods for simultaneous localization and mapping (SLAM), these methods have a circular dependency problem where the mapping relies on an accurate localization, and the localization relies on an accurate map. This section focuses on how a map M can be created given a position estimate $\boldsymbol{p}$. As a case study, we next consider an occupancy grid map in 2D, that can easily

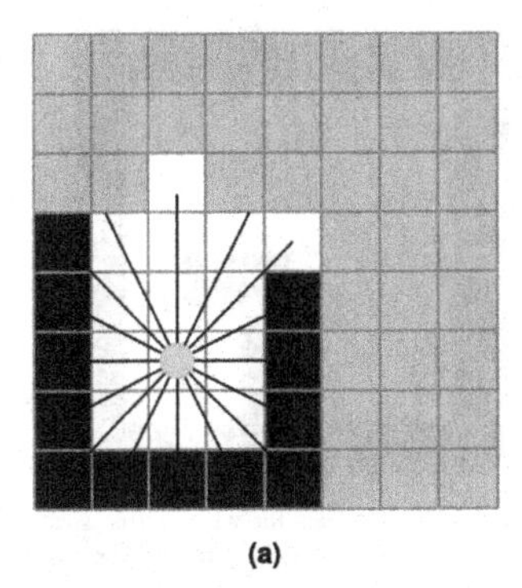

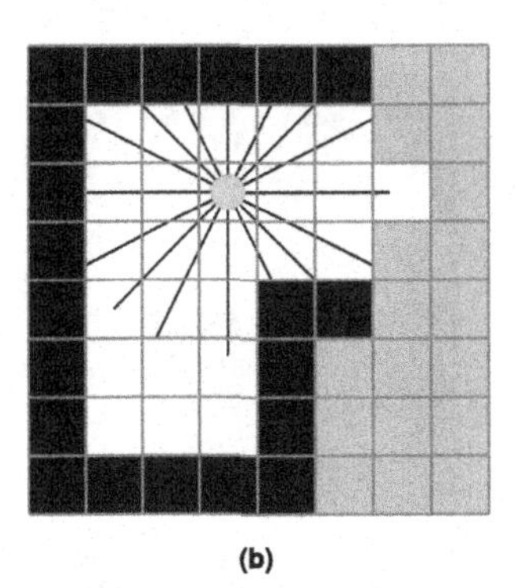

FIGURE 5.2 Examples of a mapping process. The green circle indicates the position ($\boldsymbol{p}$) of the ARW in a grid map. Black cells are occupied ($P(c) = 1$), white are free ($P(c) = 0$), and gray are unexplored. The lidar rays (ξ) emanating from the ARW are visualized with black lines. In 5.2a, an early stage of a mapping process is depicted, and the status of most areas is still unknown. As the ARW moves, the map evolves, and the map, after some exploration, is shown in 5.2b.

be extended to 3D, and the fundamental principles that are part of this case can also be utilized in other types of maps.

An occupancy grid is a grid where each cell (c) has a value corresponding to the probability $P(c)$ that the cell is occupied. In a perfect world where $\boldsymbol{p}$ is exact, and the sensor input ξ (point cloud from lidar in this example) is noise-free and precise, the creation of a map is easy. By transforming ξ from the body frame $\{\mathcal{B}\}$ to the global frame $\{\mathcal{G}\}$ and mapping the point to its corresponding cell, all cells with a point mapped to it will have $P(c) = 1$. And by doing a ray cast between $\boldsymbol{p}$ and each point in ξ, the passed through cells can be given an occupancy probability $P(c) = 0$ (as visualized in Fig. 5.2). In reality, $\boldsymbol{p}$ is not exact, and the sensor has noise, so there is a chance that the above method would register false positives and false negatives, leading to an inaccurate M. By assuming that the errors in $\boldsymbol{p}$ and ξ are Gaussian distributed each point in ξ can add some probability to the corresponding c and its surroundings. With multiple ξ some c will have a $P(c) \to 1$ and others $P(c) \to 0$ wiles other will be floating in-between. Using limits, each c can be given a binary value occupied or free to create a binary grid map M (Fig. 5.3 illustrates this process).

Using a low limit value will create a conservative M where it is highly unlikely to have a false negative, c that is noted as free in M but is occupied. The downside is that it is likely to have false positive, c that are noted as occupied but are free, this can unnecessarily restrict the ARW's ability to navigate.

Dynamic environments, environments with moving obstacles create a challenge. An obstacle that moves will change what the ARW preserve of the environment. Some areas will become blocked from view, and others will start to be visible. If the old ξ are considered as trustworthy as the latest ξ, the map will be slow at adjusting for dynamic obstacles movement. Old measurements will suggest that the map is in a certain way, and it will take more measurements of

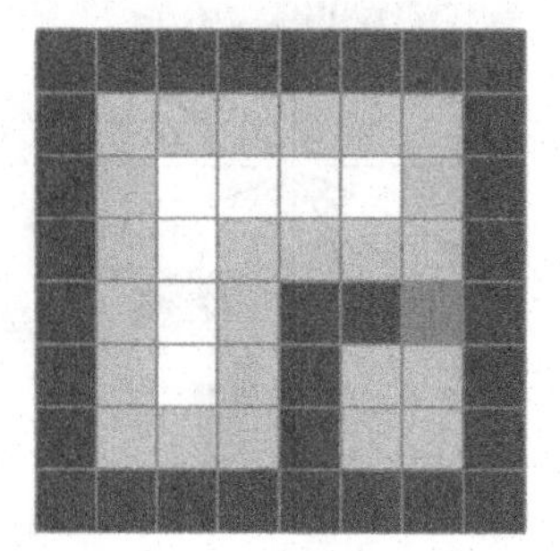

(a) A grid map with a built map where $P(c)$ are indicated with darker for $P(c) \to 1$ and lighter where $P(c) \to 0$

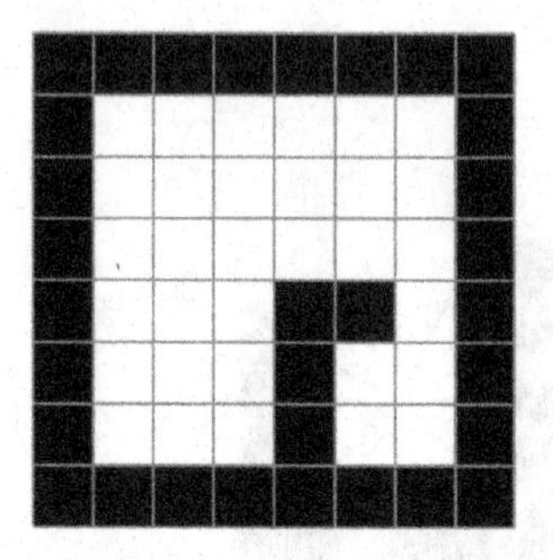

(b) By setting limit values for what is free space and what is occupied can the map in figure 1.3a be converted to a binary map.

FIGURE 5.3

a new layout that there are of the old layout to change M to the new layout. One way to deal with a dynamic obstacle is to trust the current measurements more than the old ones. Trusting the latest ξ, trust that ξ and thus subject M to the noise in that ξ. Multiple ξ can remove some noise with a simple filter, median filter, for example. One benefit of grid maps is that it naturally deals with small noises in ξ because of the discreet nature of c since the input is mapped to the same c, even if it is slightly noisy.

The size of c in M will impact the overall performance, larger c will reduce the amount of c needed to represent the environment and thus saving memory and computation time. But smaller feature in the environment risk being absorbed as a single occupied c or missed completely, and the precision will be reduced (most operation on a grid map will have the discreet precision of the c size). A too small c size will not be useful because the uncertainty from $\boldsymbol{p}$ and noise in ξ will cause too large uncertainties in M, i.e., $P(c)$ will float in between 0 and 1 in too many cases.

5.3 State estimation

5.3.1 Visual odometry

5.3.1.1 Feature extraction and matching

Visual odometry (VO), an alternative to wheel odometry, concerns estimating the pose and orientation of a mobile robot by observing the apparent motion of the static world captured by cameras. The advantage of VO compared to wheel odometry is that VO is not affected by wheel slip in uneven terrain, and it has been demonstrated that VO provides accurate trajectory estimates with error of roughly 1% [4].

The basic principle of VO is to find some representative points in images, so-called features, that will remain unchanged after minor view of the camera changes and estimate camera motion based on these features. Features such as corner features (Harris [5], FAST [6]), and BLOB features (SIFT [7], SURF [8]) are commonly used. Once a feature is detected, the region around the feature is encoded using a descriptor that can be matched against other descriptors. SIFT descriptor is a classical one that performs well in the presence of scale, rotation, and illumination changes, but the costs associated with computation are high. Some simple descriptors such as BRIEF [9] and ORB [10] have been developed for real-time feature extraction. By matching the descriptors between images, the correspondences between two images of the same scene are established, and as such, we can estimate the camera pose. Consider the features $x_t^m, m = 1, 2, \ldots, M$ extracted in image I_t and features $x_{t+1}^n, n = 1, 2, \ldots, N$ extracted in image I_{t+1}. The simplest matching approach is **Brute-Force Matcher**, calculating the Hamming distance (binary descriptors) or Euclidean distance (floating descriptors) between each x_t^m and all x_{t+1}^n, followed by sorting, where the one with the minimum distance is considered the matching feature. However, limited by the computational complexity, brute force matcher is not suitable for the real-time calculations. FLANN (Fast Library for Approximate Nearest Neighbors) [11] contains a collection of algorithms optimized for fast nearest neighbor search in large datasets and high dimensional features. It works faster than brute force matcher for large datasets and is well implemented into OpenCV.

5.3.1.2 Pose estimation

In this part, we will briefly introduce how the transformation between image I_t and image I_{t+1} can be computed given two sets of corresponding features f_t and f_{t+1}. Depending on the dimensions of features, the feature matching and corresponding motion estimation can be classified as follows:

1. 2D-2D: both f_t and f_{t+1} are 2D points, and the motion estimation is solved by epipolar geometry.
2. 3D-3D: f_t and f_{t+1} are two sets of 3D points, and the motion estimation is usually solved by ICP (Iterative Closest Point).
3. 3D-2D: f_t are specified in 3D, and f_{t+1} are 2D reprojections of f_t, and the motion estimation is solved by PnP (Perspective n Point).

a) Epipolar Geometry

As seen in Fig. 5.4, we assume that the rotation R and translation T represent the motion of the camera for capturing image I_1 and I_2. O_1, O_2 are the optical imaging centers. p_1 and p_2 is a pair of matching features in two images. The plane determined by O_1, O_2 and P is called the **Epipolar Plane**, the line O_1O_2 is called the **baseline**, **Epipoles** e_1 and e_2 are the intersections of the baseline and two image planes, and we call the intersection lines between epipolar plane and image plane as **Epipolar Lines**.

From a geometrical point of view, vectors $\overrightarrow{\mathbf{O_1p_1}}$, $\overrightarrow{\mathbf{O_1O_2}}$ and $\overrightarrow{\mathbf{O_2p_2}}$ are coplanar. Thus, we have

$$\overrightarrow{\mathbf{O_1p_1}} \cdot \left(\overrightarrow{\mathbf{O_1O_2}} \times \overrightarrow{\mathbf{O_2p_2}}\right) = 0. \tag{5.1}$$

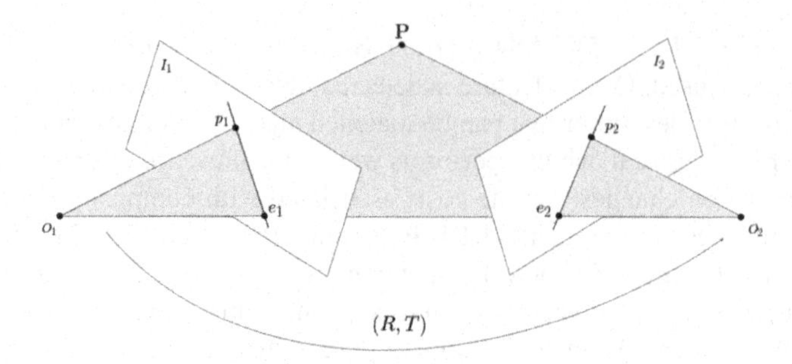

FIGURE 5.4 Coplanarity constraint of Epipolar Geometry.

Assuming the world reference system to be associated with the first camera, Eq. (5.1) can be transformed to

$$\mathbf{p_1^\top} \cdot [\mathbf{T} \times (\mathbf{R} \cdot \mathbf{p_2} + \mathbf{T})] = \mathbf{0} \tag{5.2}$$

and further simplification leads us to

$$\mathbf{p_1^\top} \cdot [\mathbf{T} \times (\mathbf{R} \cdot \mathbf{p_2})] = \mathbf{0}. \tag{5.3}$$

From linear algebra, we can represent the cross product between two vectors as a matrix-vector multiplication:

$$\mathbf{a} \times \mathbf{b} = \begin{bmatrix} 0 & -a_z & a_y \\ a_z & 0 & a_x \\ -a_y & a_x & 0 \end{bmatrix} \begin{bmatrix} b_x \\ b_y \\ b_z \end{bmatrix} = [\mathbf{a}_\times]\,\mathbf{b}. \tag{5.4}$$

Using the above notation, Eq. (5.3) can be written as

$$\mathbf{p_1^\top} \cdot [\mathbf{T}_\times]\,\mathbf{R} \cdot \mathbf{p_2} = \mathbf{0}. \tag{5.5}$$

The matrix $\mathbf{E} = [\mathbf{T}_\times]\,R$ is known as the **Essential Matrix**, which is a 3×3 matrix having 5 degrees of freedom. If the camera is not calibrated, the canonical projection of the object P to the corresponding image plane is $\mathbf{p_c} = \mathbf{K}^{-1}\mathbf{p}$, where $\mathbf{K}$ is the intrinsic matrix of the camera. Recall the coplanarity constraint,

$$\mathbf{p_{c1}^\top} \cdot [\mathbf{T}_\times]\,\mathbf{R} \cdot \mathbf{p_{c2}} = \mathbf{0}. \tag{5.6}$$

By substituting $\mathbf{p_c}$ with distorted values, we have

$$\mathbf{p_1^\top} \cdot \mathbf{K}^{-\top}\,[\mathbf{T}_\times]\,\mathbf{R}\mathbf{K}^{-1} \cdot \mathbf{p_2} = \mathbf{0}. \tag{5.7}$$

The matrix $\mathbf{F} = \mathbf{K}^{-\top}\,[\mathbf{T}_\times]\,\mathbf{R}\mathbf{K}^{-1}$ is called the **Fundamental Matrix**, which is also a 3×3 matrix but having seven degrees of freedom.

The fundamental matrix has seven degrees of freedom, in the sense that it has to be solved with seven matching features in the two images. However, from [12], we see that the computation of the fundamental matrix can be numerically unstable if only seven pairs of features are available. In [13], a classic **Eight Point Algorithm** is proposed, which solves the matrix with a set of eight or more corresponding features. After acquiring the fundamental matrix, we can use singular value decomposition (SVD) to obtain the rotation and translation of the camera.

b) Iterative Closest Point (ICP)

Assume that we match two RGB-D images and obtain a set of the corresponding 3D points in the same coordinate system: $\mathbf{P} = \{\mathbf{p_1}, \ldots, \mathbf{p_n}\}$ and $\mathbf{P}' = \{\mathbf{p}_1', \ldots, \mathbf{p}_n'\}$, scilicet, obtain the transformation matrix R. T that makes $P = R \cdot P' + T$. The Procrustes analysis provides a statistical way to solve this problem. First, we compute the center of mass of the two points sets

$$\mu_{\mathbf{p}} = \frac{1}{N_P}\left(\sum_{i=1}^{N_P} \mathbf{p_i}\right) \tag{5.8}$$

$$\mu_{\mathbf{p}}' = \frac{1}{N_P'}\left(\sum_{i=1}^{N_{\mathbf{P}'}} \mathbf{p_i}'\right) \tag{5.9}$$

and the SVD of the covariance matrix of two sets:

$$\begin{aligned} \mathbf{C}_{\mathbf{PP}}' &= \sum_{i=1}^{N_P} (\mathbf{p_i} - \mu_{\mathbf{p}}) \left(\mathbf{p_i}' - \mu_{\mathbf{p}}'\right) \\ &= \mathbf{USV}^\top \end{aligned} \tag{5.10}$$

where $\mathbf{C}_{\mathbf{PP}}'$ is a 3×3 matrix, $\mathbf{S}$ is a diagonal matrix of singular values, $\mathbf{U}$ and $\mathbf{V}$ are diagonal matrices. If $\mathbf{C}_{\mathbf{PP}}'$ has full rank, the optimal solution of transformation matrix shown in [14] has to be

$$\mathbf{R} = \mathbf{UV}^\top \tag{5.11}$$

$$\mathbf{T} = \mu_{\mathbf{p}} - \mathbf{R}\mu_{\mathbf{p}}'. \tag{5.12}$$

If correct matching features are not known, we can match the closest point in the reference point cloud to each point in the source point cloud, iteratively estimate the rotation and translation based on Procrustes analysis and transform the source point using the obtained rotation and translation until the convergence criterion is met.

c) Perspective n Point (PnP)

It is also possible to estimate the pose of the camera if some 3D points and their projection positions onto the image plane are given. This problem is known as perspective n point. As pointed out in [15], motion estimation from 3D-2D is more accurate than 3D-3D, as it minimizes the image reprojection error rather than the feature position error.

The PnP problem can be formulated as follows. Given a set of homogeneous 3D points p_w in the world frame, their projections p_c in the camera frame and the calibrated intrinsic camera parameters K, determine the rotation R and translation T of the camera with respect to the world frame

$$s\mathbf{p_c} = \mathbf{K}[\mathbf{R} \mid \mathbf{T}]\mathbf{p_w} \tag{5.13}$$

which on expansion takes the form

$$s\begin{bmatrix} u \\ v \\ 1 \end{bmatrix} = \begin{bmatrix} f_x & 0 & u_0 \\ 0 & f_y & v_0 \\ 0 & 0 & 1 \end{bmatrix} \begin{bmatrix} r_{11} & r_{12} & r_{13} & t_1 \\ r_{21} & r_{22} & r_{23} & t_2 \\ r_{31} & r_{32} & r_{33} & t_3 \end{bmatrix} \begin{bmatrix} x \\ y \\ z \\ 1 \end{bmatrix} \tag{5.14}$$

where s is a distance scale factor.

There are several solutions for PnP problem, for instance, P3P [16], Direct Linear Transform (DLT) [17], Efficient PnP (EPnP) [18], etc.

5.4 Object detection and tracking

Object detection is a fundamental pillar of computer vision and can be relevant to a variety of application scenarios for inspection and monitoring requirements. The fundamental aim of the object detection module is to use the visual sensor for classifying and localizing objects in the image plane. Classical variations of object detection include edge detection, face detection, shape detection, color detection, salient object detection, etc. The recent advances on both fronts of hardware and software brought Deep learning in the forefront of current object detection efforts. During the years, the performance of object detectors has been substantially improved, where, for example, such methods have been deployed in the recent DARPA SubT challenge.[1] Deep learning object detectors can be divided into two main categories, single-stage and two-stage detectors. The two-stage detectors include two parts, namely i) the region proposal network that provides candidates for bounding boxes and ii) feature extraction for bounding box regression tasks. This methodology allows for high accuracy of the result while operating at slower rates. The single-state detectors provide a prediction of bounding boxes directly from the input images, making them faster during inference but less accurate. In this Chapter, the state-of-the-art Convolutional

[1] https://www.darpa.mil/program/darpa-subterranean-challenge.

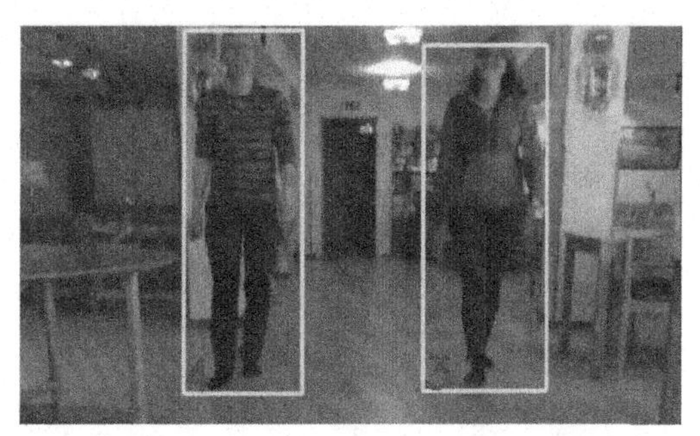

FIGURE 5.5 An example of YOLO detecting two people in an indoor environment and the Hungarian algorithm tracking them.

Neural Network (CNN) object detector YOLO (You Only Look Once) [19] is used. YOLO is based on a single CNN that models the detection problem as a regression function, which divides the input image into regions, and then it predicts the boundary boxes and probabilities for each region. A merit of the method is that it allows for predicting multiple bounding boxes and probabilities in different parts of the image. Classification networks like YOLO are capable of detecting multiple objects in an image, labeling the objects (from among a class of labels used for training), locating the objects within the image and providing an estimate of their dimension as in Fig. 5.5.

A human can track any object through time intuitively using a combination of visual resemblance and proximity to the last known position. There is generally no way for a computer to know if the objects in two different frames are the same. But a computer can use the same technique as humans do. By assuming that objects have a relatively small position change between two consecutive frames, the detected object can be paired between frames and thus tracked. Using the position information from a classification network, a combinatorial optimization algorithm like the Hungarian algorithm [20] can be used to track the detected objects.

Let us assume that five objects are detected in two consecutive frames $(a_1, ..., e_1)$ and $(a_2, ..., e_2)$. By comparing each of $a_1 ... e_1$ positions against all of $a_2 .. e_2$ positions, the best fits can be calculated. The fit can be calculated as a percentage of overlapping area in the image or as the distance between centers. Deepening on expected movement of objects, different ways of calculation fit can be preferred. Using the fits, the objects in consecutive frames can be matched by matching the best fit first and then matching the second best, and so on. Considering a fitness score on how bad fit can be considered the same object, new objects will be detected, e.g., an object exists in the new frame that do not match any in the old frame, while objects in the previous frame that do not match any in the new frame will be identified as objects outside the field of view and therefore removed from the tracked objects. This algorithm is visualized in Fig. 5.6, where one object disappears and a new one appears. To increase

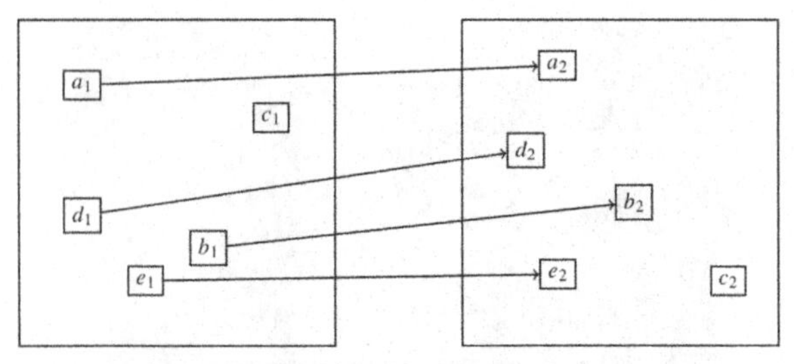

(a) Detected object in two consecutive frames are matched using the Hungarian algorithm and then tracked.

	a_2	b_2	c_2	d_2	e_2
a_1	0.9	0.2	0.5	0.7	0.1
b_1	0.2	0.92	0.8	0.78	0.82
c_1	0.5	0.75	0.23	0.4	0.3
d_1	0.65	0.85	0.45	0.89	0.86
e_1	0.2	0.75	0.4	0.45	0.96

(b) The Hungarian algorithm example matching table for figure 1.6a. The numbers are matching percentage, higher matching percentage is more likely to be a the same object. The green values are selected as best match in this example.

FIGURE 5.6 An example illustration of the Hungarian algorithm tracking process. In this example, there is no good match for c_1, and it appears a new object c_2.

the robustness of the algorithm in situations where the object comes outside the field of view of the sensor, short temporal memory mechanisms can be introduced for those objects in case they become inside the field of view. This can be combined with a module that will predict the position of the object of interest in the upcoming frames to assist the re-detection task in the case the object comes inside the Field of View of the visual sensor. A popular way of predicting the next position of the object is using a Kalman filter. The Kalman filter has the added benefit of filtering noise from the measurements, and it is also capable of estimating unmeasured states (for example, the object's velocity).

With the help of depth information of an object, the global 3D position of the object can be calculated. With the depth information, we obtain an obstacle's 3D position $p = [x, y, z]^T$ in the image frame $\{\mathcal{I}\}$. To transform p_p pixels to metric p_m, the camera intrinsic matrix $\mathbf{K}$ is used

$$\mathbf{K}^{-1} p_p = p_m. \tag{5.15}$$

The ARW's current position can be transformed into the global frame $\{\mathcal{G}\}$ through the operation

$$p^{\mathcal{I}\mathcal{G}}\mathbf{T}_{\mathcal{I}} = p^{\mathcal{G}} \tag{5.16}$$

where $\mathbf{T}_{\mathcal{I}}$ is the transformation matrix from $\{\mathcal{I}\}$ to $\{\mathcal{G}\}$.

References

[1] Velodyne, VLP-16 User Manual, Velodyne LiDAR, Inc., 2019.

[2] A. Bochkovskiy, C.-Y. Wang, H.-Y.M. Liao, YOLOv4: Optimal speed and accuracy of object detection, arXiv preprint, 2020, pp. 1–17.

[3] R. Collins, A space-sweep approach to true multi-image matching, in: Proceedings CVPR IEEE Computer Society Conference on Computer Vision and Pattern Recognition, 1996, pp. 358–363.

[4] D. Scaramuzza, F. Fraundorfer, Visual odometry [tutorial], IEEE Robotics & Automation Magazine 18 (4) (2011) 80–92.

[5] C. Harris, M. Stephens, et al., A combined corner and edge detector, Alvey Vision Conference 15 (50) (1988) 10–5244.

[6] E. Rosten, T. Drummond, Machine learning for high-speed corner detection, in: European Conference on Computer Vision, Springer, 2006, pp. 430–443.

[7] D.G. Lowe, Object recognition from local scale-invariant features, Proceedings of the Seventh IEEE International Conference on Computer Vision, vol. 2, IEEE, 1999, pp. 1150–1157.

[8] H. Bay, A. Ess, T. Tuytelaars, L. Van Gool, Speeded-up robust features (surf), Computer Vision and Image Understanding 110 (3) (2008) 346–359.

[9] M. Calonder, V. Lepetit, C. Strecha, P. Fua, Brief: Binary robust independent elementary features, in: European Conference on Computer Vision, Springer, 2010, pp. 778–792.

[10] E. Rublee, V. Rabaud, K. Konolige, G. Bradski, Orb: An efficient alternative to sift or surf, in: 2011 International Conference on Computer Vision, IEEE, 2011, pp. 2564–2571.

[11] M. Muja, D.G. Lowe, Fast approximate nearest neighbors with automatic algorithm configuration, in: VISAPP (1), vol. 2, 2009, pp. 331–340.

[12] O. Faugeras, O.A. Faugeras, Three-Dimensional Computer Vision: A Geometric Viewpoint, MIT Press, 1993.

[13] R.I. Hartley, In defense of the eight-point algorithm, IEEE Transactions on Pattern Analysis and Machine Intelligence 19 (6) (1997) 580–593.

[14] P.J. Besl, N.D. McKay, Method for registration of 3-d shapes, in: Sensor Fusion IV: Control Paradigms and Data Structures, in: Proc. SPIE, vol. 1611, SPIE, 1992, pp. 586–606.

[15] D. Nistér, O. Naroditsky, J. Bergen, Visual odometry, in: Proceedings of the 2004 IEEE Computer Society Conference on Computer Vision and Pattern Recognition, CVPR 2004, vol. 1, IEEE, 2004, p. I.

[16] X.-S. Gao, X.-R. Hou, J. Tang, H.-F. Cheng, Complete solution classification for the perspective-three-point problem, IEEE Transactions on Pattern Analysis and Machine Intelligence 25 (8) (2003) 930–943.

[17] Y. Abdel-Aziz, H. Karara, M. Hauck, Direct linear transformation from comparator coordinates into object space coordinates in close-range photogrammetry, Photogrammetric Engineering and Remote Sensing 81 (2) (2015) 103–107 [Online]. Available: https://www.sciencedirect.com/science/article/pii/S0099111215303086.

[18] V. Lepetit, F. Moreno-Noguer, P. Fua, EPnP: An accurate o(n) solution to the PnP problem, International Journal of Computer Vision 81 (Feb. 2009).

[19] J. Redmon, A. Farhadi, Yolov3: An incremental improvement, arXiv preprint, arXiv:1804.02767, 2018.

[20] H.W. Kuhn, The Hungarian method for the assignment problem, in: Naval Research Logistic, 2004.

Chapter 6

Navigation for ARWs

Björn Lindqvist, Vignesh Kottayam Viswanathan, Samuel Karlsson, and Sumeet Gajanan Satpute
Department of Computer, Electrical and Space Engineering, Luleå University of Technology, Luleå, Sweden

6.1 Introduction

The problems of navigation and path planning are some of the most well-studied and fundamental problems in Robotics [1,2]. Almost all autonomous missions require that the robot can move from some initial position to a desired one, avoiding any obstacles encountered on the way. Inspection tasks require a path planner that can query a control system to move the robot along the desired inspection path. Exploring unknown areas requires navigation methods that can continuously update as new information is acquired, while reactively avoiding any collisions with the environment. Any robot working in densely populated areas occupied by objects, other robotic agents, or, most importantly, humans, needs fast and effective obstacle avoidance schemes to ensure the safety of the robots and human workers in the area. All of these problems fall under the general class of navigation problems, in solving which safe and optimal paths are developed to take the ARW from one position to the next in order to complete the overall mission and task.

From these examples, it should be clear that the navigation task is tightly coupled with robotic perception, as there is no way to avoid an obstacle without first perceiving it. Unless we assume a completely static and fully known environment, the path planning and obstacle avoidance problems are entirely dependent on sensor information about the robot's surroundings. This information usually comes either from a constructed occupancy map of the environment (continually updated or previously known) or from the raw sensor data provided by onboard sensor equipment, such as cameras or LiDAR scanners.

In this chapter, we will separate the navigation problem into two subcategories. Reactive local planners or obstacle avoidance ensures that the ARW maintains a safe distance from any obstacle, wall, surface, etc., in the immediate surroundings. These planners should be able to operate at high run-time requirements to quickly re-plan if new obstacles are detected and should be coupled with the control framework to ensure effective avoidance maneuvers. Global path planners ensure that the robot moves through the environment in

Aerial Robotic Workers. https://doi.org/10.1016/B978-0-12-814909-6.00012-3

order to complete the desired task by calculating the path that takes the robot from an initial position to the desired one. There is a wide range of global path planners used for ARWs, mainly centered around either methods of ensuring the optimal and complete coverage during inspection tasks or occupancy map-based iterative planners that can be queried to find the path between any two points in the occupancy map. We will also discuss how simple but effective heading regulation techniques can stand in for more complex global planning architectures in specific environments.

6.2 Navigation architecture

In general, ARW missions require both a local and a global path planner. For example, in [3], a local artificial potential field is combined with a RRT-based global path planner, and in [4,5], both artificial potential fields and NMPC-based obstacle avoidance are combined with a heading regulation technique that allows the ARW to align its heading to follow subterranean tunnels while the avoidance layer maintains a safe distance to the walls. This is in general the case, since even with very efficient and fast map-based global path planners, maps and localization frameworks can suffer from drifts, uncertainties, inaccuracies or simply produce a path that would lead to collisions, and thus a local obstacle avoidance scheme that relies only on the direct sensor information is recommended as an inner safety layer to ensure there are no collisions with the environment or other robots. A general example architecture for an ARW mission is presented in Fig. 6.1. From a mission/exploration architecture or from an ARW operator, a desired goal point to reach, denoted as p_g, is selected. A global map-based path planner generates a series of way-points $\boldsymbol{wp}$ that define the path from initial point p_0 (based on the current estimated position of the ARW) to the goal p_g. The first waypoint in $\boldsymbol{wp}$, let us call it wp_1, is sent to a local avoidance scheme that calculates the relevant controller reference x_{ref} that commands the robot to move toward wp_1 while maintaining a pre-defined safety distance to any obstacles based on the onboard perception. The next step is either recalculating $\boldsymbol{wp}$ based on new map information and initial position p_0 or query the local planner once the ARW has reached wp_1 in order to move to the next waypoint wp_2 and so on until p_g is reached.

6.3 Reactive local path planners

Reactive local path planners, also commonly referred to as obstacle avoidance act as an inner safety layer whose purpose is to guide the ARW through the local environment in such a way as to maintain a specified safety distance from any obstacle encountered on the way. To be able to quickly react to new environment information, these local planners have to operate in real time, which for a quickly maneuvering platform with high run-time requirements like the ARW, implies re-calculations at 10–20 Hz or faster. This is a large restriction, but with

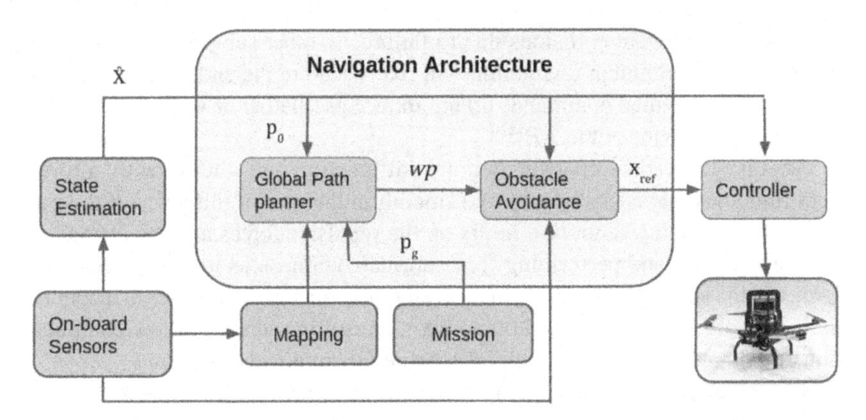

FIGURE 6.1 General example navigation architecture for ARW.

modern computational devices and smarter algorithms, wider ranges of applicable schemes for local planning and re-planning become available. Additionally, the better the pairing is between the controller and the avoidance layer, the better and smoother the avoidance maneuvers will be.

Due to multiple benefits (discussed in Section 6.3.1), the APF (Artificial Potential Field) [6,7] is the current most common approach utilized for reactive obstacle avoidance and has been used for the ARW application use cases for both mapping [8] and exploration tasks [9], as an extra reactive control layer for obstacle avoidance. Novel nonlinear MPC based avoidance schemes [10], where obstacle avoidance is directly integrated into the reference tracking controller, will also be discussed in Section 6.3.2, with the advantage of directly considering the nonlinear dynamics of the ARW platform during the avoidance maneuvers while keeping computation time low.

6.3.1 Artificial potential field

Artificial Potential Fields in simple terms operate by generating attractive and repulsive "forces fields" that operate on the ARW. Attractive forces propel the ARW toward a goal, like a desired velocity or a goal position, while repulsive forces are generated from environment information and repel the ARW away from any detected obstacle within a volume surrounding the platform. Repulsive forces can be calculated not only from a generated representation of the environment (occupancy map or other methods approximating the general shape of the local environment) but also directly from sensor data from onboard sensors on the robot, which is a major attraction toward using the APF. Let us assume an ARW equipped with a 3D LiDAR that can be utilized for both obstacle avoidance and for position state estimation.

ARWs pose a set of unique challenges to obstacle avoidance algorithms, such as high levels of reactiveness and low computations, being independent of mappers as to prevent collisions due to failures in other submodules (since for ARWs, any environment interaction will likely lead to the end of the mission), and that the generated commands do not induce oscillating or wobbling ("back-and-forth") behavior of the ARW.

This is achieved by choosing a potential function (or rather directly a force field function) that is continuous and smooth in the area of influence of the potential field, placing saturation limits on the repulsive forces and the change in repulsive forces, and performing force-normalization, so as to always generate forces of the same magnitude (which a reference-tracking controller then can be optimally tuned to follow). Additionally, the resulting scheme is low computation and directly uses the pointcloud provided by onboard sensors.

Let us denote the local point cloud generated by a 3D LiDAR as $\{P\}$, where each point is described by a relative position to the LiDAR frame as $\rho = [\rho_x, \rho_y, \rho_z]$. Let us also denote the repulsive force as $F_r = [F_{r,x}, F_{r,y}, F_{r,z}]$, the attractive force as $F_a = [F_{a,x}, F_{a,y}, F_{a,z}]$, the radius of influence of the potential field as r_F. As we are only interested in points inside the radius of influence, when considering the repulsive force, let us denote the list of such points $\boldsymbol{\rho}_r \subset \{P\}$, where $|| \rho_r^i || \leq r_F, i = 0, 1, \ldots, N_{\rho_r}$ (and as such N_{ρ_r} is the number of points to be considered for the repulsive force). Taking inspiration from the classical repulsive force function proposed by Warren [6], the repulsive force can be described by:

$$F_r = \sum_{i=0}^{N_{\rho_r}} L \left(1 - \frac{|| \rho_r^i ||}{r_F}\right)^2 \frac{-\rho_r^i}{|| \rho_r^i ||} \tag{6.1}$$

where L is the repulsive constant and represents the largest possible force-per-point inside r_F. We simply denote the attractive force as the next position waypoint $wp = [wp_x, wp_y, wp_z]$ generated by any higher-level navigation module in the local body frame $\mathcal{B}$, such that $F_a = wp^{\mathcal{B}} - \hat{p}^{\mathcal{B}}$, with p here denoting the current position-state of the robot. From an intuitive point of view, this can be seen as the attractive force being the vector from the current position to the next given waypoint with a unitary gain, while the repulsive force is the shift in the next waypoint required to avoid obstacles. Generally, the attractive and repulsive forces are summed to get the total resulting force F. But, since the requirement of a stable and smooth flight is as high for an ARW as with multiple other modules (localization, etc.) running from onboard sensors, we propose a force-normalization and saturation on the repulsive force and change in the repulsive force. Let k denote the current sampled time instant and $(k-1)$ the previous sampled time instant, then:

$$F = F_k = || \frac{F_{a,k}}{|| F_{a,k} ||} + [F_{r,(k-1)} + sgn(F_{r,k} - F_{r,(k-1)})\Delta F_{max}, sgn(F_{r,k})F_{max}, F_{r,k}]_{-} ||$$

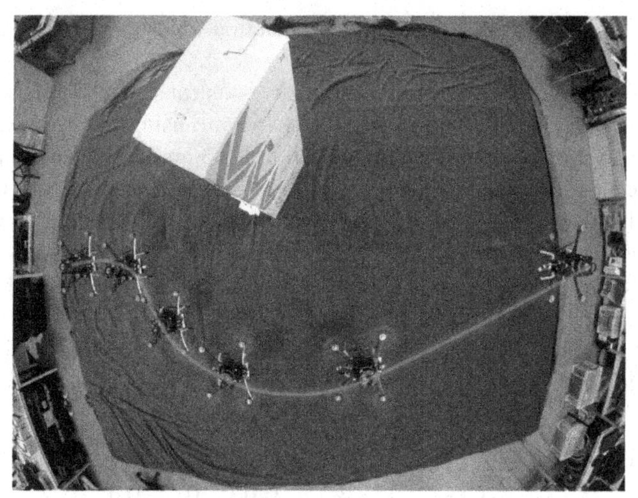

FIGURE 6.2 Artificial Potential Field navigation around a simple obstacle. The ARW is given a waypoint on the other side of the obstacle, and the avoidance algorithm continuously shifts waypoints as to provide an obstacle-free motion with a safe distance from the obstacle.

with F_{max} and ΔF_{max} being the saturation value and rate-saturation of the repulsive force and $sgn(F_r)$ denoting the sign of F_r. $[a, b]_-$ here denotes the *min(a,b)* operator and for simplicity in notation, let us consider its output the smallest-in-magnitude. The reference position, passed to the reference tracking controller (in the Micro Aerial Vehicle (MAV) body frame), with included obstacle avoidance, then becomes $p^{\mathcal{B}}_{ref} = F + \hat{p}^{\mathcal{B}}$, where $p^{\mathcal{B}}_{ref}$ are the first three elements of x_{ref}.

Since this formulation does not rely on any detection or mapping layer and can work with any position controller, it is very easy to evaluate in a realistic fully autonomous operation. Fig. 6.2 shows the path of the ARW around a simple obstacle in a laboratory environment, only relying on onboard sensors, computation, and LiDAR-based state estimation. The generated path represents a smooth and stable avoidance maneuver without excessive actuation.

6.3.2 Nonlinear MPC with integrated obstacle avoidance

We have already discussed nonlinear MPC in Chapter 4 as a nonlinear control scheme for the ARW. In the following works [10–14], obstacle avoidance was integrated directly in the control layer via constraints posed in the NMPC formulation. The general method consists of constraining the available position space for the ARW, and as such a nonlinear optimizer can be tasked with finding the NMPC trajectory that tracks the reference optimally while also avoiding obstacles. Using modern optimization software, NMPC formulations can allow large enough prediction horizons as to functionally plan a path around obstacles, while maintaining low enough computation time to satisfy the run-time requirements for control of an ARW.

As the name implies, any model predictive control scheme starts with a dynamic actuator model of the system (note: we are using the same simplified ARW model presented in Chapter 3, Section 3.3.3 and cost function as previously discussed in Chapter 4, Section 4.6 on NMPC, and as such parts related to the model and cost function are identical as for the pure controller case). The six degrees of freedom (DoF) of ARW are defined by the set of equations (6.2). The same model has been successfully used in previous works, such as in [10,12,14].

$$\dot{p}(t) = v(t) \tag{6.2a}$$

$$\dot{v}(t) = R(\phi, \theta) \begin{bmatrix} 0 \\ 0 \\ T_{\text{ref}} \end{bmatrix} + \begin{bmatrix} 0 \\ 0 \\ -g \end{bmatrix} - \begin{bmatrix} A_x & 0 & 0 \\ 0 & A_y & 0 \\ 0 & 0 & A_z \end{bmatrix} v(t), \tag{6.2b}$$

$$\dot{\phi}(t) = {}^1/_{\tau_\phi}(K_\phi \phi_{\text{ref}}(t) - \phi(t)), \tag{6.2c}$$

$$\dot{\theta}(t) = {}^1/_{\tau_\theta}(K_\theta \theta_{\text{ref}}(t) - \theta(t)), \tag{6.2d}$$

where $p = [p_x, p_y, p_z]^\top$ is the position, $v = [v_x, v_y, v_z]^\top$ is the linear velocity in the global frame of reference, and ϕ and $\theta \in [-\pi, \pi]$ are the roll and pitch angles along the x^{W} and y^{W} axes, respectively. Moreover, $R(\phi(t), \theta(t)) \in \mathrm{SO}(3)$ is a rotation matrix that describes the attitude in the Euler form, with $\phi_{\text{ref}} \in \mathbb{R}$, $\theta_{\text{ref}} \in \mathbb{R}$ and $T_{\text{ref}} \geq 0$ to be the references in roll, pitch, and total mass-less thrust generated by the four rotors, respectively. The above model assumes that the acceleration depends only on the magnitude and angle of the thrust vector, produced by the motors, as well as the linear damping terms $A_x, A_y, A_z \in \mathbb{R}$ and the gravitational acceleration g.

The attitude terms are modeled as a first-order system between the attitude (roll/pitch) and the references $\phi_{\text{ref}} \in \mathbb{R}$, $\theta_{\text{ref}} \in \mathbb{R}$, with gains $K_\phi, K_\theta \in \mathbb{R}$ and time constants $\tau_\phi, \tau_\theta \in \mathbb{R}$. The aforementioned terms model the closed-loop behavior of a low-level controller tracking ϕ_{ref} and θ_{ref}, which also implies that the UAV is equipped with a lower-level attitude controller that takes thrust, roll, and pitch commands and provides motor commands for the UAV, such as in [15].

Let the state vector be denoted by $x = [p, v, \phi, \theta]^\top$, and the control action as $u = [T, \phi_{\text{ref}}, \theta_{\text{ref}}]^\top$. The system dynamics of the ARW is discretized with a sampling time T_s using the forward Euler method to obtain

$$x_{k+1} = \zeta(x_k, u_k). \tag{6.3}$$

This discrete model is used as the *prediction model* of the NMPC. This prediction is done with receding horizon, i.e., the prediction considers a set number of steps into the future. We denote this as the *prediction horizon*, $N \in \mathbb{N}$, of the NMPC. In some applications, the *control horizon* is distinct from N, but in this chapter, we will only consider the case where they are the same, meaning both control inputs and predicted states are computed in the same horizon without loss of generality. By associating a cost to a configuration of states and inputs, at the current time and in the prediction, a nonlinear optimizer can be tasked with finding an optimal set of control actions, defined by the cost minimum of this *cost function*.

Let $x_{k+j|k}$ denote the predicted state at time step $k+j$, produced at the time step k. Also denote the control action as $u_{k+j|k}$. Let the full vectors of predicted states and inputs along N be denoted as $\boldsymbol{x}_k = (x_{k+j|k})_j$ and $\boldsymbol{u}_k = (u_{k+j|k})_j$. The controller aims to make the states reach the prescribed set points, while delivering smooth control inputs. To that end, we formulate the following cost function as:

$$J(\boldsymbol{x}_k, \boldsymbol{u}_k; u_{k-1|k}) = \sum_{j=0}^{N} \underbrace{\|x_{\text{ref}} - x_{k+j|k}\|^2_{Q_x}}_{\text{State cost}} + \underbrace{\|u_{\text{ref}} - u_{k+j|k}\|^2_{Q_u}}_{\text{Input cost}} + \underbrace{\|u_{k+j|k} - u_{k+j-1|k}\|^2_{Q_{\Delta u}}}_{\text{Input change cost}}, \tag{6.4}$$

where $Q_x \in \mathbb{R}^{8\times 8}$, Q_u, $Q_{\Delta u} \in \mathbb{R}^{3\times 3}$ are symmetric positive definite weight matrices for the states, inputs, and input rates, respectively. In (6.4), the first term denotes the *state cost*, which penalizes deviating from a certain state reference x_{ref}. The second term denotes the *input cost* that penalizes a deviation from the steady-state input $u_{\text{ref}} = [g, 0, 0]$, i.e., the inputs that describe hovering in place. Finally, to enforce smooth control actions, the third term is added that penalizes changes in successive inputs, the *input change cost*. It should be noted that for the first time step in the prediction, this cost depends on the previous control action $u_{k-1|k} = u_{k-1}$. In the aforementioned works, the method for constraining position space is based on the $[h]_+ = max(0, h)$ function defined by

$$[h]_+ = \begin{cases} 0 & \text{if } h \leq 0 \\ h & \text{otherwise.} \end{cases} \tag{6.5}$$

We can choose h to represent an expression that describes the desired obstacle geometry. The h_+ output will then be zero outside the obstacle implying that the constraint holds, and nonzero inside the obstacle, implying that the constraint is violated. Let us define $p = [p_x, p_y, p_z]$ as the position coordinates of the ARW. For a circular obstacle, we can thus impose an equality constraint as

$$h_{\text{circle}}(p, \xi^{\text{c}}) := [r_{\text{c}}^2 - (p_x - p_x^{\text{c}})^2 - (p_y - p_y^{\text{c}})^2]_+ = 0, \tag{6.6}$$

where $\xi^{\text{c}} = [p_x^c, p_y^c, r_c]$ define the x,y position coordinates of the center and the radius of the obstacle. More precisely, in a realistic context, r_c includes the radius of the obstacle, the radius of the ARW, and an additional safety distance

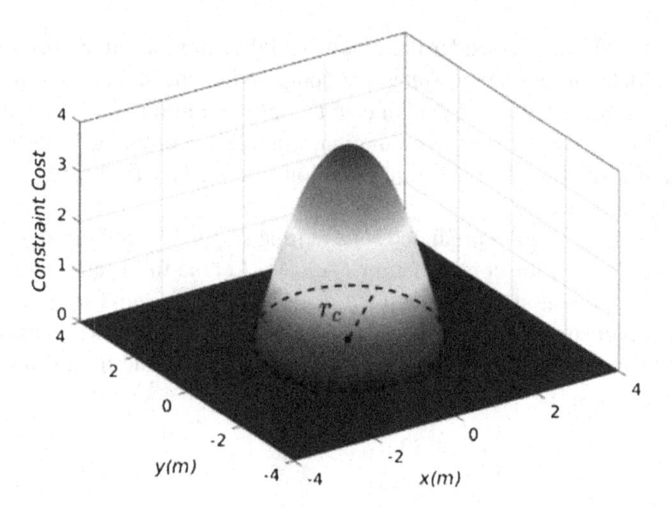

FIGURE 6.3 Cost map of cylinder with radius 2 m.

as $r_c = r_{obs} + r_{arw} + d_s$ to ensure no collisions with the environment. To maintain the equality constraint as $h_{\text{circle}}(p, \xi^c) := 0$ the position coordinates of the ARW are required to lie outside the circle, both at the current time step and at all predicted time steps within the prediction horizon.

We can visualize the obstacle in the form of a cost map, a 2D slice of the 3D space where the new z-axis represents the cost related to violating the constraint (the value of h at that point) as in Fig. 6.3.

We also directly apply constraints on the control inputs. As we are designing the NMPC to be utilized on a real ARW, hard bounds must be posed on the roll and pitch control references as the low-level attitude controller will only be able to operate with the desired flight behavior within a certain range of angles. Additionally, as the thrust of the ARW is limited, a similar hard bound must be posed on the thrust values. As such, constraints on the control inputs are posed as:

$$u_{\min} \leq u_{k+j|k} \leq u_{\max}. \tag{6.7}$$

Following the NMPC formulation, with model and cost function as previously described, the resulting NMPC problem becomes:

$$\underset{\boldsymbol{u}_k, \boldsymbol{x}_k}{\text{Minimize}}\, J(\boldsymbol{x}_k, \boldsymbol{u}_k, u_{k-1|k}) \tag{6.8a}$$

$$\text{s.t.: } x_{k+j+1|k} = \zeta(x_{k+j|k}, u_{k+j|k}),\ j = 0, \ldots, N-1, \tag{6.8b}$$

$$u_{\min} \leq u_{k+j|k} \leq u_{\max},\ j = 0, \ldots, N, \tag{6.8c}$$

$$h_{\text{circle}}(p_{k+j|k}, \xi_i^c) = 0,\ j = 0, \ldots, N,\ i = 1, \ldots, N_c, \tag{6.8d}$$

$$x_{k|k} = \hat{x}_k, \tag{6.8e}$$

where N_c denotes the number of circular obstacles. Such a nonlinear parametric optimization problem can be solved by the Optimization Engine [16,17] or OpEn for short using a penalty method [18] for dealing with the nonlinear constraints. Start by performing a *single-shooting* of the cost function via the decision variable $z = \boldsymbol{u}_k$ and define Z by the input constraints defined in the optimization problem above. We also define F to cast the circular equality constraints and f as the single-shooting of the cost function. Using a quadratic penalty method in the Optimization Engine framework, the resulting optimization becomes

$$\underset{z \in Z}{\text{Minimize}}\, f(z) + q\,\|F(z)\|^2, \tag{6.9}$$

with q as a positive penalty parameter. The penalty method in simple terms means re-solving the optimization problem with an increasing penalty parameter using the previous solution as an initial guess. This gradually pushes the cost-minima out of the obstacle until the trajectory is obstacle-free. A visualization of the penalty method concept can be seen in Fig. 6.4, where, as the q increases the trajectory moves out of the constrained position space represented by the circle.

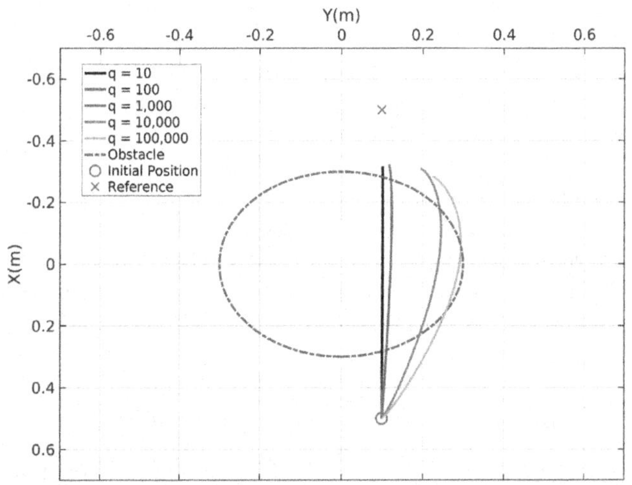

FIGURE 6.4 NMPC optimized trajectories with $N = 40$ using one to five penalty method iterations for a circular obstacle with radius $r = 0.3$ m. The cost-minima is gradually pushed out as the cost associated with violating the constraint is increased.

The above described method is applied in a simulation environment with four circular obstacles in Fig. 6.5. The prediction horizon is set to 40, with a sampling time of 0.05 s. The simulation is run in real time where only the first entry of actuator-trajectory $\boldsymbol{u}_k$ is applied to the ARW simulated system and the trajectory is then re-calculated for a full real time reactive planning and re-planning. The most major benefit of this methodology is the merging of the control system with local reactive planning in the form of the NMPC reference tracking combined with obstacle avoidance constraints, while still operating purely in the control layer and at a high run time frequency. Fig. 6.5 shows the ARW navigating past the obstacles while precisely satisfying the obstacle constraints, choosing the best local path to the references while maintaining a safe distance to the obstacle.

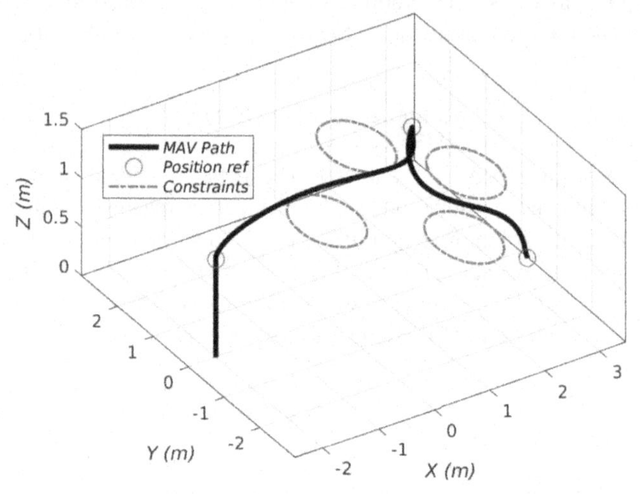

FIGURE 6.5 ARW path through the constrained environment with circular obstacles using NMPC-based combined control and obstacle avoidance.

Furthermore, more complex obstacles can be defined by taking the product of multiple $[h]_+$ terms, as based on (6.5) the product will be zero when any of the terms are zero. For example, instead of as in (6.6) where we defined an area the ARW should not go into, we can define a constrained entrance or "hoop" as the only allowed space for the ARW to navigate through in the y-z-plane by three terms, as:

$$h_{\text{entrance}}(p, \xi^{\text{e}}) := [-(r_{\text{e}}^2 - (p_y - p_y^{\text{e}})^2 - (p_z - p_z^{\text{e}})^2)]_+ \tag{6.10}$$

$$[p_x - p_x^{\text{w}}]_+ [-(p_x - p_x^{\text{w}}) + d_e]_+ = 0, \tag{6.11}$$

with $\xi^{\text{e}} = [r_{\text{e}}, p_y^{\text{e}}, p_z^{\text{e}}, p_x^{\text{w}}, d_e]$ representing the radius of the entrance, the y,z-position of the entrance, the x-position of the corresponding wall with the entrance, and lastly the depth of the entrance (here as the width of the wall). It should be noted that this is simply an additional example of how obstacles can be constructed, and how freely the optimization framework allows the definition of obstacles. This type of obstacle was also evaluated in a simulation environment in Fig. 6.6, where the ARW is tasked to move to the other side of the wall and correctly moves through the constrained entrances. This is, in general, a difficult navigation problem to handle (moving through very tight entrances), but due to the local planning capabilities and precise constraint definitions, the NMPC-based navigation can easily compute obstacle-free trajectories in real time.

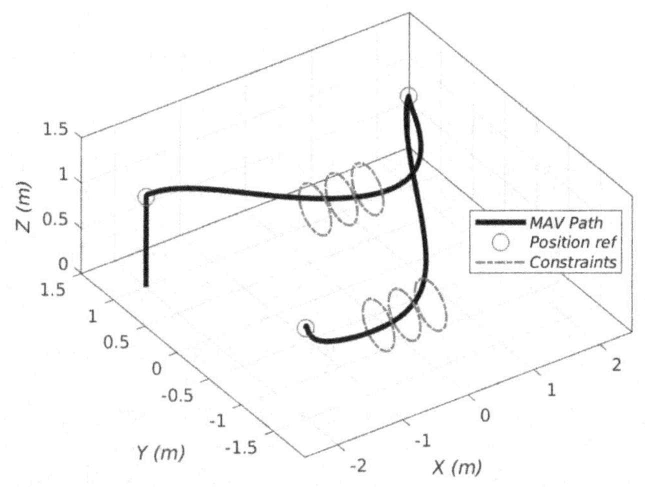

FIGURE 6.6 ARW path through the constrained environment with two constrained entrances using NMPC-based combined control and obstacle avoidance. The ARW correctly moves through the constrained entrance on both passes through the environment.

Another perfect fit for NMPC-based obstacle avoidance is the consideration of dynamic, or moving, obstacles. In [12], a trajectory classification and prediction scheme is combined with the same optimization framework as discussed above to create an obstacle avoidance scheme capable of dealing with moving obstacles. This addition comes very naturally to the NMPC formulation, as constraints in (6.6)–(6.10) are already solved for so that to satisfy the constraint along the full prediction horizon of the NMPC. Here, we consider only a spherical obstacle, representing some general obstacle moving or falling toward or being thrown at, the ARW as:

$$h_{\text{sphere}}(p, \xi^{\text{s}}) := [r_{\text{s}}^2 - (p_x - p_x^{\text{s}})^2 - (p_y - p_y^{\text{s}})^2 - (p_z - p_z^{\text{s}})^2]_+ = 0, \quad (6.12)$$

with $\xi^{\text{s}} = [p_x^s, p_y^s, p_z^s, r_s]$ representing the x,y,z-position and radius of the obstacle. The main difference is providing not just the initial positions of the obstacle ($\xi^{\text{s}} = [p_x^s, p_y^s, p_z^s, r_s]$, etc.) but just as the NMPC predicts future states of the ARW as $\boldsymbol{x}_k = (x_{k+j|k})_j$, we provide as input to the NMPC the full vector of discrete predicted obstacle positions, e.g., as $\boldsymbol{\xi}_k^s = (\xi_{k+j|k})_j$ with the same sampling time as the NMPC as to match each predicted state with a predicted obstacle position. This requires some mathematical model to predict future obstacle positions, for example, a spherical obstacle falling or thrown at the ARW follows the general projectile motion equations, with obstacle states

$x^s = [p_x^s, p_y^s, p_z^s, v_x^s, v_y^s, v_z^s]$, as

$$\dot{p}^s(t) = v^s(t), \tag{6.13a}$$

$$\dot{v}^s(t) = \begin{bmatrix} 0 \\ 0 \\ -g \end{bmatrix} - \begin{bmatrix} B_x & 0 & 0 \\ 0 & B_y & 0 \\ 0 & 0 & B_z \end{bmatrix} v^s(t), \tag{6.13b}$$

with B as linear aerodynamic drag coefficients. Discretizing this model, we can predict any $x^s_{k+j|k}$ from an initial measurement of velocity and position to form $\boldsymbol{\xi}^s$. The only real difference between this and the static obstacles from the optimization point of view, is that there is now a different obstacle position associated with the obstacle constraint at each predicted time step. This method was evaluated in a laboratory environment with a micro-quadcopter and a motion-capture system to track obstacle and ARW states. Fig. 6.7 shows an obstacle thrown at the micro-quad, and the resulting avoidance maneuver. Due to the nature of multiple moving parts, it is very hard to visualize the real performance of the ARW in the figure format. As such in https://youtu.be/vO3xjvMMNJ4, we demonstrate the dynamic obstacle NMPC (with an additional layer of trajectory classification and prediction, see [12]) in multiple scenarios also showing how the classic potential field and static obstacle NMPC are incapable of handling the fast-moving obstacles. This can be attributed to the *pro-active* component of NMPC-based obstacle avoidance. As the NMPC predicts future states within the prediction horizon, the avoidance maneuver can start as soon as the predicted obstacle positions are within the prediction horizon, in this example two seconds beforehand. The ARW-obstacle distances in the examples showcased in the video results can be seen in Fig. 6.8 and as can be seen the dynamic obstacle NMPC maintains the desired safety distances.

A natural continuation of this work is considered in [19], where instead of trying to predict future obstacle positions, the obstacle avoidance integrated NMPC considers an area densely occupied by multiple robotic agents that all share their intended predicted motions via their computed NMPC trajectories. In this situation, each ARW broadcasts its computed trajectory $\boldsymbol{u}_k$ and estimated state $\hat{x}$ to all nearby agents, and as such instead of using a discrete prediction model as in (6.13), we directly know the model-based trajectories of all agents and can formulate the agent to agent collision avoidance problem accordingly. This work also incorporates the consideration of when there are more nearby obstacles/ARWs than obstacle constraints in the NMPC formulation, e.g., when the number of agents $N_a - 1 > N_{\text{obs}}$, and, in general, a method for reducing computational complexity by *obstacle prioritization*. The presented obstacle prioritization scheme is directly based on the shared motion intentions of the aerial agents and is compactly summarized in Algorithm 1 with $\hat{x}_k^{\text{obs},i}$ and $\boldsymbol{u}_{k-1}^{\text{obs},i}$ denoting the state and shared trajectory of the ith non-ego agent, and the output is $(\xi^{\text{obs},i}_{\text{prio},j})_{i,j}$, e.g., the trajectories and radii of the prioritized obstacles that are the input parameters to the NMPC related to obstacle avoidance.

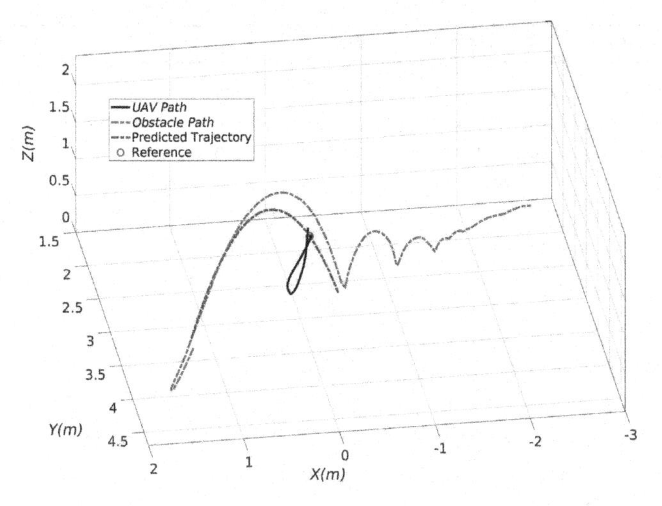

FIGURE 6.7 ARW avoidance maneuver and the path of an obstacle thrown at the AWR.

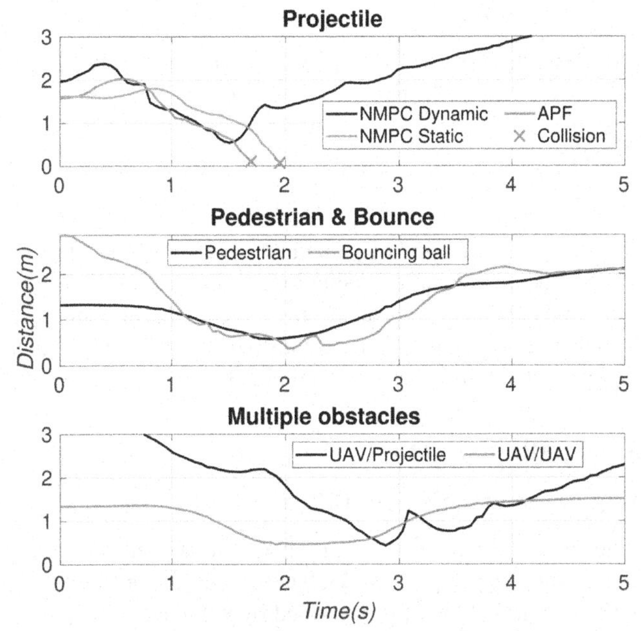

FIGURE 6.8 Distances between the obstacle and the ARW from video results of dynamic obstacle avoidance and comparisons (top). Without obstacle prediction the schemes fail, while the dynamic obstacle NMPC handles multiple moving obstacles without collisions.

This scheme was extensively evaluated in dense aerial swarms for up to ten aerial agents in a laboratory environment, with a high-level description and experiment video provided at https://www.youtube.com/watch?v=3kyiL6MZaag,

Algorithm 1: Obstacle prioritization.

1 **Inputs:** $\hat{x}_k, \boldsymbol{u}_{k-1}, \hat{x}_k^{\text{obs},i}, \boldsymbol{u}_{k-1}^{\text{obs},i}, N_a, N_{\text{obs}}, \boldsymbol{r}^{\text{obs}}, d_s$
Result: From the shared trajectories and measured states, decide which N_{obs} agents should be considered as obstacles
2 **for** $i = 1, N_a - 1$ **do**
3 　**for** $j = 0, N$ **do**
4 　　Compute $p_{k-1+j|k-1}$, $p_{k-1+j|k-1}^{\text{obs},i}$ and $v_{k-1+j|k-1}^{\text{obs},i}$
5 　　$d \leftarrow \|p_{k-1+j|k-1} - p_{k-1+j|k-1}^{\text{obs},i}\|$
6 　　$v_{\text{m}} \leftarrow \|v_{k-1+j|k-1}^{\text{obs},i}\|$
7 　　**if** $(d \leq r^{\text{obs},i})$ *and* $(j = 0)$ **then**
8 　　　$w_i \leftarrow w_i + M$
9 　　**else if** $d \leq r^{\text{obs},i} + d_s$ **then**
10 　　　$w_i \leftarrow w_i + (1 - \frac{d}{r^{\text{obs}}+d_s})^2 v_{\text{m}} \frac{N}{(j+1)^a}$
11 　　**end**
12 　**end**
13 **end**
14 Sort in descending order: $(\xi_j^{\text{obs},i})_{i,j}$ by corresponding element in w_i
15 $(\xi_{\text{prio},j}^{\text{obs},i})_{i,j} \leftarrow [(\xi_j^{\text{obs},1})_j, (\xi_j^{\text{obs},2})_j, \ldots, (\xi_j^{\text{obs},N_{\text{obs}}})_j]$
16 **Output:** $(\xi_{\text{prio},j}^{\text{obs},i})_{i,j}$

while Fig. 6.9 shows examples of generated trajectories, and Fig. 6.10 again presents the key parameter in an evaluation of obstacle avoidance: that being the minimum distance to any obstacle, in this case, the minimum agent-agent distance throughout the evaluation, here considering an obstacle radius of 0.4 m while the safety-critical distance, was around 0.3 m.

6.4 Global path planners

6.4.1 Map-based planning with D_+^*

Probably the most commonly used group of global path planners are grid-based planners such as A* [20] or D*. These planners operate on a map that is constructed as a grid (G) that is described by volumes (V) containing the information whether that particular volume is free (V_f) or occupied (V_o). Such volumetric representations of the map space are commonly known as voxels (volumetric pixels). In the literature there exists multiple mapping algorithms that can construct such Occupancy Maps from onboard sensors, one example being OctoMaps [21]. By assigning an entry cost to each voxel corresponding to if it is free $e = 1 \iff V_f$ or occupied $e = \infty \iff V_o$ so that traversing from one V to its neighbor V' is $C(V, V') = e_{V'}$ or $C(V, V') = \sqrt{2e_{V'}^2}$

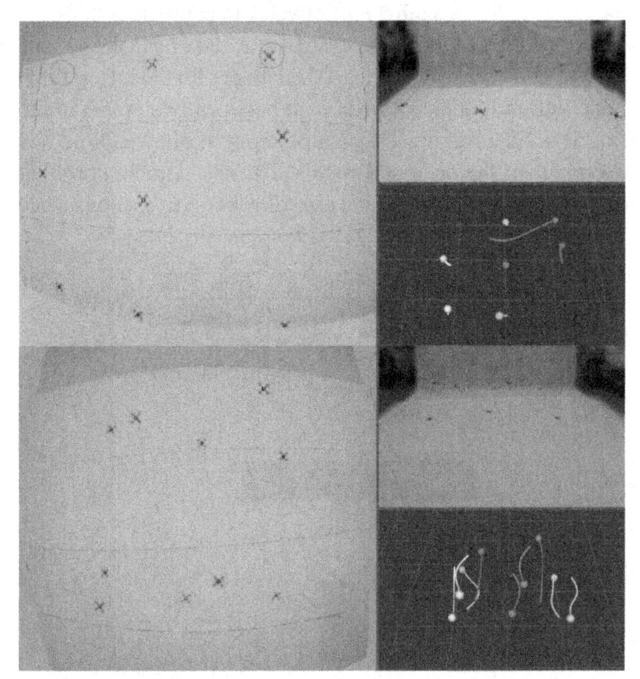

FIGURE 6.9 Showing distributed NMPC trajectories in two experimental evaluations: (top) each agent moves one-by-one and (bottom) all agents move at once.

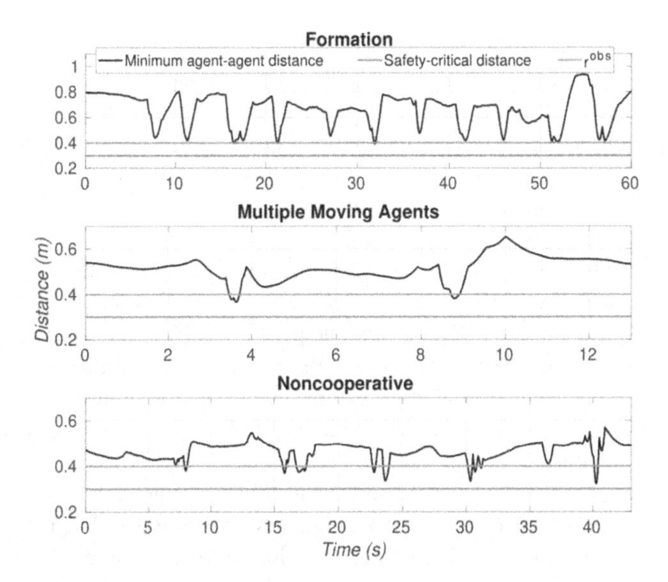

FIGURE 6.10 The minimum agent-agent distances throughout the experiments presented in [19].

if V' is diagonal to V. A path (P) between two points consists of a chain of V to V' transitions and thus the cost of traversing P can be calculated by $(C(P_{p \to w}) = \sum \forall C(V, V') \in P_{p \to w}$. D* planners find the P with the smallest cost $C(P)$ and thus that is the shortest path between two points in the grid map G. In Fig. 6.11, a 2D case of a D* path planning is shown. Note how the path touches the corners of the occupied voxels. This is a common problem for grid-based planners as they do not take the size of robot into account when planning the path.

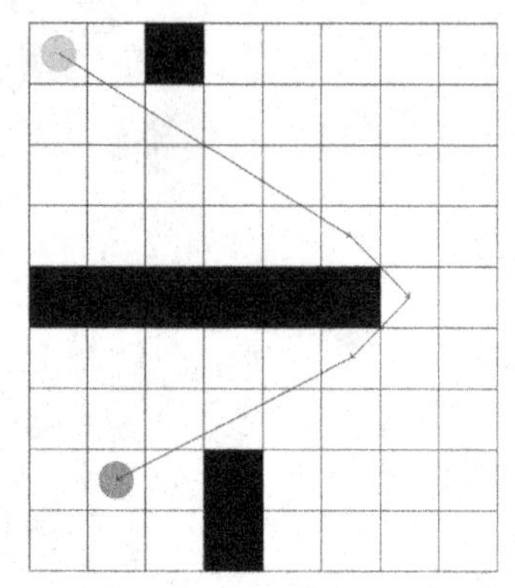

FIGURE 6.11 D* lite plans a path from the green (light gray in print) circle to the red (dark gray) circle in the free white spaces. The path planed is the shortest path, but it has no marginal to the black occupied spaces.

To deal with the corner touches, a risk layer is added to G; for V close to V_f, an e is added so that $e_{V_o} > e_r > e_{V_f}$. This creates a layer of higher-cost paths close to obstacles, thus creating a cheaper path further out from the occupied spaces. While longer, this path can now compensate for the size of the ARW and can in general compensate for risks associated with moving close to an obstacle. Fig. 6.12 depicts the outcome of the risk-aware planned path from the described methodology.

Another challenge with the D* family of path planners is the dependency on well-defined maps. More specifically, holes in a wall cased by poor sensor information can cause a path like the blue path in Fig. 6.13. By adding a representation of unknown voxels V_u to G with its one e_{V_u} so that $e_{V_o} > e_{V_u} > e_{V_f}$, this issue can be addressed. A path planner with both the representation of V_u and e_r is D^*_+. D^*_+ is built from a D* lite library [22] and can handle both 2D and 3D path planning.

D^*_+[1] have been tested during some real-life experiments where a reliable ability to plan a safe path in challenging environments has been shown. In Fig. 6.14, a path was planned and safely followed by a drone, although G had

[1] https://github.com/LTU-RAI/Dsp.

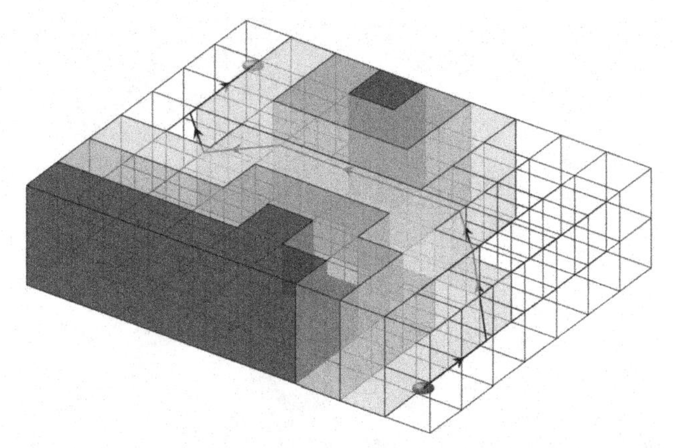

FIGURE 6.12 An illustration of how D_+^* plans a path from the blue (black in print) to the red (gray) ball with regards to the risk layer, red (dark gray) voxels is higher risk, green (light gray) voxels is lower risk, and the black voxels are occupied.

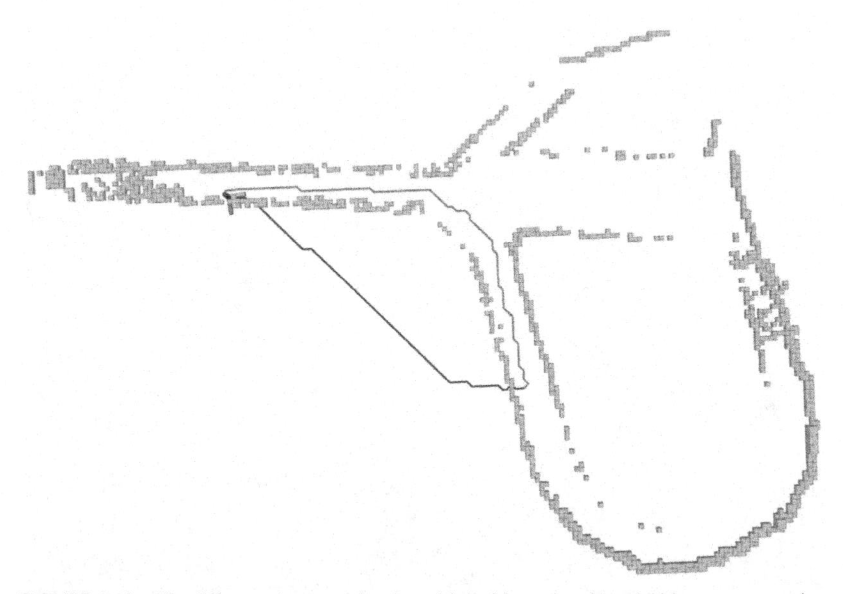

FIGURE 6.13 The difference between planning with (red (upper) path) and without representation of V_u (blue (lower) path).

some imperfections due to the narrow environment. In another experiment, the drone was tasked to go to the opposite side of a pillar-obstacle in an open space and then back to the starting position. In this case, D_+^* successfully plans a path around the pillar with suitable safety marginal without taking a longer path than necessary (e.g. the shortest safe path), see Fig. 6.15.

FIGURE 6.14 D_+^* planner plans a path from beyond the line of sight along the green (light gray in print) line.

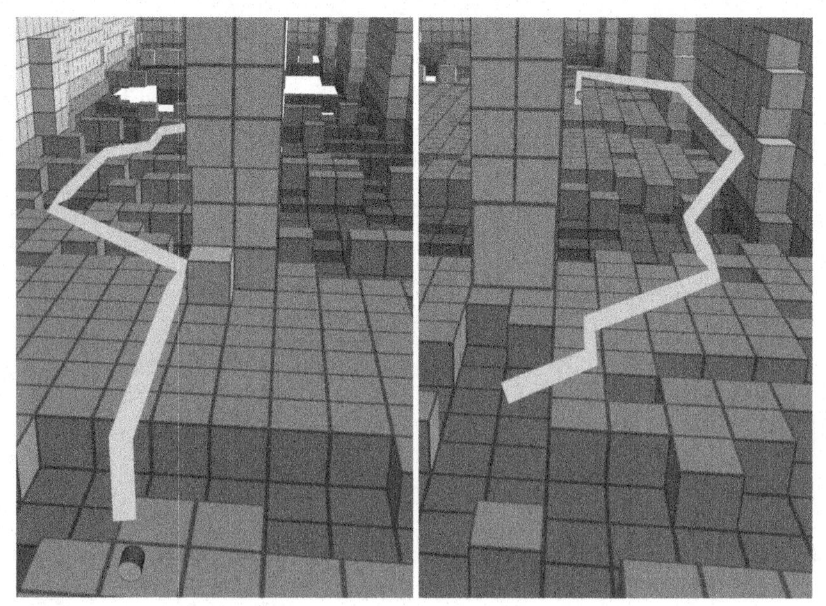

FIGURE 6.15 D_+^* planner plans a path (green (light gray in print) line) around the pillar with a suitable safety marginal. The drones position is visualized with a red (dark gray) dot.

6.4.2 Coverage path planner

Coverage path planning (CPP) is a branch of robotic motion planning that refers to the task of planning an inspection path while avoiding obstacles and ensuring complete coverage of the required volume or area of interest. Coverage planning can be classified as off-line or as an online approach. In an off-line approach, full *a-priori* knowledge of the operating environment, geometric information of the structure being inspected, and knowledge of obstacles is known while in an online version, information from onboard sensors is utilized to perform coverage tasks in real-time. Toward the task of autonomous inspection of remote infrastructures, the work presented in [23] details a coverage approach in which multiple ARWs behave cooperatively, utilizing their onboard sensors and computational resources to fulfill the task of visual inspection of a remote wind turbine.

6.4.2.1 Cooperative coverage path planner

Given a point cloud model representation of the infrastructure $S_{(x,y,z)} \subset \mathbb{R}^3$, based on the vertical camera footprint distance $\Delta\lambda \in \mathbb{R}$ from the distance $\Omega \in \mathbb{R}$ from the surface of the structure, a set of points $P_{(x,y,z)} \subset \mathbb{R}^3$ lying on the point of intersection of the horizontally sliced interval of planes $\lambda_i \in \mathbb{R}$ with $i \in \mathbb{N}$ and the structure can be obtained. Applying graph theory methodology for clustering [24], more specifically using $k - means$ method [25], m clusters can be extracted $\forall$ λ. After adding an offset distance to the clusters, for a two agent inspection scenario, the path is assigned with the two agents maintaining 180° difference. Fig. 6.16 presents an indoor scenario of coverage planning using cooperative aerial vehicles. The point cloud representation of the target structure taken by onboard sensors during the mission is given on the right. The trajectories followed by the respective vehicles during the inspection mission is presented in Fig. 6.17. A more comprehensive information on the methodology is presented in the corresponding chapter for aerial infrastructure inspection (Chapter 11).

FIGURE 6.16 On the left is the simple indoor structure to be reconstructed and on the right the cooperative pointcloud of the structure.

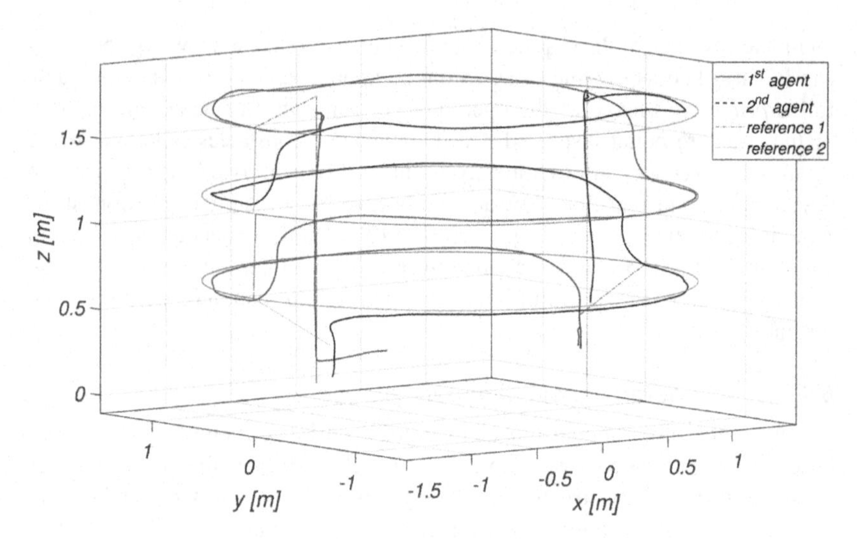

FIGURE 6.17 Trajectories that are followed in indoor experiment.

6.4.2.2 Landmark-based path planner

A landmark represents a point or a small area of interest within the surface of an object to inspect. Let $l_i \in \mathcal{L} \subset \mathbb{R}^3, i = 1, 2, \ldots, N$ where $N \in \mathbb{N}$ be the candidate landmark positions on the object surface. The set of landmarks offers an abstract and approximate representation of the external surface of a body. The higher the number of landmarks, the higher is the approximation accuracy. A landmark set representation of the surface of a 3D structure can be obtained, for example, from a point cloud or from a 3D model, using opportune sampling algorithms.

Let $\mathcal{X} \subset \mathbb{R}^n$ represent the n-dimensional region, which is sufficient to find the path for the ARW in the proximity of an object. The initial state of the ARW is denoted as $x_0 = [r_s^0, \dot{r}_s^0]^\top \in \mathcal{X}$, while $x_d^i = [r_s^i, \dot{r}_s^i]^\top \in \mathcal{X}^d \subset \mathcal{X}$ is the desired set of target states of the ARW to observe the respective i^{th} landmark. This target state x_d^i is defined by the position $r_s^i \in \mathbb{R}^3$ along the surface normal vector p^i at a distance of $r_g \in \mathbb{R}$ from the object's surface, and $\dot{r}_s^i \in \mathbb{R}^3$ is the velocity of the ARW. Finally, we denote the ARW trajectory by $x(t) \in \mathcal{X}$ and the associated control trajectory, where $\mathcal{U}$ is the feasible range of control inputs, by $u_c(t) \in \mathcal{U} \subset \mathbb{R}^3$.

The optimal trajectory planning problem can be defined as to design an optimal trajectory denoted by $x^*(t) \in \mathcal{X}$, from the ARW's initial state, to observe all the desired N landmarks under proper visual conditions. Let the ARW's observing sequence be defined as $Q = (q_0, q_1, \ldots, q_N)$, where each element $q_k \in \mathbb{Z}^+$ denotes the index of the desired ARW position at the kth sequence in the tour. The tour always starts from the ARW's initial position ($q_0 = 0 \implies x_0$), while passing through the permutation of the remaining N target states. Hence, the problem is to determine the optimal sequence of the sites to be observed and the associated trajectories for the visual coverage scenario, which will result in observing all the desired landmarks under a proper illumination condition, while minimizing the total consumption of ARW's resources or the total distance covered. Moreover, during the entire experiment, it is also required that the ARW is maintained between some maximum and minimum distances from the object's surface. These bounds are denoted as $r_{max} \in \mathbb{R}$ and $r_{min} \in \mathbb{R}$, respectively, so that the ARW does not crash into the object or move far away from the object to avoid poor coverage quality. This inspection problem can be defined as follows:

$$\min_{u_c^k(t), Q} \quad \underbrace{\sum_{k=1}^{N} \int_{t_{d,k-1}}^{t_{d,k}} \|u_c^k(t)\|^2 dt}_{\text{total control cost}} \tag{6.14a}$$

$$\text{s.t.: } Q = (q_0, q_1, \ldots, q_N) \tag{6.14b}$$

$$x(t_0) = x_0, \text{ where, } t_0 = \text{epoch}, \; x_0 \in \mathcal{X} \tag{6.14c}$$

$$\dot{x}(t) = f(x(t), u_c(t), t) \tag{6.14d}$$

$$r_{min} \leq ||r_s(t)|| \leq r_{max} \tag{6.14e}$$

$$u_c^k(t) \in \mathcal{U}, \qquad \forall k = 1, \ldots, N \tag{6.14f}$$

In (6.14), k represents the observation order of the landmarks, while the vector Q represents the ARW's observing sequence. During the kth observation sequence, the variables q_k and $t_{d,k}$ denote the index of the candidate landmark that is being observed and the ARW's departure time from the kth viewing location. The notation u_c^k denotes the control input required to observe a certain landmark during the kth observation sequence. The main objective of this cost function is to reduce the control cost required, while observing all sites represented by (6.14a). In addition, the total distance covered by the ARW can also be added to this objective function. The constraints on the ARW states are formulated through Eqs. (6.14c), (6.14d), and (6.14e) depending on the ARW's initial position, nonlinear dynamics, and the position maintained within the safety bounds to avoid any collision, respectively. The feasible range of the control inputs are maintained through (6.14f).

Thus, for the inspection task case of observing landmarks, the problem is to determine the optimal sequence and the associated ARW trajectories to observe each landmark under proper illumination conditions. This problem is very complicated to solve as it addresses the nonlinear optimization with the continuous and integer design variables, hence, falling under the mixed integer nonlinear programming category. For such problems meta-heuristic algorithms, such as genetic algorithms, particle swarm optimization, or tabu-search algorithm, etc., can be used.

6.4.3 Effective heading regulation

In certain situations, simple algorithmic regulation techniques can stand in for both planning and exploration behavior. An example of this is the heading regulation methods presented in [4,9,26] in the context of subterranean tunnel navigation. These schemes utilize the geometric shape or low-light conditions of an underground tunnel to align the body x-axis of the ARW with the tunnel axis. The ARW can then be commanded to move forward along its body x-axis while the heading regulation technique continuously aligns it with the tunnel, while a local obstacle avoidance scheme maintains a safe distance from obstacles and walls. In general, these methods fall apart when the environment changes into something that is not a tunnel, and there is no concept of optimality or "shortest distance" to get somewhere. But, in the specific environment, they were designed for, the simple algorithmic implementations can stand in for both map-based path planning and exploration behavior while being significantly less computationally heavy, completely reactive, and most importantly do not rely on an occupancy-map representation of the environment as shown in Fig. 6.18, constituting them a fundamental component for the autonomy of resource constrained aerial platforms.

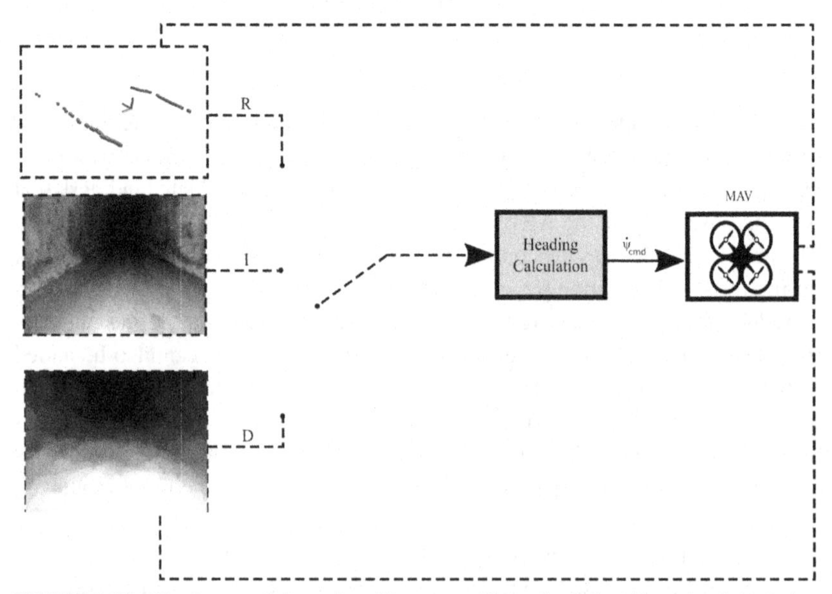

FIGURE 6.18 Heading regulation using either range (R) or visual data (I) or visual depth data (D). The dashed lines denote the sensor data from the MAV, while the solid line denotes the yaw rate command ($\dot{\psi}$) provided to the MAV.

This decouples the navigation and mapping problems completely (and in a sense, the navigation problem and the estimation of position states as well), which is a major focus for this type of navigation scheme. This Chapter presents three methods for effective heading regulation: a) using 2D LiDAR measurements to approximate the geometry (and as such the direction) of the tunnel, b) using an RGB camera to track the "darkest area" in the tunnel through contour detection, using the combination of the lack of natural illumination in the tunnel with onboard lights under the assumption that the direction of the tunnel follows the darkest area, and c) using a depth-camera to align the heading with the deepest area in the tunnel by clustering points of similar depth and heading towards the deepest cluster. The remaining parts of this section provide a description of the different heading regulation methods.

6.4.3.1 LiDAR-based heading regulation

Initially, three range measurements r_i from 2D LiDAR ranges $R = \{r_i | r_i \in \mathbf{R}^+, i \in \mathbf{Z} \cap [-\pi, \pi]\}$ are measured for the right d_r, center d_c, and left d_l distances. The $\vec{d}_c$ has a direction toward the x-axis of the body frame ($\theta_{geom,c} = 0$), $\vec{d}_r$ and $\vec{d}_l$ are right and left vectors with a constant value of $\theta_{geom,r} \in [0, \pi/2]$, and $\theta_{geom,l} \in [-\pi/2, 0]$ from the x-axis of the body frame, respectively. The values of the $\theta_{geom,r}$ and $\theta_{geom,l}$ are symmetric and depend on the design choices, and for clarity it is called θ_{geom}. The resultant distance vector of d_r, d_c, and d_l is calculated as:

$$\vec{d}_{sum} = \vec{d}_r + \vec{d}_c + \vec{d}_l \tag{6.15}$$

Based on the direction of the $\vec{d}_{sum}$, the heading can be obtained. However, achieving accurate heading especially for low-cost navigation systems is not always possible. Furthermore, most of the methods rely on information from a magnetometer and gyrocompass, multi-antenna Global Navigation Satellite Systems (GNSSs), or position information of the platform [27]. Nonetheless, magnetometer is not reliable especially in underground mines, and GNSS is not available for underground areas. Thus, in the proposed method, the obtained heading is converted to a heading rate by dividing the $\angle\vec{d}_{sum}$ by θ_{geom} as shown in (6.16). The $\dot{\psi}$ is bounded between -1 and 1, e.g., $d_r = 0$, $d_l = 0$, and $d_c > 0$ that results in $\dot{\psi} = 0$ rad/sec, $d_c = 0$, $d_l = 0$, $d_r > 0$ results in $\dot{\psi} = -1$ rad/sec, and $d_r = 0$, $d_c = 0$, and $d_l > 0$ result in $\dot{\psi} = 1$ rad/sec. However, a larger value of θ_{geom} can be selected to have slower heading rate commands.

$$\dot{\psi} = \frac{\angle\vec{d}_{sum}}{\theta_{geom}} \tag{6.16}$$

It should be highlighted that in order to avoid disturbances in the range measurements, the array of ranges can be used for $\vec{d}_c$, $\vec{d}_r$, and $\vec{d}_l$. Algorithm 2 provides a brief overview while the following part provides the outlook of the vector geometry based approach for generating heading rate commands, while the $\theta_{geom} = \pi/3$, the array of ranges from $[-\pi/4, \pi/4]$, $[\pi/4, \pi/2]$, and $[-\pi/2, \pi/4]$ are considered for $\vec{d}_c$, $\vec{d}_r$, and $\vec{d}_l$, respectively.

Algorithm 2: Heading rate based on 2D lidar.

1 **Inputs:** R
Result: Calculate heading rate command based on 2D lidar measurements
2 $\vec{d}_c = \sum_{i=-\pi/4}^{\pi/4} r_i[\cos\theta_i, \sin\theta_i]^\top$
3 $\vec{d}_r = \sum_{i=\pi/4}^{\pi/2} r_i[\cos\theta_i, \sin\theta_i]^\top$
4 $\vec{d}_l = \sum_{i=-\pi/2}^{-\pi/4} r_i[\cos\theta_i, \sin\theta_i]^\top$ //obtaining $\vec{d}_c$, $\vec{d}_r$, and $\vec{d}_l$ for arrays of ranges
5 $\vec{d}_{sum} = \vec{d}_r + \vec{d}_c + \vec{d}_l$
6 $\dot{\psi}_r = \frac{\angle \vec{d}_{sum}}{\theta_{geom}}, \theta_{geom} = \pi/3$
7 **Output:** $\dot{\psi}$

6.4.3.2 *The weighted arithmetic mean*

In the weighted arithmetic mean [28], the range value for each angle is used as a weight for the corresponding angle. The weighted arithmetic mean is calculated as:

$$\psi_k = \frac{\sum_i r_{i,k}\xi_i}{\sum_i r_{i,k}} \tag{6.17}$$

where $i \in [i_{min}, \; i_{max}]$ is the set of beam angle indexes, and r_i is the range measurement for each corresponding angle ξ of 2D lidar rotation. ξ_{min} and ξ_{max} define a range of angles, and there are angles for y^+ and y^- axes in the body frame; thus the range measurements from y^- axis up to y^+ based on the right hand rule are used. However, this result is noisy and subject to spurious changes in the environment. For improving this, the integrated z-axis of the gyro in the Inertial Measurement Unit (IMU) is used as a prior using a complementary filter as follows.

Initially, the heading is integrated from the z-axis angular rate $\omega_{z,k}$ to predict the movement as:

$$\hat{\psi}_{k|k-1} = \hat{\psi}_{k-1|k-1} + \omega_{z,k} T_s, \tag{6.18}$$

which in the sequel is corrected using the estimated angle from the 2D lidar:

$$\hat{\psi}_{k|k} = \beta\hat{\psi}_{k|k-1} - (1-\beta)\psi_k, \tag{6.19}$$

where $\beta \in (0, \; 1)$ is a classic complementary filter, which was chosen due to its simplicity in tuning for generating robust results. It should be also noted the negation in the second term, $-(1-\beta)\psi_k$, that is generated from the fact that the measured free space always rotates with the inverted angular rates relative to the MAV.

The obtained heading angle is fed to the low level proportional controller with gain of $k_p > 0$ to always have the MAV point toward the open space:

$$\dot{\psi}_r = -k_p \hat{\psi}_{k|k}. \tag{6.20}$$

6.4.3.3 Darkest contour heading regulation

This heading regulation method relies on the color camera image processing in low illuminated areas. The input to the algorithm is an image I from the on-board camera. It is considered that the images are composed by a set of physical objects (walls, pipes, lights, etc.), as well as a background. The first step deals with the separation of such physical objects from the background, which is done by finding a threshold that separates the data and uses that threshold to create a binary image [29].

The threshold t partitions the pixels of the image based on their intensity into two classes: 1) $\mathcal{T}_1 = \{0, 1, ..., t\}$, which represents the background class and 2) $\mathcal{T}_2 = \{t+1, t+2, ..., L-1\}$, which represents the foreground class where L the number of gray levels of the image. The threshold value t is found by maximizing the measure of class separability:

$$S(t) = \frac{\sigma^2_{Between}(t)}{\sigma^2_{Total}} \tag{6.21}$$

where σ^2_{Total} is the total variance of the histogram, and $\sigma^2_{Between}(t)$ is the variance between the two classes.

The optimal threshold t^* is computed as:

$$t^* = \underset{t}{\operatorname{argmax}}\, S(t) \tag{6.22}$$

After the calculation of the threshold value, the image is converted to binary, and the binary image is used to find the largest set of connected background pixels. Tracing the contour of the background is performed using the Moore-Neighborhood tracing method [30].

Let $p(x_k, z_l)$ be the pixel related to the image $I^{N \times M}$, and $k \in [0, n] \cap \mathbf{Z}$, $l \in [0, m] \cap \mathbf{Z}$ are position in the image with $N \times M$ pixels resolution. Then the Moore-Neighborhood $M(p(x_k, z_l))$ is the set of all pixels that share a vertex or an edge with $p(x_k, z_l)$ as follows:

$$M(p(x_k, z_l)) = \{p(x, y) : x = x_k + k, z = z_l + l, K = \{-1, 0, 1\}, \\ l = \{-1, 0, 1\}, (k, l) \neq (0, 0)\} \tag{6.23}$$

After the contour of each binary object has been computed, the object with maximum area is selected, and the centroid of the bounding box for this object is calculated and returned as the heading point coordinates. The method is described in Algorithm 3.

Algorithm 3: Centroid extraction using Darkness Contours.

1 **Input:** RGB image acquired by the forward looking camera.
Start;
- Step 1. Convert the RGB image to binary image using Otsu's threshold.
- Step 2. Extract contour of background objects using Moore-Neighborhood tracing algorithm.
- Step 3. Identify the darkness in the tunnel as the background object with largest area.
- Step 4. Compute a bounding rectangle for this area and return the centroid of the rectangle (s_x, s_y).

End

– Output: Pixel coordinates of the heading point (s_x, s_y).

Fig. 6.19 presents snapshots of the method evaluated in experimental runs for autonomous navigation of an aerial platform in an underground tunnel using color camera for the heading alignment. Figures from a) to c) depict the common output of the method in cases where the deepest part of the tunnel is distinctively recognized, while from d) to f) depict more challenging situations, where the illumination, dust or motion blur affect the outcome, thus without jeopardizing the stability of the mission. The method cannot provide results in well-illuminated areas.

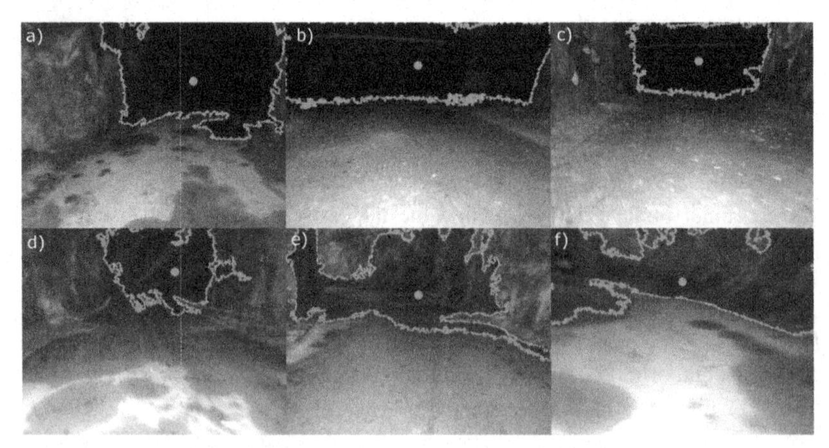

FIGURE 6.19 Extracted centroid in color images for tunnel following heading regulation. The contours denote the identified deepest areas of the tunnel.

6.4.3.4 Deepest point heading regulation

This method considers the processing of depth images. Initially, the depth images are pre-filtered using a gray scale morphological close operation [31] to remove noise and enhance the tunnel opening information in the depth frame. The filtered depth images are then processed with a clustering algorithm to identify regions of common depth and essentially extract the region that captures the deepest part of the tunnel. The clustering algorithm considered in this method is k-means, which extracts a fixed number of clusters $C_i, [i = 1, 2, 3, \ldots, N_{clusters}]$, where the $N_{clusters} = 10$ value has been selected based on the tunnel environment morphology. The number of clusters is a critical parameter and should be selected according to the geometry and morphology of the subterranean area. Once the clusters are calculated, each cluster region is evaluated based on their intensity to select the cluster with the maximum intensity, since the method relies on this assumption for segmenting the depth of the tunnel. The x-axis pixel coordinate of the cluster centroid is calculated as $s_x = \frac{1}{|C_m|}\sum_{(x,y)\in C_m} x$, where $|C_m|$ represents the number of pixels.

Fig. 6.20 shows an application of such a navigation technique in a real subterranean tunnel area, where the Deepest-Point technique aligns the ARW with the tunnel direction, and the Artificial Potential Field presented in Section 6.3.1 keeps the ARW in the middle of the tunnel. The result is a completely reactive navigation (e.g., not based on a map of the environment) architecture for an unknown environment.

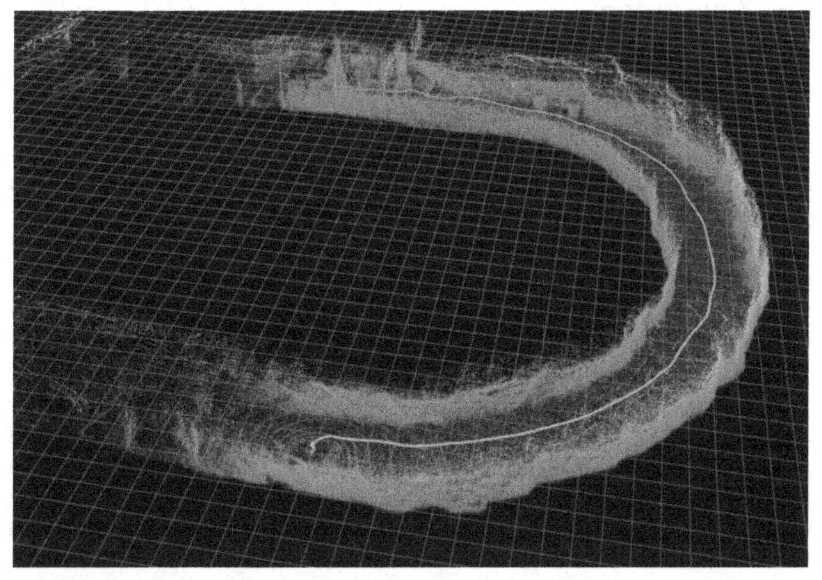

FIGURE 6.20 Exploration and Navigation through a curving subterranean tunnel environment using a combination of Heading Regulation and Artificial Potential Fields. The green (gray in print) line is the resulting exploration path of the ARW.

6.4.3.5 *Centroid to heading rate command mapping*

Methods presented in Sections 6.4.3.3 and 6.4.3.4 are used for regulating the heading of the aerial platform with the tunnel axis. These methods extract centroid coordinates and require an additional processing step to converted them to heading rate command $\dot{\psi}_d$. In this case, the x-axis coordinate is only considered. Initially, the pixel coordinate s_x is normalized to [0, 1] using the normalization parameter α equal to the width of the image. The second step considers the generation of the yaw rate command in the interval $[\dot{\psi}_{d,min}, \dot{\psi}_{d,max}]$ through the parameter β since the selection of β defines the yaw rate limits. The overall mapping scheme is described in Algorithm 4 as follows.

Algorithm 4: Heading rate based on the centroid of the open space.

---

1 **Input:** RGB image acquired by the forward looking camera.
- Input: Centroid of the heading point (s_x)
  Start;
- Step 1. Linear mapping $\overline{s}_x \to [0, 1]$;$\overline{s}_x = \frac{s_x}{\alpha}$, $\overline{s_x} \in [0, 1]$
- Step 2. Linear mapping $\overline{s}_x \to [\dot{\psi}_{d,min}, \dot{\psi}_{d,max}]$ rad/sec
  $\dot{\psi}_d = \frac{\overline{s}_x - 0.5}{\beta}$
  End
- Output: Heading rate correction $\dot{\psi}_d$.

6.5 Conclusion

This chapter has presented navigation schemes comprised of reactive local path planners and map-based global path planners that are used for ARWs to autonomously perform tasks safely and optimally.

For local reactive navigation, the chapter presented two different methodologies: the high-TRL and easily applicable Artificial Potential Field that can directly work with raw sensor data, and the modern NMPC with integrated obstacle avoidance that can account for moving obstacles and generates optimal trajectories for proactive and efficient obstacle avoidance maneuvers. A global risk-aware path planning framework was also presented in the D_+^* path planner, which uses an occupancy grid representation of its environment to find the shortest path between any two points in a global map. The chapter has also presented the idea of complete coverage path planning. The goal is to generate a trajectory that completely inspects a piece of infrastructure, a very common use-case of an ARW. The chapter was concluded by introducing of a series of application-driven heading regulation techniques that simplify the path planning problem for the use case of subterranean tunnel navigation utilizing the tunnel morphology to generate exploratory behavior for the ARW.

Finally, to support the efficacy of the presented navigation schemes, numerous application-orientated examples have also been showcased to provide real-life examples utilizing these autonomy modules. Reliable navigation architectures are an enabling technology for any autonomous mission use-case for an ARW, as being able to plan a safe path through its environment, avoid any obstacles encountered along the way, and fulfill mission criteria such as inspection coverage, are absolute necessities for real-life applications of fully autonomous operations of ARWs.

References

[1] S.M. LaValle, Planning Algorithms, Cambridge University Press, 2006.

[2] C. Goerzen, Z. Kong, B. Mettler, A survey of motion planning algorithms from the perspective of autonomous UAV guidance, Journal of Intelligent and Robotic Systems 57 (1–4) (2010) 65.

[3] P. Pharpatara, B. Hérissé, Y. Bestaoui, 3-D trajectory planning of aerial vehicles using RRT, IEEE Transactions on Control Systems Technology 25 (3) (2016) 1116–1123.

[4] S.S. Mansouri, C. Kanellakis, D. Kominiak, G. Nikolakopoulos, Deploying MAVs for autonomous navigation in dark underground mine environments, Robotics and Autonomous Systems 126 (2020) 103472.

[5] S.S. Mansouri, C. Kanellakis, E. Fresk, B. Lindqvist, D. Kominiak, A. Koval, P. Sopasakis, G. Nikolakopoulos, Subterranean MAV navigation based on nonlinear MPC with collision avoidance constraints, arXiv preprint, arXiv:2006.04227, 2020.

[6] C.W. Warren, Global path planning using artificial potential fields, in: 1989 IEEE International Conference on Robotics and Automation, IEEE Computer Society, 1989, pp. 316–317.

[7] E. Rimon, D.E. Koditschek, Exact robot navigation using artificial potential functions, Departmental Papers (ESE), 1992, p. 323.

[8] D. Droeschel, M. Nieuwenhuisen, M. Beul, D. Holz, J. Stückler, S. Behnke, Multilayered mapping and navigation for autonomous micro aerial vehicles, Journal of Field Robotics 33 (4) (2016) 451–475.

[9] C. Kanellakis, S.S. Mansouri, G. Georgoulas, G. Nikolakopoulos, Towards autonomous surveying of underground mine using MAVs, in: International Conference on Robotics in Alpe-Adria Danube Region, Springer, 2018, pp. 173–180.

[10] E. Small, P. Sopasakis, E. Fresk, P. Patrinos, G. Nikolakopoulos, Aerial navigation in obstructed environments with embedded nonlinear model predictive control, in: 2019 18th European Control Conference (ECC), IEEE, 2019, pp. 3556–3563.

[11] B. Lindqvist, S.S. Mansouri, G. Nikolakopoulos, Non-linear MPC based navigation for micro aerial vehicles in constrained environments, in: 2020 European Control Conference (ECC), May 2020.

[12] B. Lindqvist, S.S. Mansouri, A.-a. Agha-mohammadi, G. Nikolakopoulos, Nonlinear MPC for collision avoidance and control of UAVs with dynamic obstacles, IEEE Robotics and Automation Letters 5 (4) (2020) 6001–6008.

[13] S.S. Mansouri, C. Kanellakis, B. Lindqvist, F. Pourkamali-Anaraki, A.-A. Agha-Mohammadi, J. Burdick, G. Nikolakopoulos, A unified NMPC scheme for MAVs navigation with 3D collision avoidance under position uncertainty, IEEE Robotics and Automation Letters 5 (4) (2020) 5740–5747.

[14] S. Sharif Mansouri, C. Kanellakis, E. Fresk, B. Lindqvist, D. Kominiak, A. Koval, P. Sopasakis, G. Nikolakopoulos, Subterranean MAV navigation based on nonlinear MPC with collision avoidance constraints, International Federation of Automatic Control, 2020.

[15] J. Jackson, G. Ellingson, T. McLain, ROSflight: A lightweight, inexpensive MAV research and development tool, in: 2016 International Conference on Unmanned Aircraft Systems (ICUAS), June 2016, pp. 758–762.

[16] P. Sopasakis, E. Fresk, P. Patrinos, Optimization engine [Online]. Available: http://doc.optimization-engine.xyz/, 2019.

[17] P. Sopasakis, E. Fresk, P. Patrinos, Open: Code generation for embedded nonconvex optimization, International Federation of Automatic Control, 2020.

[18] B. Hermans, P. Patrinos, G. Pipeleers, A penalty method based approach for autonomous navigation using nonlinear model predictive control, IFAC-PapersOnLine 51 (20) (2018) 234–240.

[19] B. Lindqvist, P. Sopasakis, G. Nikolakopoulos, A scalable distributed collision avoidance scheme for multi-agent UAV systems, arXiv preprint, arXiv:2104.03783, 2021.

[20] F. Duchoň, A. Babinec, M. Kajan, P. Beňo, M. Florek, T. Fico, L. Jurišica, Path planning with modified a star algorithm for a mobile robot, Procedia Engineering 96 (2014) 59–69.

[21] A. Hornung, K.M. Wurm, M. Bennewitz, C. Stachniss, W. Burgard, OctoMap: An efficient probabilistic 3D mapping framework based on octrees, Autonomous Robots 34 (3) (2013) 189–206.
[22] M. Sheckells, DSL Gridsearch: D*-lite on a uniformly spaced 3D or 2D grid [Online]. Available: https://github.com/jhu-asco/dsl_gridsearch, 2018. (Accessed February 2021).
[23] S.S. Mansouri, C. Kanellakis, E. Fresk, D. Kominiak, G. Nikolakopoulos, Cooperative coverage path planning for visual inspection, Control Engineering Practice 74 (2018) 118–131.
[24] D.B. West, et al., Introduction to Graph Theory, vol. 2, Prentice Hall, Upper Saddle River, 2001.
[25] S. Lloyd, Least squares quantization in PCM, IEEE Transactions on Information Theory 28 (2) (1982) 129–137.
[26] B. Lindqvist, C. Kanellakis, S.S. Mansouri, A.-A. Agha-Mohammadi, G. Nikolakopoulos, Compra: A compact reactive autonomy framework for subterranean MAV based search-and-rescue operations, arXiv preprint, arXiv:2108.13105, 2021.
[27] K. Gade, The seven ways to find heading, Journal of Navigation 69 (5) (2016) 955–970.
[28] A. Madansky, H. Alexander, Weighted standard error and its impact on significance testing, The Analytical Group, Inc., 2017.
[29] N. Otsu, A threshold selection method from gray-level histograms, IEEE Transactions on Systems, Man and Cybernetics 9 (1) (1979) 62–66.
[30] G. Blanchet, M. Charbit, Digital Signal and Image Processing Using MATLAB, Wiley Online Library, vol. 4, 2006.
[31] P. Soille, Morphological Image Analysis: Principles and Applications, 2nd ed., Springer-Verlag, Berlin, Heidelberg, 2003.

Chapter 7

Exploration with ARWs

Akash Patel, Samuel Karlsson, Björn Lindqvist, and Anton Koval
Department of Computer, Electrical and Space Engineering, Luleå University of Technology, Luleå, Sweden

7.1 Introduction

The continuous development of robotic technologies is constantly increasing the number of real-life and mission-oriented deployments in a variety of environments with different complexity and time-varying characteristics. Among the latest major trends in the robotics community, the focus is on demonstrating resilient and robust autonomy in large-scale subterranean (Sub-T) environments, for example, the case of the DARPA Sub-T competition [1]. In such applications, one of the main challenges and real requirements for the robotic autonomy frameworks is the ability to ensure safe and robust navigation of the robotic platforms while operating in a-priori known, spatially, or completely unknown environments. Except the fundamental problem of localization, the problem of navigation in these environments and the planning of safe and optimal trajectories for reaching goal points in a 3D space have been in the focus of robotics research for many decades [2,3].

The representation of the real world in a 3D map form is an important aspect of path planning, while a common approach to generating these maps is to store geometric properties of the environment in a form of a 3D point cloud. However, with the dynamically increasing size of the environment, its representation with a point cloud becomes computationally expensive and not memory efficient due to data redundancy, which makes the overall approach unfeasible to be utilized in large-scale scenarios [4]. Furthermore, point clouds do not provide differentiation between free and occupied spaces, thus is very common in robotics to represent an arbitrary environment in various approaches, such as the OctoMap [5], the Elevation map [6], the Signed Distance Field (SDF) map [7], and the multi-level surface map [8]. From all these approaches, the OctoMap framework has become one of the most frequently utilized occupancy grid generators, which is based on octotrees [9] and provides the capability to dynamically adapt the map to be efficiently used in path planning.

From a path planning perspective, the most classical approaches with re-planning capabilities in dynamic environments are the D* Lite [10], the Field D* [11], and the Theta* [12]. These algorithms are capable of fast re-planning

Aerial Robotic Workers. https://doi.org/10.1016/B978-0-12-814909-6.00013-5

based on the local information and can identify situations when the path is blocked. For these cases, these algorithms store the computed path after the obstacle and try to find the new path, which can lead either to he stored path after the obstacle or to the goal if a shorter path exists. However, these algorithms need to be extended for a full operation in 3D environments and especially integrated into more generic and combined exploration frameworks. Related works like [13], [14], and [15] extended these three algorithms to be applied in the case of 3D environments; however, none of them was tested in dynamically expanded 3D environments and at a large scale, while with the term large scale, we consider situations where the local surroundings environmental perception is multiple and sequentially expanded, for example, the case of a robot equipped with a 3D laser scanner and the merging of a progressive 3D dense point cloud registrations during the advancement of the exploration task.

To evaluate the performance of the path planning algorithms and in realistic scenarios, there is a need to integrate this with related 3D mapping frameworks like [5–8] that are capable of handling dynamic map updates. For example, in [16], the authors introduced a heuristic angular search method that utilizes a direct way-point search method on the global map, which is created by streaming the point cloud of obstacles from onboard sensors to OctoMap framework. In [17], the authors proposed a perception-aware trajectory re-planning method for ARW, which uses a gradient-based approach for re-planning and is integrated with a volumetric mapping framework. In [18], an online path planning method for autonomous underwater vehicles has been presented. In their approach, the transition-based RRT method was used to obtain the path, while the volumetric representation of the environment was modeled by means of an OctoMap framework. In [19], the frontier cell-based exploration was integrated with the Lazy Theta* path planning algorithm and adopted to OctoMap, which was built on the point cloud raw data. However, there was no performance evaluation of the algorithm throughout a mission with the increasing amount of information, while they considered a very limited small 3D world.

On the basis of the state-of-the-art literature, this chapter proposes first an integrated exploration and path planning framework, suitably configured for real-time implementations for expanding large-scale maps. The presented scheme is demonstrated in an expanding map handling scenario under continuous obstacle-free path planning. Then we introduce a full 3D D* algorithm suitable for path planning over Octomaps and propose an algorithmic revision of generated Octomaps that is able to handle false opening/holes in the considered map expansions. We also evaluate the overall algorithmic enhancements of the combined frontier exploration and D* path planning regarding the computational efficiency, the updating rates and the overall update time in a large-scale online navigation simulated scenario. Secondly, this chapter introduces the Exploration Rapidly Exploring Random Tree (ERRT)-based methodology for exploration of the unknown environment. Unlike the frontier-based method, ERRT uses sampling-based exploration approach. Each pseudo-random goal is computed while subject to minimizing actuation effort and maximizing the information gained to continue the exploration efficiently. The details of this approach will be explained briefly, along with an example in Section 7.3.2.

7.2 The overall framework for exploration and path planning

An overview of the proposed framework and its architecture is shown in Fig. 7.1, where the combined exploration and path planning framework has as a baseline a robotic platform equipped with a sensor that provides ranging measurements of the environment in the form of 3D point cloud. As will be demonstrated below, the point cloud is stored in a volumetric form utilizing the OctoMap mapping framework [5]. It builds and maintains the global representation of the environment and provides OctoMap updates O to an occupancy grid that creates the 3D occupancy grid G, which is essential for path planning and re-planning in a dynamically expanding environment. In the sequel, the D* lite algorithm computes the path based on the Occupancy grid updates G, while receiving information about the current position $p(x, y, z)$ from the robot and the next position or way-point $w(x, y, z)$ from the frontiers exploration algorithm [20]. On the basis of this information, the D* lite computes the path $P_{p \to w}$, which afterward is provided to a nonlinear model predictive controller [21] to move the robot to the next desired way-point. The overall system design has a modular structure that can be linked with any ARW and software packages based on Robot Operating System (ROS) [22], and it can provide the current position p of a robot, the next way-point $w(x, y, z)$, and the volumetric occupancy information.

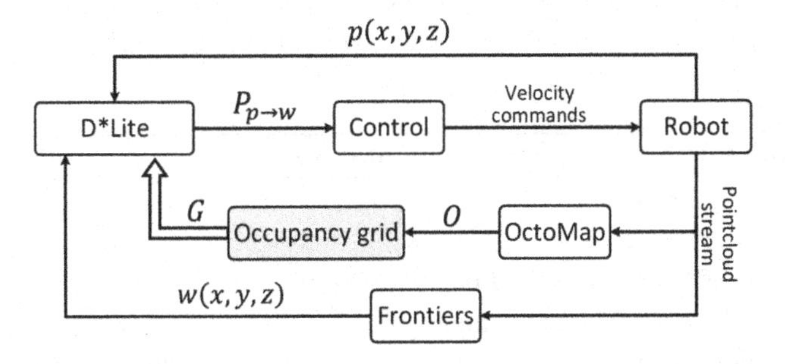

FIGURE 7.1 Software architecture.

7.2.1 Environment representation

To generate a global map of the surrounding environment from an instantaneous 3D point cloud, the OctoMap framework [5] has been utilized providing the Octomap O as depicted in Fig. 7.2. O is built from voxels $v(x, y, z)$ with center coordinates at x, y, z and rib length l, while each voxel can be either occupied or free and is denoted by v_o and v_f, respectively. Additionally, O contains the map dimensions and the overall resolution information needed to reconstruct the environment to be explored.

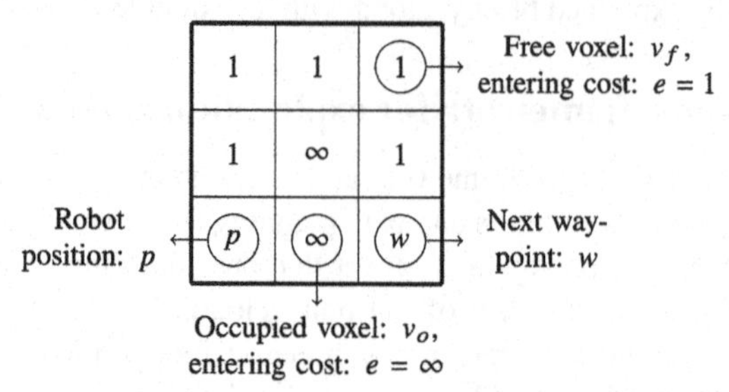

FIGURE 7.2 A schematic view of one layer of the 3D OctoMap O, with hidden layers in front or behind the visualized layer. In this example, the map size is 3 × 3, while here, only a 2D projection is visualized. At the same time, each of the squares represents a voxel v and with the corresponding voxel's value to represent the entrance cost e, 1 for free space and ∞ for the occupied space. The robot's current position p and the next way-point w in this representation are the hiding voxels entrance cost from this view.

The OctoMap updates are used to build the 3D occupancy grid G, where $\forall v \in O : \exists v \in G$, with an entry-cost $e = \infty \iff v_o$ and $e = 1 \iff v_f$, for path planning. In this approach, the grid G is connected in such a way that the movement cost from any given v to its neighbor v' (diagonal included) is $C(v, v') = e_{v'}$. In Algorithms 1–3, the steps required to transform the OctoMap O to 3D grid G are shown.

The creation of a 3D occupancy grid G is presented in Algorithm 2. This is a slow and time-consuming process, and it is desirable to update it with new information instead of recreating it as often as possible. Updates of the 3D occupancy grid G are carried out by the proposed Algorithm 3, requiring the OctoMap O has the constant dimensions.

Algorithm 1: Decide if G will be updated or recreated.

```
1 Input: New OctoMap ⁿO
2 Old OctoMap ˡO
3 Output: 3D occupancy grid G
4 if size of ⁿO == size of ˡO then
5 |  Algorithm 3                    ▷ If the same update existing G
6 else
7 |  Algorithm 2                    ▷ Else create a new G
```

To decide whether the 3D occupancy grid G can be updated or needs to be recreated, Algorithm 1 is utilized. This algorithm checks if the incoming OctoMap O size is the same as the last received OctoMap O. If the size is the same, Algorithm 3 will be called to update the 3D occupancy grid G; otherwise, Algorithm 2 will be called to create a new 3D occupancy grid G. The update

Algorithm 3 is looping through the OctoMap to check which voxels v have changed and then update those voxels in the 3D occupancy grid G. To create a new 3D occupancy grid G, multiple conversion steps are needed. First, the map is mapped to the array that is used to create the grid. Subsequently, the grid is used as a base to create a connected 3D occupancy grid, which is the resulting G on which path planning is performed.

Algorithm 2: Create 3D occupancy grid G.

1 **Input:** New OctoMap ^{n}O
2 **Output:** New 3D occupancy grid G
3 $m \leftarrow$ new array[size of O : 1]
4 **for** ***each*** v ***in*** ^{n}O **do**
5 | $m[v(x, y, z)] \leftarrow e_v$
6 OctoGrid $\leftarrow$ form m
7 ▷ Connect and search the OctoGrid
8 $G \leftarrow$ from OctoGrid

Algorithm 3: Update 3D occupancy grid G.

1 **Input:** New OctoMap ^{n}O
2 Old OctoMap ^{l}O
3 **Output:** Updated 3D occupancy grid G
4 **for** ***each*** ^{n}v ***in*** ^{n}O **do**
5 | $^{l}v \leftarrow {}^{n}v$'s corresponding $v \in {}^{l}O$
6 | **if** $e_{nv} \neq e_{lv}$ **then**
7 | | ▷ Update G's corresponding v
8 | | $G({}^{n}v(x, y, z)) \leftarrow e_{nv}$

7.2.2 Frontier based exploration

The fundamentals of frontier-based exploration technique were initially published by Yamauchi [23]. In this approach, the sensor readings were obtained by a laser-limited sonar. As a next step, a probability was assigned to each cell of the created evidence grid. The ability to fuse information from multiple different sensors in order to update the map is an advantage of using evidence grids. The main challenge associated with sonar is degraded accuracy when a beam is reflected from an uneven surface. Due to this uncertainty, the cell either seems free, or if it is occupied, the cell is sensed to be further away from its actual position. The standard evidence grid assumes that each sensor reading is independent of every other sensor reading. Therefore, if the laser scanner reads the distance of the obstacle to be less than that of sonar, they update the evidence grid as if sonar had returned the value of laser scanner while marking the cell returned by laser to be occupied. As a result, the evidence grid has fewer errors caused by specular reflections, and obstacles are also detected above and below the plane of the laser scanner. Frontier detection is done by grouping and labeling the cells based on their occupancy probabilities.

The only case where the occupancy probability will be equal is when the occupancy information about the particular node has not been updated or registered with the sensor data in the octree. The process of defining a cell to be a frontier is divided into three steps. In the first step, an open cell adjacent to an unknown cell is labeled as a frontier edge cell. In the second step, all adjacent frontier edge cells are grouped into frontier regions. In the third step, a frontier region that is roughly of the same size as the robot is defined as a frontier. Now once the frontier has been detected, the planner uses Depth-First-Search on the occupancy grid starting from robot's current cell and attempting the shortest obstacle-free path to the closest frontier, which has not been visited before. For path planning in [23], the evidence grid is considered as a graph of nodes, where the center of each open cell represents a node, and each node is connected to all adjacent open cells. Reactive obstacle avoidance is added to avoid dynamic obstacles, which were not present when the grid map was constructed. When the robot reaches the goal position, the location is added to the list of previously visited frontiers, upon which the robot performs a 360-degree sensor sweep to update the occupancy grid. The robot then attempts to reach the next closest new frontier in the updated grid.

Fig. 7.3 describes frontiers as a concept with respect to the field application of using aerial robots for exploration purposes. This technique of single robot frontier exploration is adapted to a group of robots to achieve the exploration task in less time. For this multi-robot case, each robot has its own occupancy grid. When a robot reaches a new frontier, it does a sensor sweep and stores the information about its surrounding in a local grid. The local grid is then updated in the global grid, and the local grid is also sent to other robots. The other robots update this local grid in their own global grid. A method of log odds representation is used to integrate two evidence grids in real-time. Applying this representation, independent probabilities can be combined using addition rather than multiplying them. The log odd of each pair of cells is added, and the sum is stored in the corresponding cell of the new grid. The probabilities are normalized so that the log prior probability is zero. As a result, the cells with no prior information in either grid remain equal to the prior probability in the new grid. The main advantage of this approach is that it is cooperative and decentralized. It also allows getting data from another robot to decide where to navigate. Another advantage is that a robot can identify which areas are being explored by another robot and then choose to navigate to an unexplored area. However, there are also some limitations of this method. As the navigation is independent, it may not be optimally efficient for the overall task in some cases. Reactive navigation helps avoid obstacles, but it can be blocked by other robots. Based on the current approach to choosing which frontier to go to, a frontier might sometimes

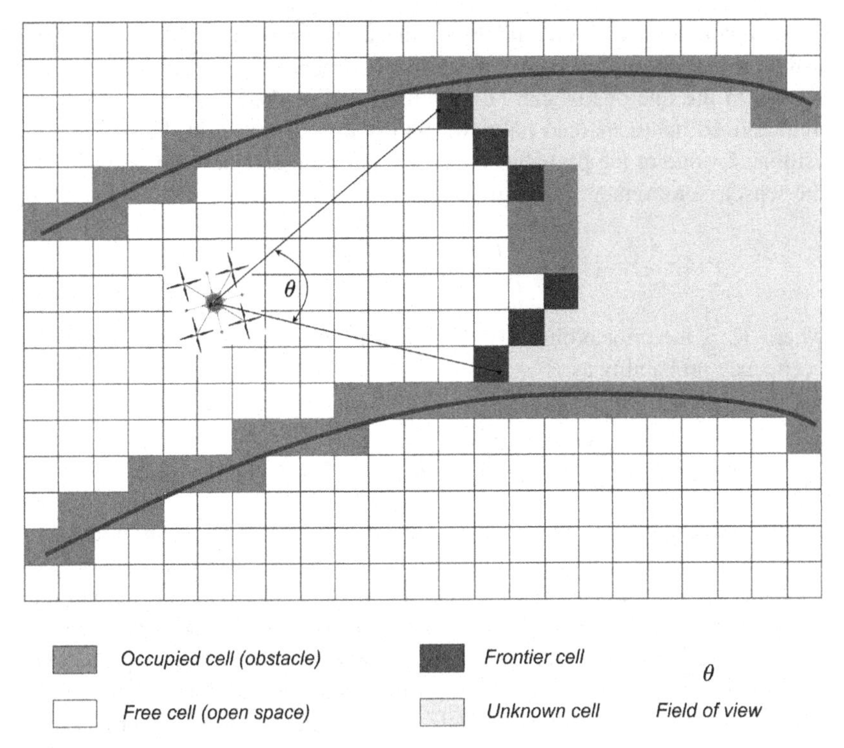

FIGURE 7.3 Description of frontiers.

be inaccessible. In this case, the robot will mark the place of the blocked robot as inaccessible. However, the other robot can still navigate to a new frontier.

Frontier based method uses occupancy grid maps as a mapping algorithm that can generate a 2D or 3D probabilistic map. A binary value of occupancy is assigned to each cell that represents a cell to be either free or occupied in the grid. Suppose $m_{x,y}$ is the occupancy of a cell at (x, y) and $p(m_{x,y} \mid z^t, x^t)$ is the numerical probability, then using a Bayes filter [24], the odds of the cell to be occupied can be denoted as:

$$\frac{p(m_{x,y} \mid z^t, s^t)}{1 - p(m_{x,y} \mid z^t, s^t)} = \frac{1 - p(m_{x,y})}{p(m_{x,y}} \times \frac{p(m_{x,y} \mid z^{t-1}, s^{t-1})}{1 - p(m_{x,y} \mid z^{t-1}, s^{t-1})}$$

where z^t represents a single measurement taken at the location s^t. To construct a 3D occupancy grid, an existing framework OctoMap [5] based on octree is used in this work. In this case, each voxel is subdivided into eight voxels until a minimum volume defined as the resolution of the octree is reached. In the octree, if a certain volume of a voxel is measured, and if it is occupied, then the node containing that voxel can be initialized and marked occupied. Similarly, using the ray casting operation, the nodes between the occupied node and the sensor, in the line of ray, can be initialized and marked as free. It leaves the uninitialized nodes marked unknown until the next update in the octree. The estimated value of the probability $P(n \mid z_{1:t})$ of the node n to be occupied for the sensor measurement $z_{1:t}$ is given by:

$$P(n|z_{1:t}) = [1 + \frac{1 - P(n|z_t)}{P(n|z_t)} \frac{1 - P(n|z_{1:t-1})}{P(n|z_{1:t-1})} \frac{P(n)}{1 - P(n)}]^{-1}$$

where P_n is the prior probability of node n to be occupied. Let us denote the occupancy probability as P^o.

$$p^{voxelstate} = \begin{cases} Free, & \text{if } P^o < P_n \\ Occupied, & \text{if } P^o > P_n \end{cases}$$

The presented basic frontier algorithm is made up of three essential modules, namely the frontier generation, the optimal frontier selection, and the goal publisher as presented in Fig. 7.4. The first module converts a point cloud into a voxel grid V defined as $V = \{\vec{x}\}$ based in the octomap framework discussed earlier. The next module generates frontiers based on the occupancy probability function for each voxel corresponding to each sensor measurement. The frontier points are fed into the next module, which evaluates each frontier based on a cost function and selects an optimal frontier point. This point is published to the ARW controller as a temporary goal point to visit, and the overall process continues for each sensor scan until no frontier remains.

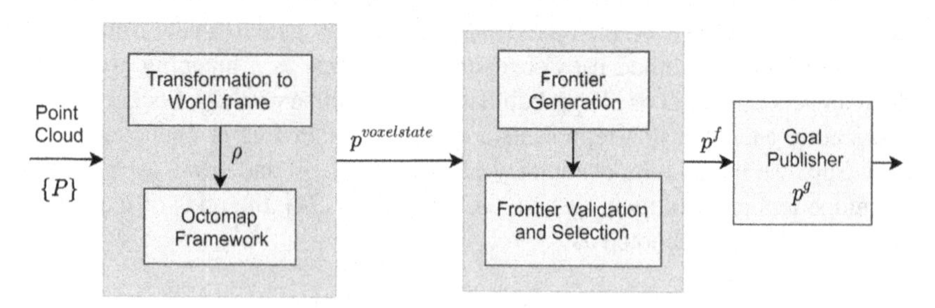

FIGURE 7.4 Frontier algorithm framework.

The classical definition of frontiers considers a cell in the voxel grid to be a frontier if at least one of its neighboring cells is marked as unknown. It is highly inefficient in the real scenario where the area to be explored is vast. Therefore, in Algorithm 4, the number of unknown neighbors can be set by the user before the start of the exploration, reducing the overall computational complexity by a significant margin. Another change compared to the classical frontier approach is made by not allowing any neighbor to be an occupied voxel. This results in frontier not being generated in extremely narrow openings and due to sensor measurement noise.

The third layer of refinement is set as if a frontier is detected within the sphere of radius r (set by the user at the start of exploration), then the cell is marked as free. This helps limit the undesired change in vertical velocity of the ARW if the frontier is detected above or below the ARW. Algorithm 4 uses a Velodyne 3D Lidar to get the point cloud, and because of the limited field of view of the 3D Lidar, the above-mentioned layer of refinement helps in an overall stable operation. The detailed implementation of the frontier approach is also presented in [25].

Algorithm 4: Frontier generation.

```
1        Input : m_octree : Currentoctree,
2                n : Number of unknowns required
3                r : radius of sphere around ARW
4        Output : List of Frontiers
5  for cell ∈ m_octree.updatedStateCells() do
6  |   if cell.is.Free() then
7  |   |   if cell.distance() < r then
8  |   |   |   i = 0
9  |   |   |   for neighbours ∈ cell.getNeighbour() do
10 |   |   |   |   if neighbour.isOccupied() then
11 |   |   |   |   |   i = 0;
12 |   |   |   |   |   break;
13 |   |   |   |   else
14 |   |   |   |   |   i = i + 1
15 |   |   if i >= n then
16 |   |   |   frontiers.add(cell);
```

The pseudo-code for the frontier generation from octomap is depicted in Algorithm 4. To reduce the number of cells to be examined in each update, only the cells whose state (free or occupied) is changed are examined. For the selection of frontier, a local and global set of frontiers is defined. For a large environment, it is computationally inefficient to go through each frontier, while evaluating the optimal frontier to visit. Therefore, in the presented method, the frontiers evaluation is done in two stages. First, the frontiers in the field of view in the direction of the ARW are added to the local set of frontiers and are prioritized in the evaluation process. Second, when a frontier is generated outside the field of view, it is added to the global list of frontiers. Thus, when there is no local frontier to visit, a frontier from the global set is selected based on the cost function. The pseudo-code for the frontier validation and extraction, for both local and global frontiers, is presented in Algorithm 5. A frontier from the local set is selected based on a minimum angle $\alpha \in [-\pi, \pi]$ between then frontier vector

Algorithm 5: Frontier validation and extraction.

```
1       Input : frontiers_FromCurrrentList
2               n : Number of unknowns required
3               r : radius of sphere around ARW
4               α : frontier vector angle
5       Output : frontiers_ValidFrontiers
6                frontiers_LocalFrontiers
7                frontiers_GlobalFrontiers
8  for cell ∈ Frontiers do
9      if cell_distance < r then
10         i = 0;
11         for neighbours ∈ cell.getNeighbour() do
12             if neighbour.isOccupied() then
13                 i = 0;
14                 break;
15             else
16                 i = i + 1;
17         if i < n then
18             frontiers.remove(cell);
19         for frontier ∈ frontiers_ValidFrontiers() do
20             if α < θ/2 then
21                 frontiers_LocalFrontiers.add(cell)
22             else
23                 frontiers_GlobalFrontiers.add(cell)
```

$\vec{x_f}$ and the ARW's direction of travel $\vec{v_d}$. In Fig. 7.5, $(X_{\mathcal{B}}, Y_{\mathcal{B}}, Z_{\mathcal{B}})$ represents the body-fixed coordinate frame, and $(X_{\mathcal{W}}, Y_{\mathcal{W}}, Z_{\mathcal{W}})$ represents the world coordinate frame. The frontiers from octomap are generated in the world frame, but the frontier vector $\vec{x_f}$ is calculated relative to the position of the ARW. As shown in Fig. 7.5, the angle α with respect to $\mathcal{B}$ can be defined as:

$$\alpha = cos^{-1}(\frac{\vec{x_f} \cdot \vec{v_d}}{|\vec{x_f}| \cdot |\vec{v_d}|})$$

Thus, by selecting the frontier with the lowest angle value α, the exploration can be continued with a relatively higher speed as the ARW does not have to go through a significant change in velocity. When there is no frontier in the local set, a frontier from the global set is selected. The selection of a frontier from the global set can be tuned based on a cost function as shown in Eq. (7.1)

$$C = w_\alpha \alpha + w_h \Delta H + w_d \Delta D \tag{7.1}$$

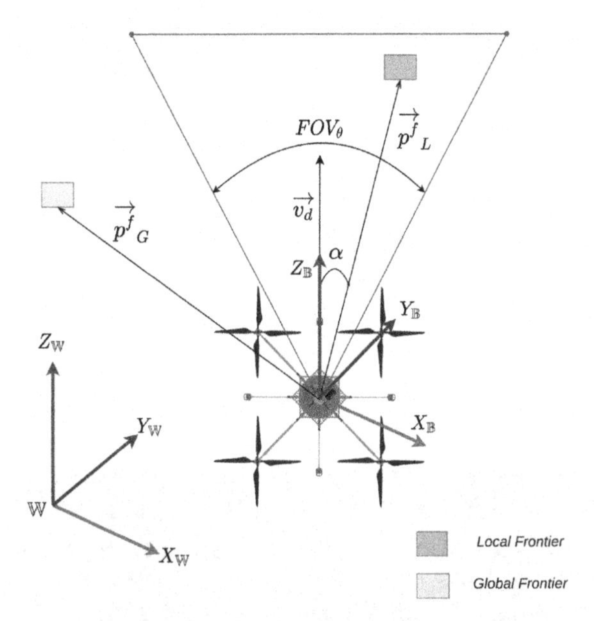

FIGURE 7.5 Local (green; dark gray in print) and Global (yellow; light gray in print) frontiers.

where ΔD is the distance between the ARW's current position and the frontier point, ΔH is the height difference between ARW and the frontier point, w_h, w_d, and w_α are weights on the height, distance, and angle, respectively. The cost for each valid frontier in the global set is calculated, and the frontier with a minimum cost value from Eq. (7.1) is selected as the next frontier to visit. As discussed previously, the weights can be set before the start of the exploration, which allows where the frontier will be selected relative to the ARW's current position in the 3D space. For example, if a high value of w_h is set, then the cost concerning a height difference is increased, resulting in a frontier higher or lower than the ARW position in the Z direction that would be less likely to be selected.

In Fig. 7.6, an example of overall Guidance, Navigation, and Control framework is presented.

In Fig. 7.10, incremental exploration using frontier based approach is presented. The resulted occupancy map shown in Fig. 7.10 is built using a reactive exploration framework, meaning the path planner is not involved in this case. In Fig. 7.7, an instance of exploration is shown when the ARW explores a cave environment, and if an inaccessible frontier is detected at the dead end of the cave channel, then the ARW selects the next frontier from the global list based on the cost formulated in Eq. (7.1) and globally re-positions itself. Similarly, in Fig. 7.8, another exploration instance is shown, where the ARW reconnects the map using the local and global frontier merging when the ARW returns to such opening in the map.

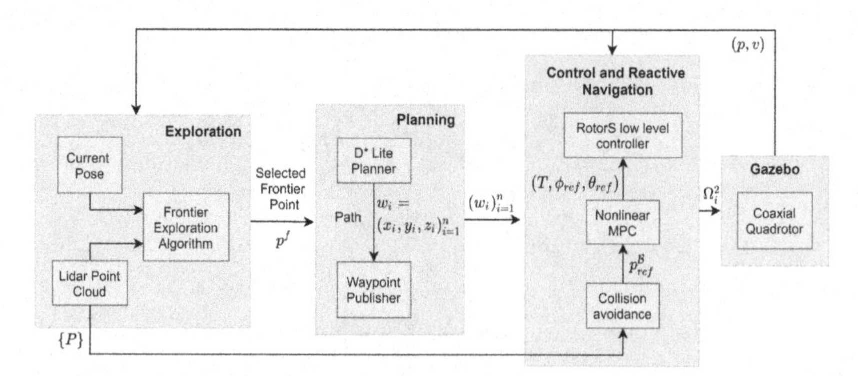

FIGURE 7.6 GNC architecture containing exploration, planning, and control.

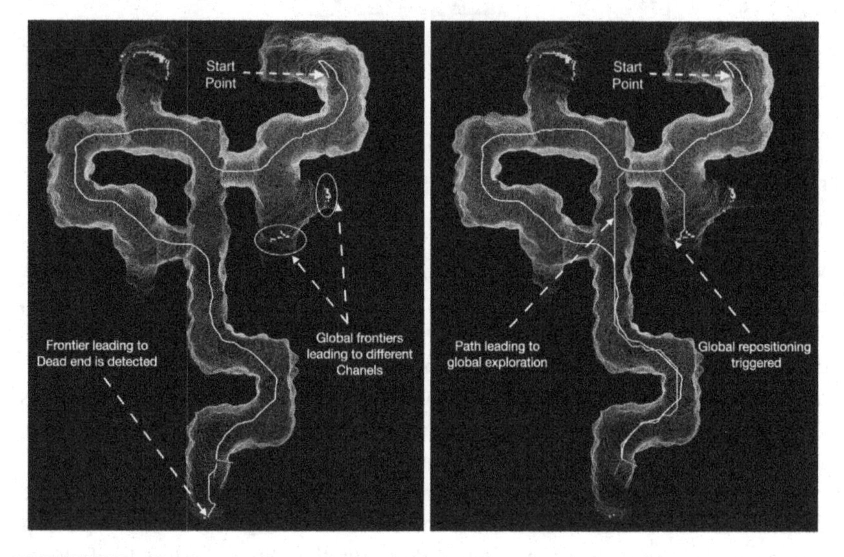

FIGURE 7.7 Global re-positioning due to in accessible frontier.

In Fig. 7.11, the reconstruction of the point cloud map of a cave environment is demonstrated using the frontier-based reactive exploration approach. The red (dark gray in print) line in Fig. 7.11 represents the robot trajectory while exploring autonomously.

In Fig. 7.10 and Fig. 7.9, the point cloud reconstruction and robot trajectory are shown using frontier based exploration as well as D^*_+ based global path planning module. Please note that the difference in robot trajectory in Fig. 7.11 and Fig. 7.10 is because in the full exploration framework the path to the next best frontier is planned using a dedicated global planning module. As presented in Fig. 7.11, the reactive exploration always moves the robot to the nest best

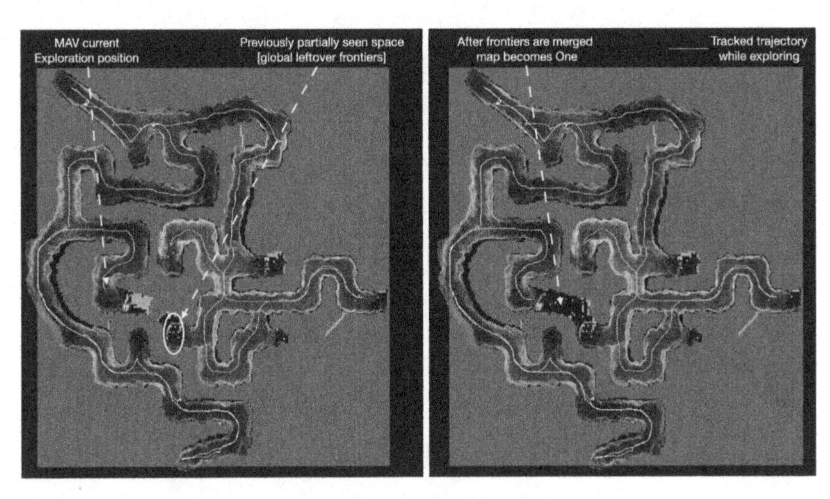

FIGURE 7.8 Global frontiers are merged when previously partially seen area is discovered again.

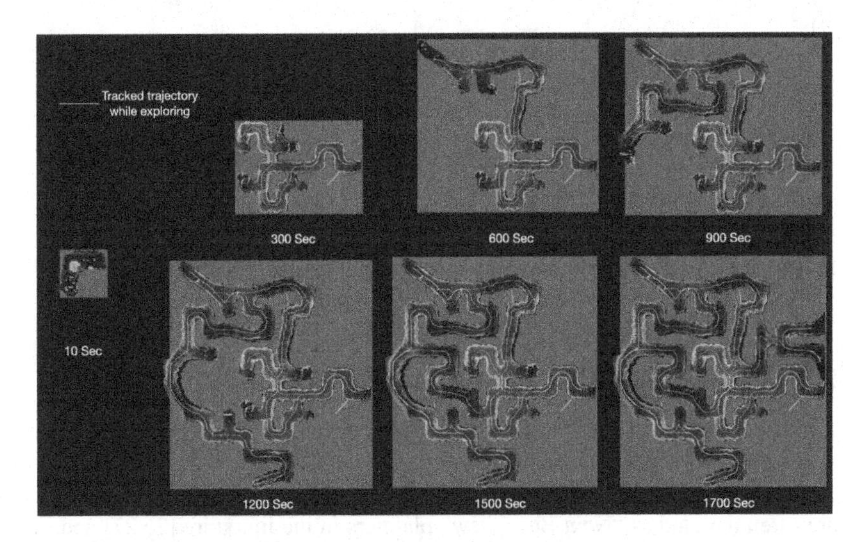

FIGURE 7.9 Incrementally built map by the ARW while exploring.

frontier in the shortest path possible, which is a straight line in open spaces. However, as presented in Fig. 7.10, the planned path also takes into account the safety margin of a such path and distance of the robot from the potential obstacle (cave walls).

In Fig. 7.12, an octomap representation of the environment is shown. The map is built incrementally as the ARW explores the environment.

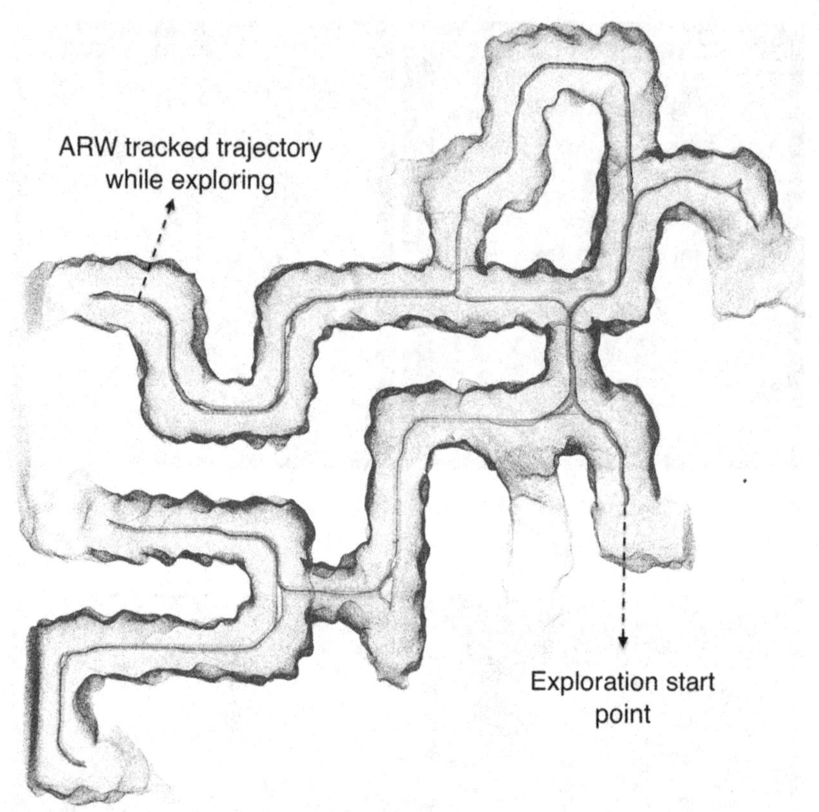

FIGURE 7.10 Point cloud reconstruction and robot trajectory using complete exploration framework, including global path planning.

7.3 Combined path planning and exploration

As an alternative to the previously discussed methods of selecting a frontier and then querying a path planner to plan the path, there are methods that include exploration behavior in the path planning problem directly. Such methods are often denoted as "Next Best View" planners in the literature [26,27] and in a very generalized description try to solve the problem of maximizing information gain while minimizing the distance traveled or other such metrics. The difference is that the path to the frontier, point of interest, or unknown area, is considered when choosing "where to go next", whereas, in standard frontier selection methods, this is not the case. Of course, to evaluate those metrics, it is necessary to compute the path to multiple different points. In the Exploration-RRT framework [28], an RRT-based [29] planner plans the path to multiple pseudo-random goals and decides which goal to go to based on a combination of the predicted information gain along that path, the distance traveled to reach that goal, and the model-based actuation required to track such a path.

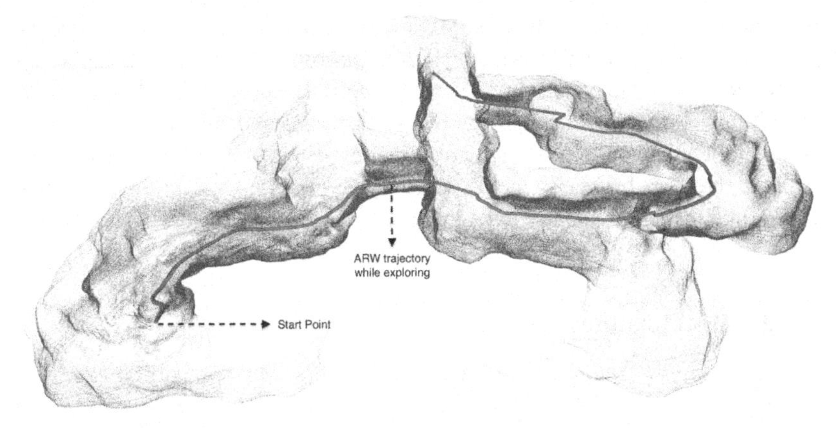

FIGURE 7.11 Point cloud reconstruction and robot trajectory using reactive frontier-based exploration.

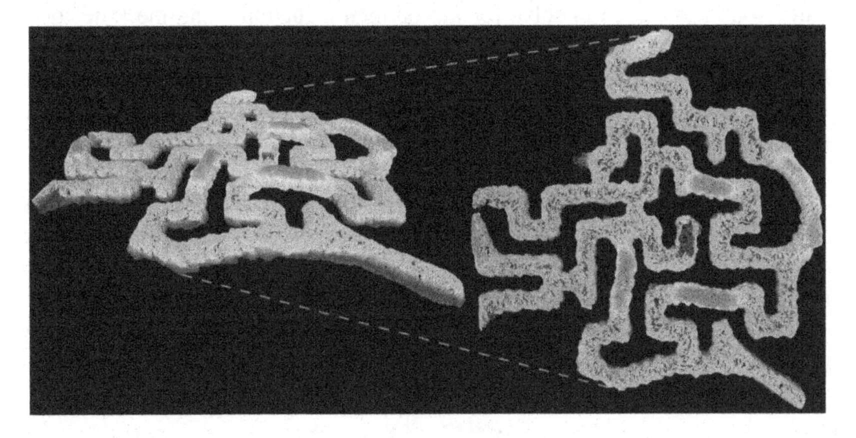

FIGURE 7.12 Complete map of a virtual cave environment built after exploration.

7.3.1 The problem

The overarching goal of coupling the path planning and exploration problem in the ERRT framework is considered as the minimization of three quantities, namely: 1) The total distance of the 3D trajectory $\boldsymbol{x}$, 2) the actuation required to track the trajectory where $\boldsymbol{u}$ denotes a series of control actions, and 3) increasing the known space here considered as a negative cost associated with the information gain $\nu \in \mathbb{R}$ along trajectory $\boldsymbol{x}$. These quantities result in the following minimization:

$$\begin{aligned} &\text{Minimize } J_a(\boldsymbol{u}) + J_d(\boldsymbol{x}) + J_e(\nu) \\ &\qquad\qquad \text{subj. to: } \boldsymbol{x} \subset V_{free} \\ &\qquad\qquad\qquad J_e(\nu) \neq 0 \end{aligned} \tag{7.2}$$

with $J_a(\boldsymbol{u})$ denoting the actuation cost, $J_d(\boldsymbol{x})$ the distance cost, $J_e(\nu)$ the exploration, or information-gain cost, which should be non-zero to expand the known space, and $V_{free} \subset V_{map}$ denoting the obstacle-free space, with $V_{map} \subset \mathbb{R}^3$ as the 3D position-space encompassed by the current map configuration. If solved completely, this would result in the optimal trajectory for the exploration task, in a compromise between quickly discovering more space, limiting actuation based on a dynamic system model and being *lazy*, e.g., moving as little as possible.

7.3.2 ERRT solution

ERRT proposes a solution composed of four components: pseudo-random goal generation based on a sensor model (under the condition that each goal should have non-zero information gain), a multi-goal RRT* planner that plans the path to each pseudo-random goal denoted ρ^r, a receding horizon optimization problem (Nonlinear MPC) to solve for the optimal actuation along the trajectory, and finally computing the total costs associated with each trajectory and choosing the minimal-cost solution. The process of generating many goals, solving the path to each of them and then computing the total costs associated with each trajectory, turns (7.2) from a true optimization problem into:

$$\arg\min(J_a(\boldsymbol{u}_i) + J_d(\boldsymbol{x}_i) + J_e(\nu_i))_i, i = 1, \ldots, n_{goals} \quad \text{where } \boldsymbol{x}_i \subset V_{free} \tag{7.3}$$

that considers the finding of $\boldsymbol{x}_{\min} \subset V_{free}$ that is the $\boldsymbol{x}_i$ trajectory having the lowest cost associated with it, and $n_{goal} \in \mathbb{N}$ is the overall number of path planner

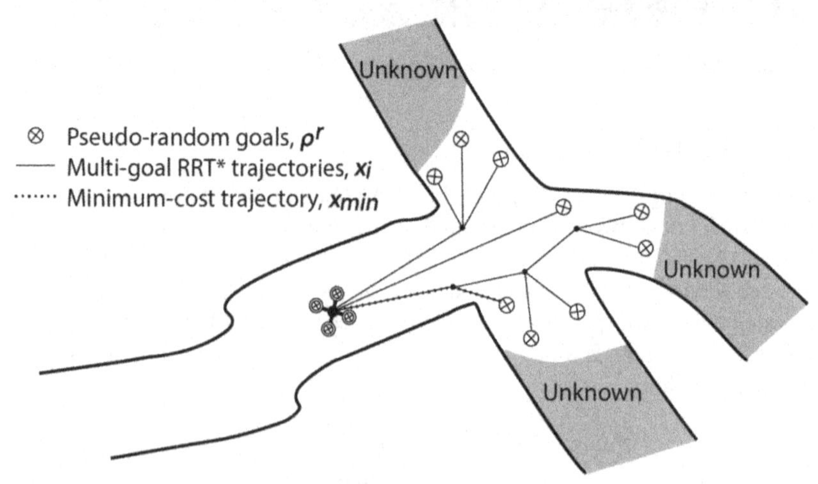

FIGURE 7.13 The ERRT concept. Multiple trajectories are calculated to pseudo-random goals, and the lowest cost trajectory is chosen.

goals. This process will converge toward approximating the complete problem in (7.2), as we increase n_{goal} and optimize the trajectory-generating algorithm (or simply by increasing the number of iterations of the RRT*) as more and more possible solutions are being investigated.

Fig. 7.13 shows a low-resolution concept of the ERRT framework, while Algorithm 6 shows the ERRT pipeline. Here, $\boldsymbol{G}$ denotes the occupancy grid map, $\{U\}$ is a set of unknown voxels (e.g., frontiers or similar), r_s, θ_s, l_s are sensor parameters, $K_d, K_v, Q_u, Q_{\Delta u}$ are weights related to cost computation, and $\boldsymbol{N}_{goal}$ is the list of graphs produced by the RRT* that reach each individual goal. An evaluation of the ERRT framework in a simulated subterranean environment using a 3D-lidar equipped ARW can be founds at https://drive.google.com/file/d/1v3vg3Z9iB2DR-Oec3MxWUg_39d3lIj1F/view?usp=sharing, while Fig. 7.14 shows the progression of such a simulated exploration run in an unknown environment.

Algorithm 6: The ERRT algorithm.

1 **Inputs:** $\boldsymbol{G}, \{U\}, n_{goal}, r_s, \theta_s, l_s, \hat{x}$
Result: Minimum-cost trajectory $\boldsymbol{x}_{min}$
2 $\boldsymbol{\rho}^r$ = generate_goals($\boldsymbol{G}, \{U\}, n_{goal}, r_s, \theta_s, l_s$)
3 $\boldsymbol{N}_{goal}$ = multigoal_rrt($\boldsymbol{\rho}^r, \boldsymbol{G}$)
4 $\boldsymbol{x}_i$ = trajectory_improvement($\boldsymbol{\rho}^r, \boldsymbol{N}_{goal}, \boldsymbol{G}$)
5 $\boldsymbol{u}_i$ = NMPC_module($\boldsymbol{x}_i, \hat{x}$)
6 $(J_a, J_d, J_e)_i$ = cost_calc($\{U\}, \boldsymbol{G}, \boldsymbol{x}_i, \boldsymbol{u}_i, K_d, K_v, Q_u, Q_{\Delta u}$)
7 $\boldsymbol{x}_{min}$ = path_selection($(J_a, J_d, J_e)_i, \boldsymbol{x}_i$)
8 **Output:** $\boldsymbol{x}_{min}$

7.4 Conclusions

This chapter presents different exploration strategies that are directly applicable for deploying autonomous Aerial Robotic Workers in subterranean exploration missions. The main motivation behind the exploration strategy is to compute where to move next so that the ARW will acquire more information about the environment and autonomously navigate to such area. Different frontier-based and sampling-based exploration methods were presented in detail to address the problem with different methodologies. Exploration results have been presented in terms of the incrementally built map as well as using custom modeled occupancy information-based simulators as presented in the combined exploration-planning module.

References

[1] DARPA, DARPA Subterranean (SubT) challenge [Online]. Available: https://www.darpa.mil/program/darpa-subterranean-challenge, 2020. (Accessed February 2021).

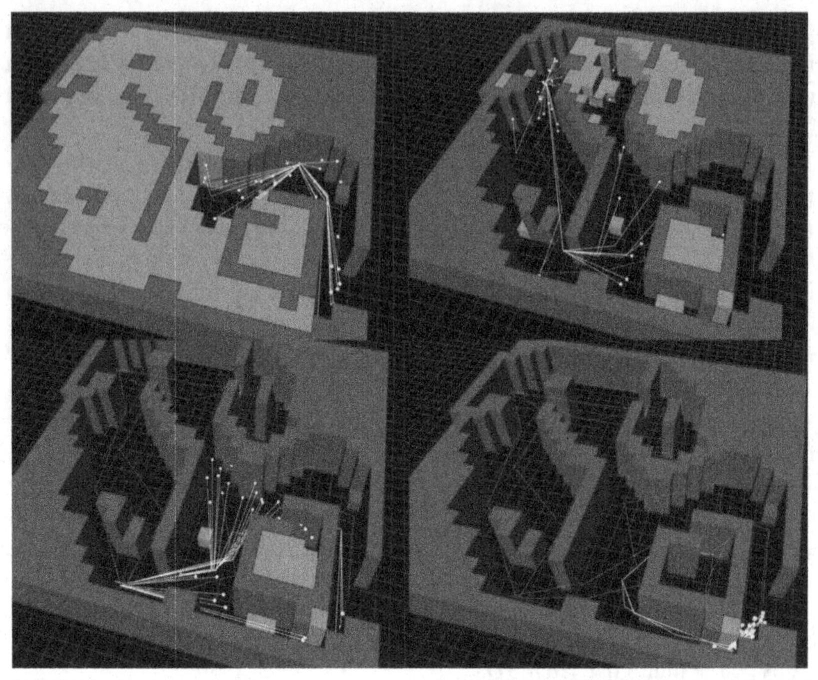

FIGURE 7.14 The exploration progression using the ERRT algorithm. Blue (dark gray in print) voxels are occupied, while green (light gray) voxels represent the currently unknown areas. White dots represent ρ^r, white lines the path to those goals, and the purple line is the selected minimum-cost trajectory.

[2] T. Sasaki, K. Otsu, R. Thakker, S. Haesaert, A.A. Agha-Mohammadi, Where to map? Iterative rover-copter path planning for mars exploration, IEEE Robotics and Automation Letters 5 (2) (2020) 2123–2130.

[3] B. Morrell, R. Thakker, G. Merewether, R. Reid, M. Rigter, T. Tzanetos, G. Chamitoff, Comparison of trajectory optimization algorithms for high-speed quadrotor flight near obstacles, IEEE Robotics and Automation Letters 3 (4) (2018) 4399–4406.

[4] Y. Kwon, D. Kim, I. An, S.-e. Yoon, Super rays and culling region for real-time updates on grid-based occupancy maps, IEEE Transactions on Robotics 35 (2) (2019) 482–497.

[5] A. Hornung, K.M. Wurm, M. Bennewitz, C. Stachniss, W. Burgard, Octomap: An efficient probabilistic 3D mapping framework based on octrees, Autonomous Robots 34 (3) (2013) 189–206.

[6] S. Choi, J. Park, E. Lim, W. Yu, Global path planning on uneven elevation maps, in: 2012 9th International Conference on Ubiquitous Robots and Ambient Intelligence (URAI), IEEE, 2012, pp. 49–54.

[7] H. Oleynikova, A. Millane, Z. Taylor, E. Galceran, J. Nieto, R. Siegwart, Signed distance fields: A natural representation for both mapping and planning, in: RSS 2016 Workshop: Geometry and Beyond—Representations, Physics, and Scene Understanding for Robotics, University of Michigan, 2016.

[8] R. Triebel, P. Pfaff, W. Burgard, Multi-level surface maps for outdoor terrain mapping and loop closing, in: 2006 IEEE/RSJ International Conference on Intelligent Robots and Systems, IEEE, 2006, pp. 2276–2282.

[9] P. Payeur, P. Hébert, D. Laurendeau, C.M. Gosselin, Probabilistic octree modeling of a 3D dynamic environment, in: Proceedings of International Conference on Robotics and Automation, vol. 2, IEEE, 1997, pp. 1289–1296.
[10] S. Koenig, M. Likhachev, D* lite, in: AAAI/IAAI, vol. 15, 2002.
[11] D. Ferguson, A. Stentz, Using interpolation to improve path planning: The field D* algorithm, Journal of Field Robotics 23 (2) (2006) 79–101.
[12] A. Nash, K. Daniel, S. Koenig, A. Felner, Theta*: Any-angle path planning on grids, in: AAAI, vol. 7, 2007, pp. 1177–1183.
[13] M. Sheckells, DSL Gridsearch: D*-lite on a uniformly spaced 3D or 2D grid [Online]. Available: https://github.com/jhu-asco/dsl_gridsearch, 2018. (Accessed February 2021).
[14] J. Carsten, D. Ferguson, A. Stentz, 3d field d: Improved path planning and replanning in three dimensions, in: 2006 IEEE/RSJ International Conference on Intelligent Robots and Systems, IEEE, 2006, pp. 3381–3386.
[15] A. Nash, S. Koenig, C. Tovey, Lazy Theta*: Any-angle path planning and path length analysis in 3D, in: Proceedings of the AAAI Conference on Artificial Intelligence, vol. 24, 2010, No. 1.
[16] H. Chen, P. Lu, Computationally efficient obstacle avoidance trajectory planner for UAVs based on heuristic angular search method, arXiv preprint, arXiv:2003.06136, 2020.
[17] B. Zhou, J. Pan, F. Gao, S. Shen, Raptor: Robust and perception-aware trajectory replanning for quadrotor fast flight, arXiv preprint, arXiv:2007.03465, 2020.
[18] J.D. Hernández, G. Vallicrosa, E. Vidal, È. Pairet, M. Carreras, P. Ridao, On-line 3D path planning for close-proximity surveying with AUVs, IFAC-PapersOnLine 48 (2) (2015) 50–55.
[19] M. Faria, I. Maza, A. Viguria, Applying frontier cells based exploration and lazy Theta* path planning over single grid-based world representation for autonomous inspection of large 3D structures with an UAS, Journal of Intelligent & Robotic Systems 93 (1) (2019) 113–133.
[20] T. Cieslewski, E. Kaufmann, D. Scaramuzza, Rapid exploration with multi-rotors: A frontier selection method for high speed flight, in: 2017 IEEE/RSJ International Conference on Intelligent Robots and Systems (IROS), IEEE, 2017, pp. 2135–2142.
[21] S.S. Mansouri, C. Kanellakis, B. Lindqvist, F. Pourkamali-Anaraki, A.A. Agha-Mohammadi, J. Burdick, G. Nikolakopoulos, A unified NMPC scheme for MAVs navigation with 3D collision avoidance under position uncertainty, IEEE Robotics and Automation Letters 5 (4) (2020) 5740–5747.
[22] M. Quigley, K. Conley, B. Gerkey, J. Faust, T. Foote, J. Leibs, R. Wheeler, A.Y. Ng, ROS: an open-source robot operating system, in: ICRA Workshop on Open Source Software, vol. 3, Kobe, Japan, 2009, p. 5, No. 3.2.
[23] B. Yamauchi, Frontier-based exploration using multiple robots, in: Proceedings of the Second International Conference on Autonomous Agents, 1998, pp. 47–53.
[24] Chen Zhe, et al., Bayesian filtering: From Kalman filters to particle filters, and beyond, 2003.
[25] C. Zhu, R. Ding, M. Lin, Y. Wu, A 3D frontier-based exploration tool for MAVs, in: 2015 IEEE 27th International Conference on Tools with Artificial Intelligence (ICTAI), IEEE, 2015, pp. 348–352.
[26] R. Pito, A solution to the next best view problem for automated surface acquisition, IEEE Transactions on Pattern Analysis and Machine Intelligence 21 (10) (1999) 1016–1030.
[27] A. Bircher, M. Kamel, K. Alexis, H. Oleynikova, R. Siegwart, Receding horizon "next-best-view" planner for 3D exploration, in: 2016 IEEE International Conference on Robotics and Automation (ICRA), IEEE, 2016, pp. 1462–1468.
[28] B. Lindqvist, A.-a. Agha-mohammadi, G. Nikolakopoulos, Exploration-RRT: A multi-objective path planning and exploration framework for unknown and unstructured environments, arXiv preprint, arXiv:2104.03724, 2021.
[29] S.M. LaValle, et al., Rapidly-exploring random trees: A new tool for path planning, 1998.

Chapter 8

External force estimation for ARWs

Andreas Papadimitriou[a] and Sina Sharif Mansouri[b]
[a]*Department of Computer, Electrical and Space Engineering, Luleå University of Technology, Luleå, Sweden,* [b]*Autonomous Driving Lab in Scania Group, Stockholm, Sweden*

8.1 Introduction

No matter how accurate the modeling of an ARW or how well-designed the ARW controller is, a common problem for the ARWs is the unwanted disturbances that might occur during a mission. Examples of such disturbances could be wind gusts or the transportation of a small payload. In these cases, the platform should react fast against the disturbances to maintain its stability. When the disturbances acting on the platform are known or measured with onboard hardware [1,2], they can be mitigated, but usually, the nature of those disturbances varies, making the compensation a problematic task. Furthermore, it is inefficient to mount additional sensors on an ARW as they will reduce its flight time, thus allowing only shorter missions, and often, it is unrealistic to use them due to their size and weight. Estimating disturbances using estimation techniques without relying on external sensors, such as weather station measurements or force meters, would positively impact the battery life. There are many methods to estimate forces, like the Extended and Unscented Kalman Filter [3–6]. However, this Chapter will focus on using Nonlinear Moving Horizon Estimation (NMHE) while considering the nonlinear dynamics of the ARW without the requirement of the platform system identification. The estimated forces can be fed to a position controller like the NMPC from Chapter 4 with minimum additions, which will provide thrust and attitude commands to the low-level controller while simultaneously accounting for the estimated disturbances. The Moving Horizon Estimation (MHE) is an optimization method that uses measurements that are affected by noise, biases, variation, and other inaccuracies and produces estimates of unknown variables or parameters [7]. Similar to MPC, it relies on the minimization of a sum of stage costs subject to a dynamic model. The NMHE can handle nonlinear models and nonlinear equality and inequality constraints, and its stability and robustness are currently being studied [8,9]. The most significant trade-off of the MHE methods, similar to other receding horizon estimation approaches, is that it comes with increased computational requirements compared to other estimation methods like the *Kalman Filter*. However,

Aerial Robotic Workers. https://doi.org/10.1016/B978-0-12-814909-6.00014-7

nowadays, the increased computational resources of ARWs, and recently developed advanced formulations for fast nonlinear solvers [10] for constrained nonlinear systems, enable ARWs to cope with the computation burden. Thus, enabling the use of MHE approaches to the field of robotics and in general applications for systems with fast dynamics.

8.1.1 External force estimation

The proposed NMHE estimates the system's states and external forces applied to the ARW. Fig. 8.1 depicts the effect of the external forces on the body frame of the ARW. The body frame forces f_x, f_y, f_z result in the position drift of the ARW in the x, y, z body axes accordingly.

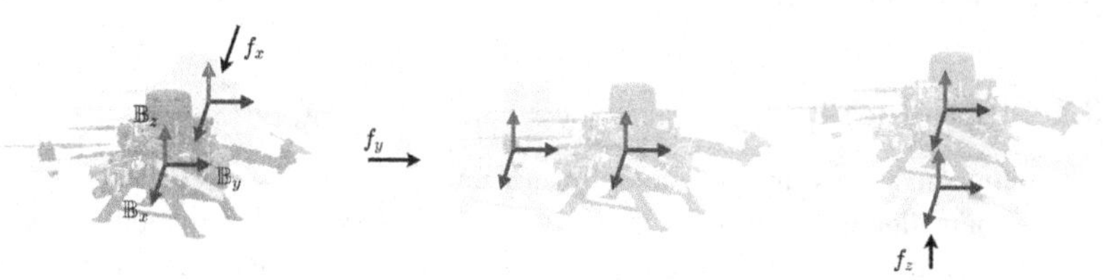

FIGURE 8.1 Schematic illustrating the effect of the external forces on the ARW body frame axes. The f_x, f_y, f_z results in the displacement of the ARW in the x, y, z body axes, respectively.

In the NMHE formulation to be presented next, we consider the following discrete-time dynamic evolution of the system as presented in (8.1):

$$\bar{\boldsymbol{x}}_{k+1} = \mathcal{F}(\bar{\boldsymbol{x}}_k, \boldsymbol{u}_k) + \boldsymbol{w}_k, \tag{8.1a}$$

$$\boldsymbol{y}_k = \mathcal{H}(\bar{\boldsymbol{x}}_k) + \boldsymbol{\Lambda}_k, \tag{8.1b}$$

where, $\bar{\boldsymbol{x}} = [\boldsymbol{x}, \boldsymbol{f}]^\top$, $\mathcal{F} : \mathbb{R}^{n_s} \times \mathbb{R}^{n_u} \to \mathbb{R}^{n_s}$ is a nonlinear function, $\mathcal{H} : \mathbb{R}^{n_s} \to \mathbb{R}^{n_m}$ is a linear vector function of the states $\bar{\boldsymbol{x}}$, and $\boldsymbol{y} = [x, y, z, v_x, v_y, v_z, \phi, \theta]^\top$ is the measured output. Furthermore, n_s, n_u, and n_m are the number of states, inputs, and measurements, respectively, and $\boldsymbol{\Lambda}_k \in \mathbb{R}^{n_m}$ and $\boldsymbol{w}_k \in \mathbb{R}^{n_s}$ represent the measurement noise and the model disturbances, respectively. Note that in the NMHE formulation, the external forces are considered in the state space of the dynamic model, that is, $\boldsymbol{f}$ is regarded as an unmeasured state. In contrast, in the NMPC formulation, $\boldsymbol{f}$ is a variable for the prediction horizon, which is updated based on the NMHE estimations. The external force $\boldsymbol{f}$ changes over time; however in, each estimation window, it is assumed that the external force is static ($\dot{\boldsymbol{f}} = 0$).

The process disturbance $\boldsymbol{w}_k$, the measurement noise $\boldsymbol{\Lambda}_k$, and the initial Probability Density Function (PDF) of the state vector are unknown, and it is assumed that they are randomly distributed according to the Gaussian PDF with covariance matrices $\boldsymbol{Q} \in \mathbb{R}^{n_s \times n_s}$, $\boldsymbol{\Omega} \in \mathbb{R}^{n_m \times n_m}$, and $\boldsymbol{\Psi} \in \mathbb{R}^{n_s \times n_s}$, respectively. Furthermore, the initial condition $\tilde{\boldsymbol{x}}_0$ is assumed to be known. Based on the information about random noises and a set of available noisy measurements

$Y = \{y_j : j = 1, ..., N_e\}$, the estimated states of system $\bar{X} = \{\bar{x}_j : j = 0, ..., N_e\}$ are obtained by solving the following optimization problem, while the N_e is the length of the fixed horizon window. Moreover, $\bar{x}_{k-j|k}$ and $y_{k-j|k}$ are the $k-j$ previous states and measurements from the current time k.

$$\min_{\bar{x}_{(k-N_e|k)}, W^{(k-1|k)}_{(k-N_e|k)}} \quad J(k) \tag{8.2a}$$

$$\text{s.t. } \bar{x}_{i+1|k} = \mathcal{F}(\bar{x}_{i|k}, u_{i|k}) + w_{i|k} \tag{8.2b}$$

$$y_{i|k} = \mathcal{H}(\bar{x}_{i|k}) + \Lambda_{i|k} \quad i = \{k - N_e, \ldots k-1\} \tag{8.2c}$$

$$w_k \in \mathbb{W}_k, \quad \Lambda_k \in \Lambda_k, \quad \bar{x}_k \in \mathbb{X}_k \tag{8.2d}$$

where,

$$\begin{aligned} J(k) = & \underbrace{\|\bar{x}_{k-N_e|k} - \tilde{x}_{k-N_e|k}\|^2_{\Psi}}_{\text{arrival cost}} + \sum_{i=k-N_e}^{i=k} \underbrace{\|y_{i|k} - \mathcal{H}(\bar{x}_{i|k})\|^2_{\Omega}}_{\text{stage cost}} \\ & + \sum_{i=k-N_e}^{i=k-1} \underbrace{\|\bar{x}_{i+1|k} - f(\bar{x}_{i|k}, u_{i|k})\|^2_{Q}}_{\text{stage cost}} \end{aligned} \tag{8.3}$$

In (8.2), $W^{(k-1|k)}_{(k-N_e|k)} = col(w_{(k-N_e|k)}, \ldots, w_{(k-1|k)})$ is the estimated process disturbance from time $k - N_e$ up to $k - 1$, which is estimated at the time k, and the estimation horizon is defined with a fixed window of size $N_e \in \mathbb{Z}^+$.

The first term of the objective function in (8.3) is the arrival cost weighted by Ψ, which describes the uncertainty in the initial state at the beginning of the horizon, considering the error between the observation model and the predicted initial state $\tilde{x}(k - N_e \mid k)$. In general, there are different approaches to transfer the arrival cost at each time. In this work, the smoothing approach is used which only uses one time-step before the window to approximate the arrival cost. The second and third terms are called stage costs. The $\|y_{i|k} - \mathcal{H}(\bar{x}_{i|k})\|^2$, weighted by Ω, is the bias between the measured output and the estimated state. The $\|\bar{x}_{i+1|k} - f(\bar{x}_{i|k}, u_{i|k})\|$, weighted by Q, is the estimated model disturbance. At every instant k, a finite-horizon optimal problem with horizon window of N_e is solved and the corresponding estimated states and external forces sequence of $\bar{x}^\star_{k-N_e|k}, \ldots \bar{x}^\star_{k|k}$ are obtained. The final estimated state $\bar{x}^\star_{k|k}$ is fed to the controller.

8.2 Disturbance rejection NMPC

Finally, we will define a method to counteract the estimated disturbances by taking an augmented version of the NMPC defined in Section 4.6. The ARW is modeled by the position of the center of mass in the inertia frame and the body's orientation around each axis with respect to the inertial frame. The ARW dynamics are defined in the body frame and are modeled by (8.4):

$$\dot{\boldsymbol{p}}(t) = \boldsymbol{v}(t), \tag{8.4a}$$

$$\dot{\boldsymbol{v}}(t) = \boldsymbol{R}_{x,y}(\theta,\phi)\begin{bmatrix}0\\0\\T\end{bmatrix} + \begin{bmatrix}0\\0\\-g\end{bmatrix} - \begin{bmatrix}A_x & 0 & 0\\0 & A_y & 0\\0 & 0 & A_z\end{bmatrix}\boldsymbol{v}(t) + \boldsymbol{f}(t), \tag{8.4b}$$

$$\dot{\phi}(t) = {}^1/_{\tau_\phi}(K_\phi \phi_d(t) - \phi(t)), \tag{8.4c}$$

$$\dot{\theta}(t) = {}^1/_{\tau_\theta}(K_\theta \theta_d(t) - \theta(t)), \tag{8.4d}$$

The main difference from the default NMPC model is that in (8.4b), we consider the forces $\boldsymbol{f} = [f_x, f_y, f_z]^\top$ acting on the body frame of the ARW. This way, the controller can account for the disturbances in a feedforward manner. This method will generate roll-pitch-thrust commands to the low-level controller, which then will consider the effect of forces that act on the body frame of the ARW. The objective function J to be minimized remains the same as in Eq. (4.18). To limit the control actions of the NMPC within a range, the control input $\boldsymbol{u}$ is bounded. The constraints are implemented to avoid aggressive behavior during maneuvers and represent the desired physical constraints of the platform. However, as will be discussed subsequently in the example section, the boundaries have to be wider to allow the ARW to compensate for disturbances, and actually, they can act as hard limiters in the maximum forces that the ARW can counteract. Based on the previous definitions, the following optimization problem is defined in (8.5):

$$\min_{\{u_{k+j|k}\}_{j=0}^{N-1}} \quad J \tag{8.5a}$$

$$\text{s.t.} \quad \boldsymbol{x}_{k+j+1|k} = f(\boldsymbol{x}_{k+j|k}, \boldsymbol{u}_{k+j|k}), \tag{8.5b}$$

$$0 \leq T \leq T_{max}, \tag{8.5c}$$

$$\phi_{min} \leq \phi_d \leq \phi_{max}, \tag{8.5d}$$

$$\theta_{min} \leq \theta_d \leq \theta_{max}. \tag{8.5e}$$

In the proposed framework, the NMHE and NMPC optimization problems are solved online by the utilization of PANOC [10,11] that is a fast solver for nonlinear optimal control problems and guarantees real-time performance, a key component for embedded applications. Finally, both developed modules are evaluated in different experimental scenarios with different platforms.

8.3 Examples

The rest of this chapter will focus on the presentation of a few experimental examples using the NMHE-NMPC formulation for various scenarios to compensate the effect of external forces. In addition, the NMHE-NMPC formulation will be compared with the standard NMPC without force estimation that is blind to external disturbances. For the demonstration of the examples, two custom designed ARWs are used. These are a standard quadrotor (Fig. 8.2a) and a re-configurable quadrotor (Fig. 8.2b). The NMPC and NMHE parameters for all the presented examples are included in Table 8.1. The constraints used for the NMPC are linked with the maximum roll and pitch angles described below as they vary between the examples. To evaluate the waypoint tracking performance, a motion capture system is used to provide accurate localization information.

(a) Standard ARW

(b) Re-configurable ARW

FIGURE 8.2 Experimental quadrotor platforms used for the evaluation of the NMHE-NMPC disturbance estimation and compensation.

TABLE 8.1 NMPC and NMHE tuning parameters.

Q_x	Q_u	$Q_{\Delta u}$
$[5, 5, 5, 5, 5, 5, 1, 1]^\top$	$[10, 10, 10]^\top$	$[20, 20, 20]^\top$
Ψ	Q	Ω
$\mathbf{I}_{11}$	$\mathbf{I}_8$	$\mathbf{I}_{11}$

8.3.1 MAV subject to wind-wall disturbances

Fig. 8.3 depicts the flying arena, where the wind-wall is located on the left side and generates wind toward x-axis of the ARW (due to the reference heading of the platform). The wind-wall dimensions are 2 m and 2.1 m for the width and height, with a total number of 18 fan modules. The modules can produce a maximum wind speed of 16 m/s.

The ARW is commanded to hover at the position $[0.5, -1.6, 2.0]^\top$ (m) in front the wind-wall, which generates wind toward x-axis of the platform. The NMPC constraints for the roll and pitch angles are $|\phi_d|, |\theta_d| \leq 0.6$ rad. Fig. 8.4a depicts the estimated and measured position of the ARW, while the Root Mean Square Error (RMSE) between the measured and estimated values is 0.1 m. The RMSE between the $\boldsymbol{x}$ and $\boldsymbol{x}_r$ is 0.2 m, 0.18 m, and 0.25 m for x, y, and z axes, respectively. The estimated external forces of the wind-wall are depicted in Fig. 8.4b. The mean and absolute max value of the forces for each axis x, y, and z are $(-1.2, 2.2)$ N, $(-0.2, 0.9)$ N, and $(0.2, 0.55)$ N, respectively.

Fig. 8.5a shows the power percentage of the wind-wall. The operator increases the power of the wind-wall from zero to 55%, which generates airflow

FIGURE 8.3 Flying arena in CAST laboratory at the California Institute of Technology. In the left side of the illustration is the wind-wall, and in the middle is the flying ARW.

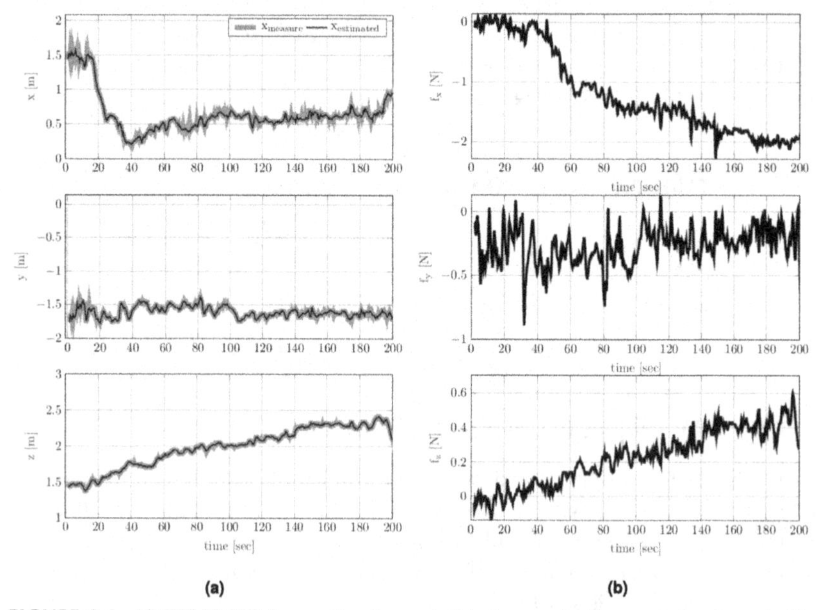

FIGURE 8.4 NMHE-NMPC force estimation and disturbance rejection example when wind is generated by a wind-wall. Position states $[x, y, z]^\top$, the measured and estimated values are shown by gray and black colors (left). Estimated external forces $[f_x, f_y, f_z]^\top$ by the NMHE when wind is generated against ARW x-axis (right).

of up to 8.8 m/s. As the wind-wall faces toward the ARW x-axis, the estimated force f_x is gradually decreasing following the same trend of the increasing wind-wall power percentage.

Moreover, the ARW is evaluated with the use of the NMPC module and without the NMHE module, thus the external forces are not estimated. In this case, the ARW is commanded to hover at $[0.5, -1.6, 1.3]^\top$ [m] and same tuning

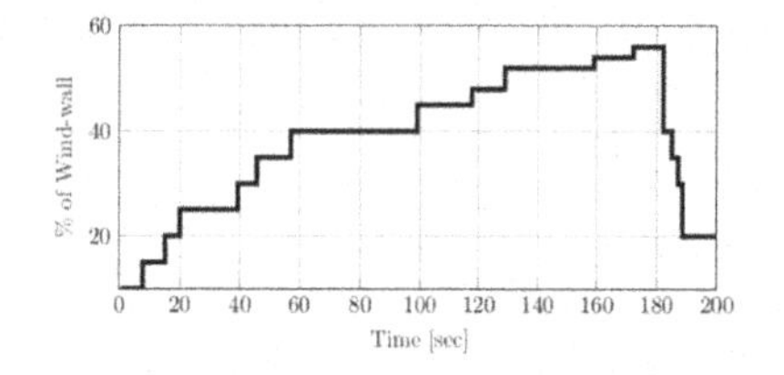

(a) Wind-wall power percentage with NMHE enabled.

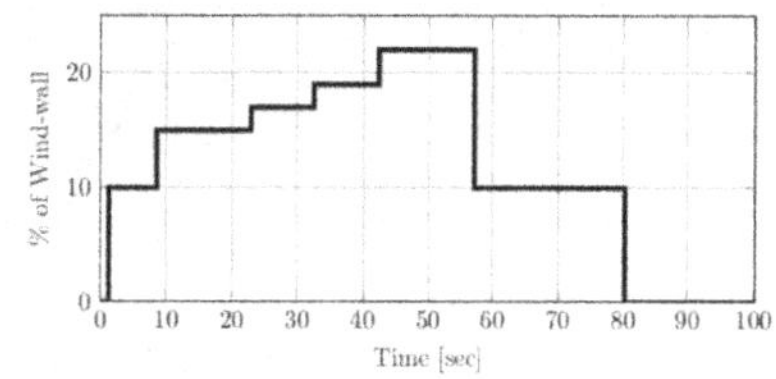

(b) Wind-wall power percentage with NMHE disabled.

FIGURE 8.5 (a) Wind-wall power percentage for the case of the ARW subject to winds estimated up to 8.8 m/sec. (b) Wind-wall power percentage for the case of the ARW subject to winds estimated up to 3.5 m/sec.

of the NMPC is used. Fig. 8.6 depicts the position of the ARW. The RMSE of the position and the waypoint for x, y, and z axes is 0.46 m, 0.35 m, and 0.40 m, respectively, while the maximum absolute error in x-axis is 1.58 m. This is due to the generated wind toward the x-axis of the ARW. Moreover, Fig. 8.5b shows the power percentage of the wind-wall that approximately reaches up to 3.5 m/s. It is observed that the NMPC tracks the desired waypoint with a high error when the external forces are not estimated. In addition, for this scenario, the maximum wind speed is approximately 2.5 times lower compared to the previous case. Thus, the proposed NMPC with NMHE modules provide at all times improved tracking error even when the wind flow is high.

8.3.2 Reducing the effect of a tethered payload

In this example, an external payload of 0.25 kg is tethered to the ARW as depicted in Fig. 8.7. The tether length is 0.68 m, resulting in a period of motion of 1.65 s. The scope of the proposed method for this example is to reduce the effect of the swinging load, while an alternative method would be to augment the states of the system with the pendulum equations of motion to dampen the swinging of the pendulum. The ARW is commanded to change continuously its position between two $\boldsymbol{x}_r$: $[-0.7, -1.6, 1.5]^\top$ and $[0.0, -1.6, 1.5]^\top$ [m] along the global X axis. Fig. 8.8a depicts the position and estimated position of the ARW. The Root Mean Square Error (RMSE) between the position and waypoint is 0.5 m, 0.4 m, and 0.2 m for x, y, and z-axes, respectively. Fig. 8.8b presents the estimated forces on the tether experiment. The oscillatory motion of the pendulum is evident in the estimated forces along the x and y axes of the platform. Note that when the ARW is approaching the landing set-point, a positive value for f_z is estimated. That occurs as the tether payload touches the ground, and the force is omitted from the ARW. It is worth highlighting that the stand-alone NMHE is not able to stabilize the aerial platform, and there is a need for compensation on the low-level controller; however, the NMHE estimation still provides collision-free navigation.

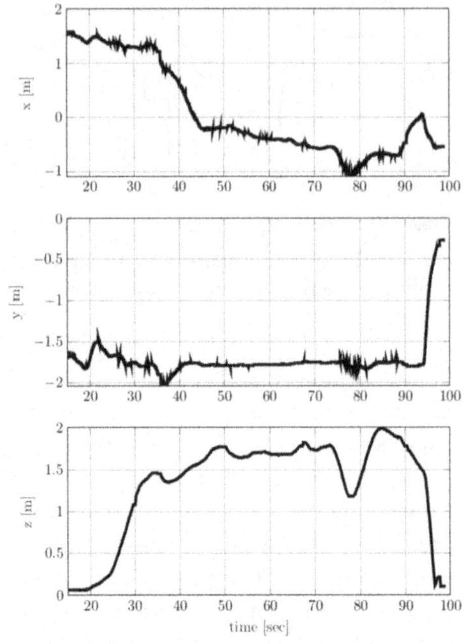

FIGURE 8.6 Measured x, y, z positions of the ARW without external forces estimation and compensation for the scenario where the wind is generated from a wind-wall.

FIGURE 8.7 Photographic still of the experimental scenario where the ARW has a tethered payload.

8.3.2.1 Example of center of gravity compensation

In this example, we test a re-configurable ARW [12], like the one presented in Chapters 3 and 4, which can fold its brackets to alter its formation to H, X, Y, and T shapes as depicted in Fig. 8.9. The different ARW configurations have a direct impact on the moment of inertia of the platform and for asymmetric

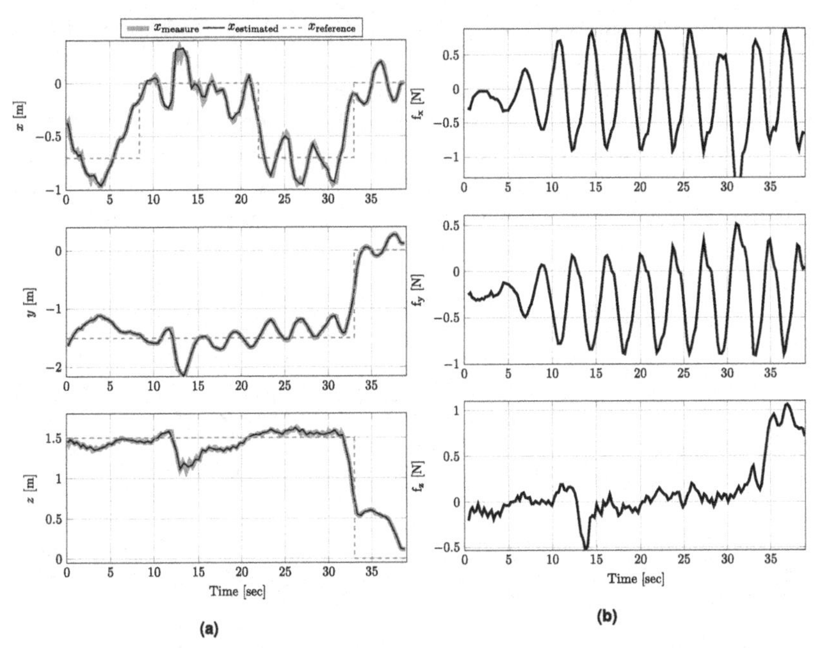

FIGURE 8.8 NMHE-NMPC force estimation and disturbance rejection for the scenario where a payload is attached on the ARW. The measured, estimated, and reference values of the position states $[x, y, z]^\top$ are shown by red (dark gray in print), black, and dashed gray lines, respectively (left). Estimated forces $[f_x, f_y, f_z]^\top$ for the scenario where a payload is attached on the ARW (right).

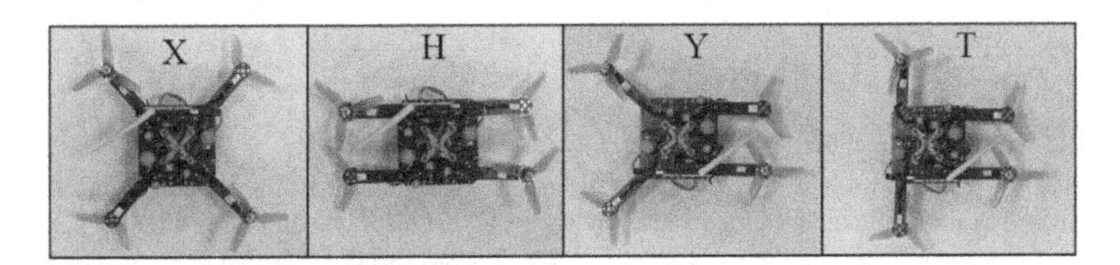

FIGURE 8.9 Different formations *X*, *H*, *Y*, and *T* based on re-configurable ARW's arms position.

configurations, like *Y* and *T*, on its center of gravity. In contrast to Chapter 4, where we compensate the effect of the different shapes on the attitude controller level, here, a classic PID-based low-level controller without compensation for the different formations is used. The selection of this low-level controller is to emphasize the NMHE force estimation capabilities and disturbance rejection when the platform dynamics are affected by major changes like its structural re-configuration.

The ARW is commanded to hover at $\boldsymbol{x}_r = [0, 0, 0.6]^\top$ [m] while the input angles are constrained in $|\phi_d|, |\theta_d| \leq 1$ rad. Fig. 8.10a shows the actual and estimated position of the ARW. The initial RMSE, while the platform maintains the H and X formations for the first 35 s, stays at 0.11 m, 0.13 m, and 0.23 m for x, y, and z-axes, respectively. The higher altitude fluctuations due to the aerodynamic effects and loss of energy among the propellers occurred since there is overlap between them in H-formation. When the platform changes to the Y-formation and later on to the T-formation, the RMSE between the position and the waypoint is 0.7 m, 0.45 m, and 0.6 m for $x-y-z$ axes. The increased RMSE is expected as the asymmetry nature of those two formations results in a major shift of the platform's center of gravity.

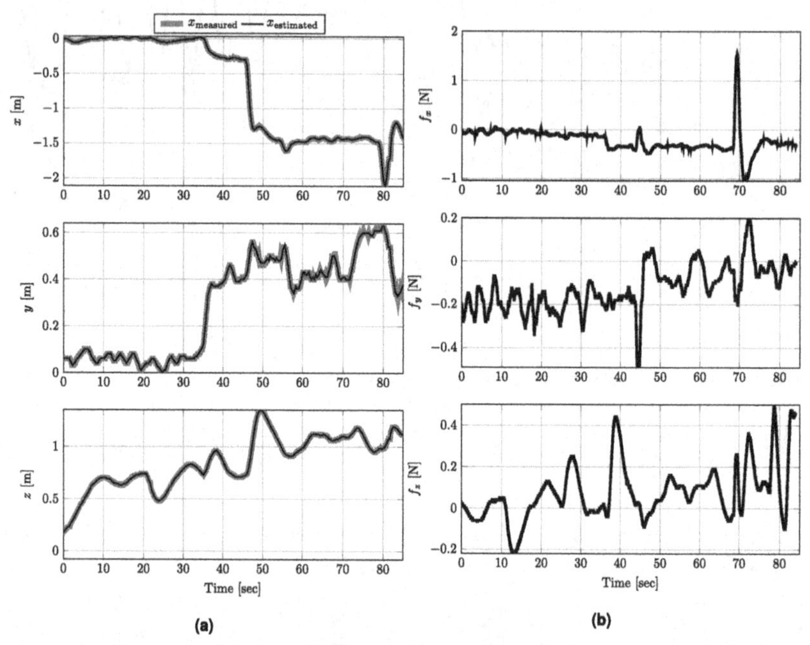

FIGURE 8.10 Position states $[x, y, z]^\top$ of the experimental evaluation with the re-configurable ARW. The measured and estimated values are shown by red (dark gray in print) and black colors, respectively. Estimated forces $[f_x, f_y, f_z]^\top$ for the experimental evaluation of the proposed methodology with the re-configurable ARW.

Figs. 8.10b and 8.11 present the estimated forces of the NMHE and the generated control commands of the NMPC. The mean and absolute force levels are $(-0.18, 1.53)$ N, $(-0.14, 0.48)$ N, and $(0.06, 0.51)$ N for each axis x, y, and z. From Fig. 8.11, one can observe that the re-configurable drone reaches the input constraints, which are already expanded compared to the other experimental scenarios. More specifically, when the ARW changes to the T configuration, the input θ_d reaches closer to 1 rad, which is the NMPC bound for the pitch angle. It should be highlighted that the same experiment is performed without the NMHE module, and the stand-alone NMPC cannot compensate for the arm re-configuration; thus, the experiment results in the collision of the platform with the protection net. As already emphasized, there is no model-based compensation for the low-level controller of the re-configurable ARW. Thus the high RMSE values are expected, while the ARW avoids the collision in contrast to the scenario where the NMHE module is suspended.

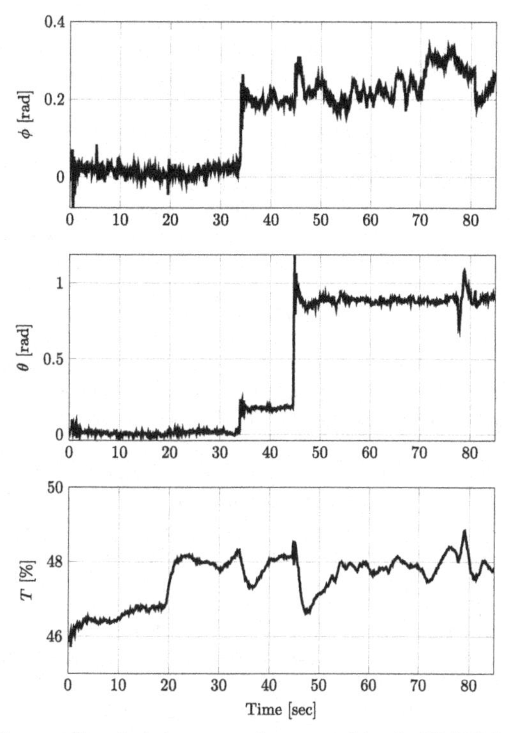

FIGURE 8.11 Thrust, roll, and pitch commands generated by the NMPC for the experimental evaluation of the proposed methodology with the re-configurable ARW.

8.4 Conclusion

In this chapter, an NMHE method is presented to estimate external disturbances that can occur via various sources, and they affect the ARWs flight performance. The estimated disturbances are assumed as forces that act on the body frame of the ARW, and an augmented NMPC is presented that compensates for them. Three example cases were used to showcase the performance of the estimator and disturbance rejection, including 1) Wind-wall, 2) tethered payload, and 3) an ARW with re-configurable arms. The NMHE successfully estimated the disturbances, and the NMPC was used to compensate for them. The actuation constraints were tuned for each example as the boundaries increased depending on the disturbance levels. The same three example cases were tested again with no estimation module, and the NMPC did not have the augmented states to compensate for these forces. The controller fails, and the flight is either terminated immediately or results in a collision with the protection net.

References

[1] D. Hollenbeck, G. Nunez, L.E. Christensen, Y. Chen, Wind measurement and estimation with small unmanned aerial systems (SUAS) using on-board mini ultrasonic anemometers, in: 2018 International Conference on Unmanned Aircraft Systems (ICUAS), IEEE, 2018, pp. 285–292.

[2] C.A. Wolf, R.P. Hardis, S.D. Woodrum, R.S. Galan, H.S. Wichelt, M.C. Metzger, N. Bezzo, G.C. Lewin, S.F. de Wekker, Wind data collection techniques on a multi-rotor platform, in: 2017 Systems and Information Engineering Design Symposium (SIEDS), IEEE, 2017, pp. 32–37.

[3] D. Hentzen, T. Stastny, R. Siegwart, R. Brockers, Disturbance estimation and rejection for high-precision multirotor position control, in: 2019 IEEE/RSJ International Conference on Intelligent Robots and Systems (IROS), 2019, pp. 2797–2804.

[4] A. Cho, J. Kim, S. Lee, C. Kee, Wind estimation and airspeed calibration using a UAV with a single-antenna GPS receiver and Pitot tube, IEEE Transactions on Aerospace and Electronic Systems 47 (1) (2011) 109–117.

[5] S. Sun, C. de Visser, Aerodynamic model identification of a quadrotor subjected to rotor failures in the high-speed flight regime, IEEE Robotics and Automation Letters 4 (4) (2019) 3868–3875.

[6] External force estimation and disturbance rejection for micro aerial vehicles, Expert Systems with Applications 200 (2022) 116883 [Online]. Available: https://www.sciencedirect.com/science/article/pii/S095741742200327X.

[7] J. Rawlings, Moving Horizon Estimation, 2013, pp. 1–7.

[8] C.V. Rao, J.B. Rawlings, Nonlinear moving horizon state estimation, in: F. Allgöwer, A. Zheng (Eds.), Nonlinear Model Predictive Control, Birkhäuser Basel, Basel, 2000, pp. 45–69.

[9] K.R. Muske, J.B. Rawlings, Nonlinear Moving Horizon State Estimation, Springer Netherlands, Dordrecht, 1995, pp. 349–365. Available: https://doi.org/10.1007/978-94-011-0135-6_14.

[10] P. Sopasakis, E. Fresk, P. Patrinos, Open: Code generation for embedded nonconvex optimization, International Federation of Automatic Control, 2020.

[11] A. Sathya, P. Sopasakis, R. Van Parys, A. Themelis, G. Pipeleers, P. Patrinos, Embedded nonlinear model predictive control for obstacle avoidance using PANOC, in: 2018 European Control Conference (ECC), IEEE, 2018, pp. 1523–1528.

[12] A. Papadimitriou, S.S. Mansouri, C. Kanellakis, G. Nikolakopoulos, Geometry aware NMPC scheme for morphing quadrotor navigation in restricted entrances, in: 2021 European Control Conference (ECC), 2021, pp. 1597–1603.

Chapter 9

Perception driven guidance modules for aerial manipulation

Samuel Karlsson and Christoforos Kanellakis
Department of Computer, Electrical and Space Engineering, Luleå University of Technology, Luleå, Sweden

9.1 Introduction

MAVs are characterized by simple mechanical design and versatile movement, which makes them a popular platform within the robotics community. These capabilities define missions that are impossible or dangerous for the human operator to perform. Some examples of up-to-date efforts to employ MAVs include the fields of infrastructure inspection [1,2], public safety, such as surveillance [3], and search and rescue missions [4].

In general, MAVs are agile platforms with various navigation modes, from hovering over a target to aggressive maneuvering in $\mathbb{R}^3$, making it possible to access remote and distant places, compared to mobile robots. MAVs are distinguished mainly for their payload capabilities with a trade in their size. Usually, they are equipped with sensors like cameras, sonars, lasers, and GPS to provide useful information to their operators. Thus, a novel trend currently emerging with fast pace includes the interaction capabilities of such platforms with the surrounding environment.

Taking advantage of the MAV sensor payload concept, MAVs can also carry other type of hardware mechanisms, e.g., lightweight dexterous robotic arms, as depicted in Fig. 9.1. In such way, Aerial Robotic Workers (ARWs) expand their operational workspace, which can be deployed in applications that require payload transportation, infrastructure maintenance, precision farming, and environment sampling.

Within the related literature, studies on MAV's with aerial manipulators have been focusing to object picking, transportation [5], and assembling [6] tasks. Nowadays, the research community in aerial robotics is pursuing dexterity regarding the manipulation capabilities of aerial vehicles [7–9]. A fundamental feature that could assist these platforms in complex missions is the development of a guidance system for the manipulator based on robust perception modules. The ability to guide precisely the end-effector, relative to targets in the surrounding workspace, introduces a wide span of novel applications for the aerial platform.

Aerial Robotic Workers. https://doi.org/10.1016/B978-0-12-814909-6.00015-9

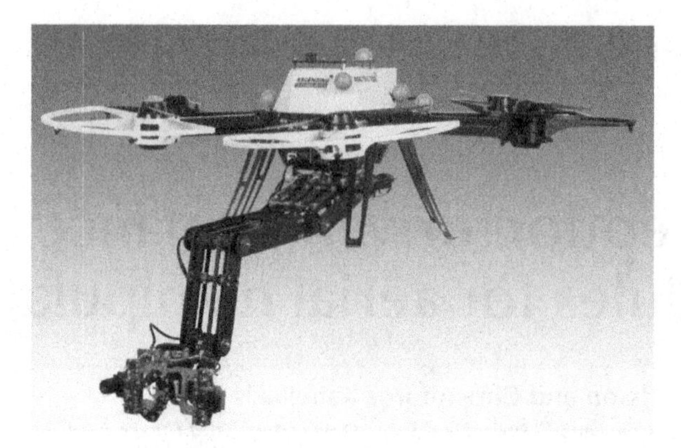

FIGURE 9.1 Example of an aerial robotic worker with endowed manipulator.

Searching the literature within this research field, several works have considered studying and developing the visual guidance system as a means to assist the manipulation task. More specifically, in [10], an image-based impedance control strategy for force tracking of an unmanned aerial manipulator is presented. The visual part is performed with planar image features that have decoupling property to drive the end-effector perpendicular to the object plane. In [11], a vision-based guidance system for a 3DoF manipulator has been developed. This work presented an Image-Based Visual Servoing (IBVS) scheme using image moments to derive the velocity references for commanding the coupled system (MAV and manipulator), while the object detection was based on color thresholding. An adaptive controller was designed to switch between position and IBVS control, while the authors of [11] extended their work on manipulation in [12] by proposing a guidance system for cylindrical objects, where the detection has been performed using RANSAC ellipse detection. Regarding the visual servoing task, a stochastic MPC has been employed to handle x and y rotational velocities as stochastic variables. In [13], an aerial manipulator guidance system has been presented, where the novelty of this work stems from the designed hierarchical control law that prioritizes the tasks like collision avoidance, visual servoing, the center of gravity compensation, and joint limit avoidance during a flight, where the visual servo control considers an uncalibrated camera and drives the manipulator to the desired pose. In [14], a tree cavity inspection system has been presented based on the depth image analysis and image processing (contour extraction, ellipse fitting), while the overall goal was to drive the end-effector inside the cavity. In [15], a stereo vision system for object grasping has been proposed with a detection algorithm to learn a feature-based model in an of-line stage and then use it online to detect the targeted object and estimate its relative pose using the information from the 3D model.

The main aim of the Chapter is to expand on visual processing modules for guidance of aerial manipulators, focusing on object detection, tracking, and localization techniques. Finally, the Chapter ends with the presentation of an aerial manipulator design and experimental examples on the topic guidance for aerial manipulation.

9.2 Perception architecture for ARW guidance

The basis of the perception architecture is the definition of coordinate frames as depicted in Fig. 9.2. The world frame $\mathcal{W}$ is fixed inside the workspace of the robotic platform, the body frame of the vehicle $\mathcal{B}$ is attached on its base, while the manipulator's frame $\mathcal{M}$ is fixed on the base of the manipulator. Finally, the stereo camera frame C origins on the left camera and is firmly attached to the end-effector frame $\mathcal{E}$. The transformation of the point p^C to the frame $\mathcal{E}$ is expressed through the homogeneous transformation matrix $T_C^{\mathcal{E}}$ ($p^{\mathcal{E}} = T_C^{\mathcal{E}}\, p^C$). For the rest of this Chapter, the superscript denotes the reference frame. Accordingly, $p^{\mathcal{E}}$ can be expressed in the manipulator's frame $\mathcal{M}$ using the forward kinematics. More specifically, $p^{\mathcal{M}} = T_{\mathcal{E}}^{\mathcal{M}}(q)\, p^{\mathcal{E}}$, where $T_{\mathcal{E}}^{\mathcal{M}}(q)$ is the homogeneous transformation matrix from the end effector's frame to the base frame, which depends on the current manipulator joint configuration $q = [q_1, \cdots, q_n]$. Finally, the manipulator is firmly attached to the MAV, thereafter the transformation matrix $T_{\mathcal{M}}^{\mathcal{B}}$ is constant, expressing the relative pose between the vehicle and the manipulator base. The pose of the target $p^{\mathcal{B}}$, relative to the multirotor base frame, is calculated through $p^{\mathcal{B}} = T_{\mathcal{M}}^{\mathcal{B}}\, p^{\mathcal{M}}$.

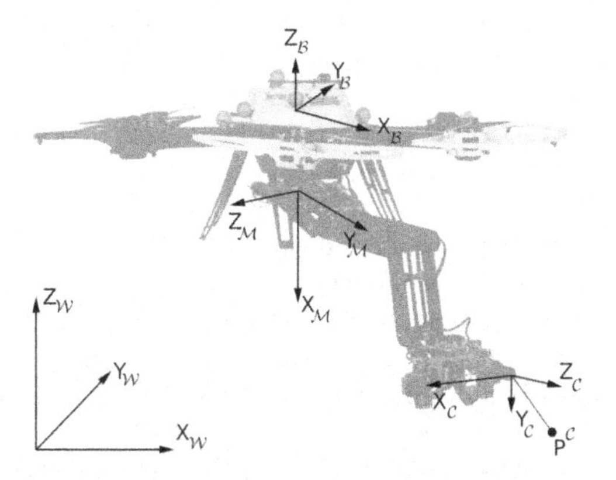

FIGURE 9.2 Coordinate frames of the aerial platform.

Fig. 9.3 depicts the perception architecture to support the guidance of the aerial manipulator. The I_1, I_2 are camera frames, P^C and $P^C_{bounded}$ are 3D point-clouds, B is the bounding box, and x^C, y^C, z^C are the centroid coordinates and waypoints p^W, ϕ. The waypoints are afterward transferred to the motion planning and control modules of the ARW for performing the guidance maneuver.

The guidance components consist of an integrated stereo-based system. The target is identified within the sequential received frames in the form of a bounding box using the combined detection and tracking scheme (described in Sec-

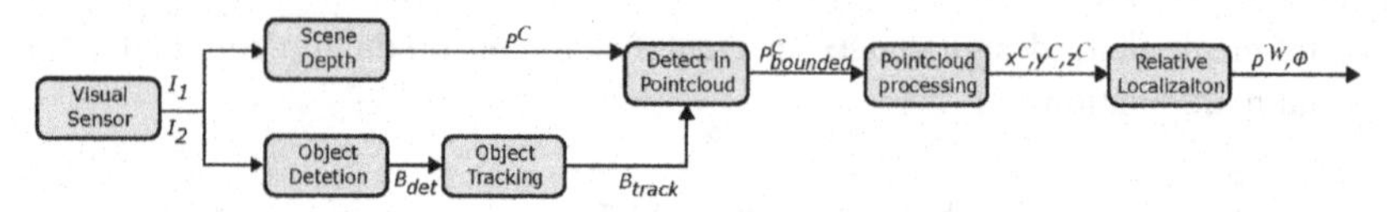

FIGURE 9.3 Perception system architecture of the guidance system.

tions 5.4, 9.3). The object detection information is integrated with a pointcloud processing to extract the centroid of the manipulated object, compute its relative configuration with respect to the MAV, generate a proper trajectory, and align the end-effector properly with the grasping point. All computations regarding the detection and tracking components are executed onboard the MAV to avoid communication latency issues.

9.3 Object tracking

The object detection module has been presented in Section 5.4. This part describes an approach for object tracking. During the years, multiple efficient tracking algorithms [16] have been proposed, but many algorithms are not suitable for MAV applications, since they require high computational resources. A tracking category that could address these challenges is the tracking-by-detection algorithms. Briefly, these tracking algorithms are treated as binary classification methods since they constantly try to discriminate between the target and the background using decision boundaries. The tracking mechanism is online using patches of both target and background captured in recent and past frames. In this Chapter, the tracking-by-detection approach for robust tracking during manipulation guidance is based on the Kernelized Correlation Filter (KCF) [17].

In a nutshell, the KCF pipeline is divided into three parts: 1) the training, 2) the detection, and 3) the training-update at the new target position. The tracker learns a kernelized least squares classifier $f\langle \phi_{kcf}(x_{kcf_{m,n}}), w_{kcf}\rangle = y_{kcf}$ of a target from a single patch x_{kcf}, centered around the target, of a size $M \times N$. It uses a circulant matrix to learn all the possible shifts of the target, considering the cyclic shifts $x_{kcf_{m,n}}$, $(m,n) \in \{0,\dots,M-1\} \times \{0,\dots,N-1\}$ of the patch as training samples for the classifier. The classifier is trained by minimizing C_{kcf} (Eq. (9.1)) over w_{kcf}, where ϕ_{kcf} is the mapping to the feature space defined by the kernel $k(f,g) = \langle \phi_{kcf}(f), \phi(g)\rangle$, and λ is a regularization parameter. The regression target $y_{kcf}(m,n) \in [0,1]$ is generated considering a Gaussian function.

$$C_{kcf} = \sum_{m,n} |\langle \phi_{kcf}(x_{kcf_{m,n}}), w_{kcf}\rangle - y_{kcf}(m,n)|^2 + \lambda \langle w_{kcf}, w_{kcf}\rangle \tag{9.1}$$

By denoting $\mathcal{F}$ the Discrete Fourier Transform (DFT) operator, the cost function is minimized by $w_{kcf} = \sum_{m,n} \alpha_{kcf}(m,n)\phi_{kcf}(x_{kcf_{m,n}})$, where the coefficients α_{kcf} are calculated from $A = \mathcal{F}(\alpha_{kcf}) = \frac{\mathcal{F}(y_{kcf})}{\mathcal{F}(u_x)+\lambda}$, where $u_x(m,n) = k(x_{kcf\,m,n}, x_{kcf})$ is the output of the kernel function.

During the detection step, a patch z_{kcf} of size $M \times N$ is cropped out in the new frame. The algorithm computes the detection scores, considering the cyclic shift alterations of z_{kcf}, as $\hat{y_{kcf}} = \mathcal{F}^{-1}\{AU_z\}$, where U_z is the Fourier transformed kernel output of the example patch z_{kcf}. The target position in the new frame is estimated by finding the translation that maximizes the score $\hat{y_{kcf}}$.

Finally, the target model is updated over time. In the KCF tracker, the model consists of the learned target appearance model $\hat{x_{kcf}}$ and the transformed classifier dual space coefficients a_{kcf}. Within the classifier update of the target, appearance model $\{x^j_{kcf} : j = 1, \ldots, p\}$ includes information from the previous frames till the current frame. The model is updated in each new frame with a fixed learning rate, which is not adaptive to appearance changes.

9.4 Object localization

The outcome of this process results in a 2D bounding box, defined as a set B with x_b and y_b coordinates (Eq. (9.2)).

$$B = \{(x_b, y_b) \in \mathbb{R}^2 |\ x_{min} < x_b < x_{max}\ ,\ y_{min} < y_b < y_{max}\} \tag{9.2}$$

where $\{(x_{min}, y_{min}), (x_{min}, y_{max}), (x_{max}, y_{min}), (x_{max}, y_{max})\}$ are the four corners of the bounding box in the image plane.

A major part of the proposed system includes the guidance layer based on stereo vision during the exploration phase of the MAV. This part is used when the target of interest lies within the depth range of the stereo camera. The goal is to bring the aerial platform in the proximity of the target by following a simple but efficient strategy.

The basis of the 3D perception of the system is structured around the reconstruction capabilities of the stereo sensor. The overall process is initiated by calculating the 3D structure of the area perceived from the stereo pair using Semi Global Block Matching (SGBM) [18] method. The stereo mapping function $S(x, y)$ maps a point (x,y) from the image pixel coordinate frame to the camera frame as shown in Eq. (9.3).

$$S(x, y) = (X, Y, Z) \tag{9.3}$$

Thus, a pointcloud $\mathcal{P}$ is formulated as $\mathcal{P} = \{S(x, y)\}$.

A pointcloud filtering method is proposed to robustly isolate the region of interest, combining information from both the depth map and the object tracker presented in Section 9.3. More specifically, the points belonging to the 2D bounding box B are translated to a pointcloud $\mathcal{P}_{bounded}$ using the stereo mapping function as

$$\mathcal{P}_{bounded} = \{S(x, y)|\ x \in x_b\ ,\ y \in y_b\} \tag{9.4}$$

In the proposed system, the centroid extraction depends on the processed pointcloud; therefore, additional background parts in the model will downgrade the accuracy of the centroid. Therefore, the clustering method Region Growing Segmentation [19], part of the pointcloud processing component (Fig. 9.3) is implemented using smooth constraints, to partition $\mathcal{P}_{bounded}$ into separate regions. The clustering of the bounded 3D points into groups is selected to remove parts of $\mathcal{P}_{bounded}$ that do not belong to the desired target and are directly passed from the object tracker. Usually, the extracted bounding box does not entirely enclose the target but includes parts of the background.

The assumption in the proposed process is based on the concept that the target of interest covers the largest part of the bounding box and therefore the largest part of $\mathcal{P}_{bounded}$. The size of every cluster in $\mathcal{P}_{bounded}$ is verified by a heuristic threshold that has been designed to further merge neighboring clusters that do not meet size requirements. In this manner, the 3D centroid of the target in $\mathcal{P}_{bounded}$ lies in the cluster with the maximum area. Finally, the centroid $[x^c, y^c, z^c]$ is extracted as the average position of the point in the cluster. Overall, there is no metric information of the target provided a priori.

On top of the already described process, the pointcloud is filtered to remove invalid values with the aim of further refining the centroid position. It is also downsampled to reduce the number of points through Voxel Grid Filtering [20] for faster processing, which is critical for the aerial platform. An extra step is considered for targets that are attached in planar surfaces, where the background plane is segmented using RANSAC [21]. Fig. 9.4 provides a stepwise visualization of the pointcloud filtering process. In the clustered point cloud, the points include only the circle and cross parts of the target, while the white background is merged after the final filtering step as shown on the right.

FIGURE 9.4 Pointcloud filtering steps, on the left the original pointcloud, in the middle the clustered pointcloud and on the right the final filtered pointcloud including the whole target.

The centroid information is transferred to the body frame of the aerial vehicle $\mathcal{B}$ using the transformation from camera as well as the manipulators kinematics. The stereo guidance subsystem is finalized with the generation of the proper waypoint $Wp = [p^W, \Phi]$ using the extracted centroid location, where p^W represent the x, y, z positions in frame $\mathcal{W}$, while Φ the orientation of the MAV in frame $\mathcal{W}$. In this case, the aerial manipulator is given a predefined joint configuration q_1, q_2, q_3, q_4 according to the task requirements. The MAV waypoint is converted into a position-velocity-yaw trajectory provided to the platform high-level position controller.

9.5 Aerial manipulator system

9.5.1 Hexacopter manipulator carrier

This work employs the aerial research platform from Ascending Technologies, the NEO hexacopter, depicted in Fig. 9.1. The platform has a diameter of 0.59 m, height of 0.24 m, and propeller length of 0.28 m, while it can carry a payload of 2 kg with up to 2 kg flight time. The sensor payload is configurable based on the mission requirements and can vary, while the onboard processor is an Intel NUC i7-5557U with Ubuntu Server Operating System. The multirotor includes three main subsystems to provide autonomous flight, namely the localization system based Vicon MoCap,[1] a Multi-Sensor-Fusion Extended Kalman Filter (MSF-EKF) [22] for state estimation and finally the linear Model Predictive Control (MPC) position controller [23–25] for trajectory following.

9.5.2 The CARMA aerial manipulator

The robotic arm introduces manipulation capabilities to the multirotor, and it is a planar robotic arm with 4 revolute joints mounted underneath the aerial platform, as shown in Fig. 9.5. The manipulator weight is 500 g, while it is capable of holding various types of end-effectors like a grasper, a brusher, a camera holder, or even an electromagnet for lifting heavy objects.

FIGURE 9.5 Left – CAD design of the CARMA manipulator. Right – CARMA parts explosion view.

Some highlights on the design of the manipulator are the following:

- a robust and sturdy mechanism with belts for motion transmission;

[1] https://www.vicon.com/.

- linear potentiometers for joint angle feedback;
- multiple end-effector types.

Compact AeRial MAnipulator (CARMA) is regulated using a cascaded position-velocity Proportional Integral Derivative (PID) control scheme. More specifically, the joint positions derived from the inverse kinematics consist of the reference to four standalone PID controllers, one controller for every joint. In the developed system, the manipulator kinematics define the end-effector position relative to the manipulator base. A full description of the design and modeling of the manipulator was presented in [26].

9.5.3 Visual sensor

The guidance system of the manipulator consists of a custom-made stereo camera with a frame rate of 20 fps at 640 × 480 resolution and a baseline of 10 cm, as depicted in Fig. 9.6. The sensor has been calibrated to calculate the intrinsics and extrinsic parameters before its use.

FIGURE 9.6 Visual sensor in an eye-in-hand configuration for aerial manipulation (video assembly at [27]).

9.6 Results

This Section provides experimental results from various components covered in the Chapter to showcase the deployment of the methods in real hardware and realistic environments. The visual tracking and object detection outcome has been collected from autonomous flight datasets that were performed indoors in the Field Robotics lab flight arena located at Luleå University of Technology and at the Mjolkudden area in Luleå. The visual guidance experiments of the aerial manipulator were performed in the Field Robotics Lab flight arena as well as in a subterranean tunnel complex located in Luleå, Sweden. On the software side, the implementation was held in C++, using ROS[2] framework, OpenCV,[3] and PCL[4] libraries.

[2] http://www.ros.org/.

[3] https://opencv.org/.

[4] http://pointclouds.org/.

9.6.1 Object detection

The results presented in this Section utilized the tiny version of YOLOv4 [28]. This selection was mainly made to reduce the computational requirements and increase the inference frame rate, compared to the vanilla version of YOLO, since the aim is to deploy it in a processor that does not host any graphics processing unit (GPU) to eventually respect the payload requirements of the aerial platform. On top of that the Vision Processing Unit (VPU), Intel Neural Compute Stick 2[5] module was used to further offload the computation load from the Central Processing Unit (CPU). Compared to previous versions YOLOv4 architecture uses CSPDarknet53 [29] as the backbone, which is cascaded with Spatial Pyramid Pooling block on top, PAN [30] for neck, and YOLOv3 [31] for head.

The network was trained on 6 classes defined by the SubT competition using a custom dataset consisting of approximately 700 images for each class. The input size of the images is 416 × 416, and the output of the algorithm is the detected bounding boxes and the class probability. Moreover, the model was further modified to be executed on the VPU using the Openvino[6] framework, which is an open-source toolkit for optimizing and deploying AI inference, with FP16 precision.

The identified objects have not been part of the training dataset for the Convolutional Neural Network (CNN)-based detector, showing the applicability to identify object in those classes that have not been seen before. Fig. 9.7 depicts snapshots from the object detection module. The detection module rate varied around 8 fps. Overall, such architectures, when deployed in realistic environments, are related to the mission requirements, and the existing trade-offs should always be considered for smoother and more efficient mission accomplishment.

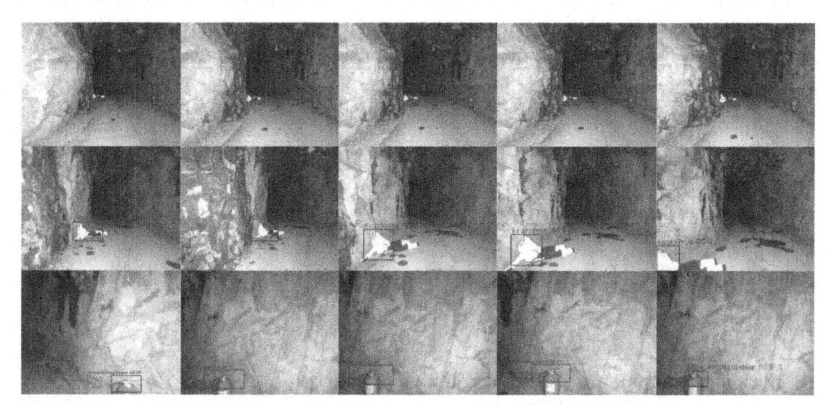

FIGURE 9.7 Visualization of true positives from detected objects (bounding boxes and class probabilities) during the full mission.

[5] https://www.intel.com/content/www/us/en/developer/tools/neural-compute-stick/overview.html.
[6] https://docs.openvino.ai/latest/index.html.

9.6.2 Visual tracking

This experimental part is designed to demonstrate the performance of the tracker, while the MAV is flying close to the target of interest. These experiments include the manual navigation of the ARW in the frontal area of the object of interest following different paths, including hovering, longitudinal, and lateral motions. The main goal is to provide an insight into the tracker capabilities to track targets with different characteristics (e.g., shape, color) during the deployment of the aerial manipulator. The trials have been performed considering three different types of objects to track: 1) a planar pattern, 2) a custom 3D printed object with a rectangular base housing a semicircle, and 3) a screwdriver tool, which are targets with incremental complexity.

Fig. 9.8 demonstrates the use of KCF in the current guidance system. More specifically, the figure provides snapshots of different instances from the on-board visual sensor of the two objects, showing the ability to continuously monitor the target that lies within the field of view of the camera.

9.6.3 Stereo based guidance

This section presents experimental trials of the guidance architecture of the aerial manipulator system for a target monitoring mission. More specifically, the end-effector of the aerial platform is autonomously guided to a desired position relative to the target in an initially unknown environment, without performing any physical interaction.

The mission behavior of the guidance system is presented in Algorithm 1. Initially, the aerial vehicle takes off and navigates to a user-defined waypoint using the high-level position control. When the MAV reaches the waypoint, the target of interest lies within the field of view of the stereo camera. The next step for the operator is to select the bounding box for the desired target, so that the tracking algorithm can learn online the target for sequential detection, as discussed in the previous section. A generic object of interest is placed on

Algorithm 1 Stereo guidance.

Take-off
Reach user commanded waypoint
Select object to track
for $\{i : 1 : \#frames\}$ **do**
  $\{x_b, y_b\} \leftarrow$ Object Tracker
  $\mathcal{P}_{bounded} \leftarrow$ Stereo Mapping(x_b, y_b)
  $x^c, y^c, z^c \leftarrow$ Pointcloud processing$(\mathcal{P}_{bounded})$
  $q_1, q_2, q_3, q_4 \leftarrow$ Manipulator Joint configuration(x^c, y^c, z^c)
  $p^W, \Phi \leftarrow$ Waypoint Extraction(x^c, y^c, z^c)
end for

FIGURE 9.8 Experimental tracking results for three different objects. In the first row, multiple snapshots of the tracking process for the object 1 have been extracted; in the second row, multiple snapshots of the tracking process for the object 2 have been extracted, while in the third row, multiple snapshots of the tracking process for the object 3 have been extracted.

the top of a bar inside the flight arena. While the aerial platform hovers at the initial waypoint, the depth from the stereo camera is converted into a pointcloud and is processed using the refining methods to extract its 3D position from the rest of the background. In this manner, the relative position between the MAV body frame and the target is calculated. In parallel, the current position of the manipulator is calculated from the forward kinematics to calculate the relative transformation between the end-effector and the MAV base. Afterward, the end-effector is driven to the final grasping configuration, based on the application requirements, using its inverse kinematics. The joint configuration for the final grasping is predefined but always considers the position of the object.

Within this work, three experimental trials have been performed to showcase the performance of the system in various situations. More specifically, experiments one and two deal with the same target but different monitoring positions, while experiment three presents the system operation with a different target.

Fig. 9.9 depicts the 3D trajectory followed by the aerial platform, while Fig. 9.10 depicts the path of the end-effector position versus the execution time of the experiment. x_{ee}, y_{ee} and z_{ee} correspond to the end-effector position measurements in the $\mathbb{W}$ frame. Fig. 9.10 shows that the proposed approach was able to perform the task and drive the end-effector close to the target approaching the reference values in all axes. The object tracking process detected and kept the object inside the cameras' field of view during all the phases of the experiment successfully. Moreover, the MAV is able to hover in front of the object at a desired distance.

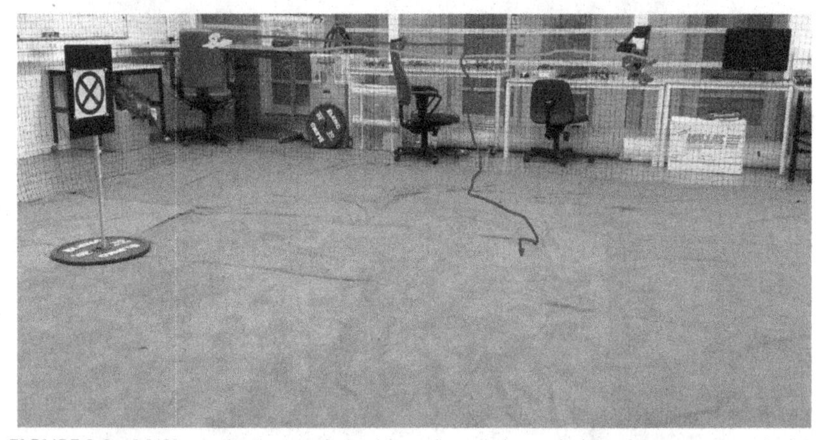

FIGURE 9.9 MAV actual trajectory derived from the experimental trials of the stereo-based guidance. Case 1: relative distance with the target 25 cm. The developed guidance system contributes to the task with the red (gray in print) and green (light gray) part of the overall trajectory. The red part is followed after the extraction of the centroid, while in the green part, the robotic platform is hovering. The blue (dark gray) parts of the trajectory constitute the initialization (hovering on a fixed position) and termination phases (landing) of the experiment.

Similarly, Fig. 9.11 depicts the 3D trajectory followed by the aerial platform, while Fig. 9.12 depicts the path of the end-effector position versus the execution time of the experiment. From the implemented tests the proposed approach was able to perform the task and drive the end-effector close to the target. The object tracking process detected and followed the object during all the phases of the experiment successfully. Additionally, the method showed satisfactory performance for extracting the target centroid position.

Finally, experiment three presents the deployment of the system to approach a target with different shape and color. Fig. 9.13 depicts the 3D trajectory followed by the aerial platform, while Fig. 9.14 depicts the path of the end-effector position versus the execution iterations of the experiment. In this case, the ob-

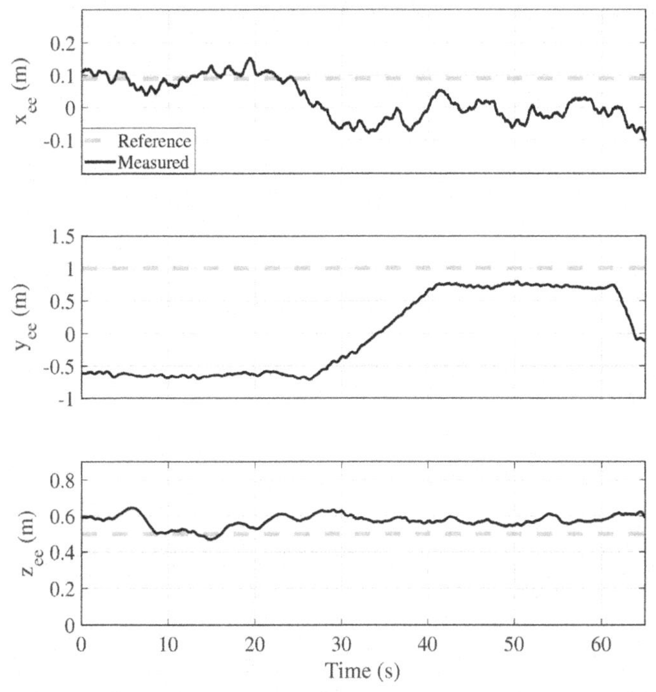

FIGURE 9.10 End-effector setpoints vs. the actual setpoints for the first experiment. The plots represent the initialization phase (centroid and waypoint calculation, the waypoint following, the hovering part relative to the target and finally the landing.

FIGURE 9.11 MAV actual trajectory derived from the experimental trials of the stereo-based guidance. Case 2: the end-effector reaches the target. The developed guidance system contributes to the task with the red (gray in print) and green (light gray) part of the overall trajectory. The red part is followed after the extraction of the centroid, while in the green part the robotic platform is hovering. The blue (dark gray) parts of the trajectory consist of the initialization (hovering on a fixed position) and termination phases (landing) of the experiment.

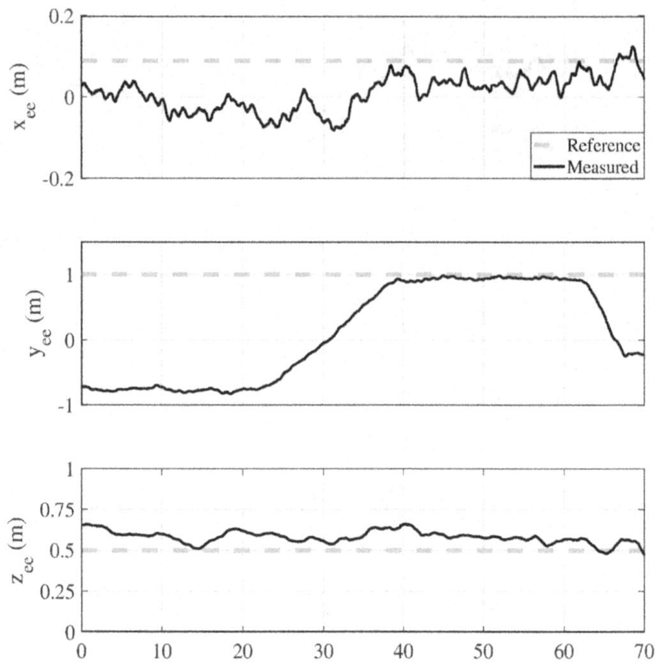

FIGURE 9.12 End-effector setpoints vs. the actual setpoints for the second experiment. The plots represent the initialization phase (centroid and waypoint calculation, the waypoint following, the hovering part relative to the target and finally the landing.

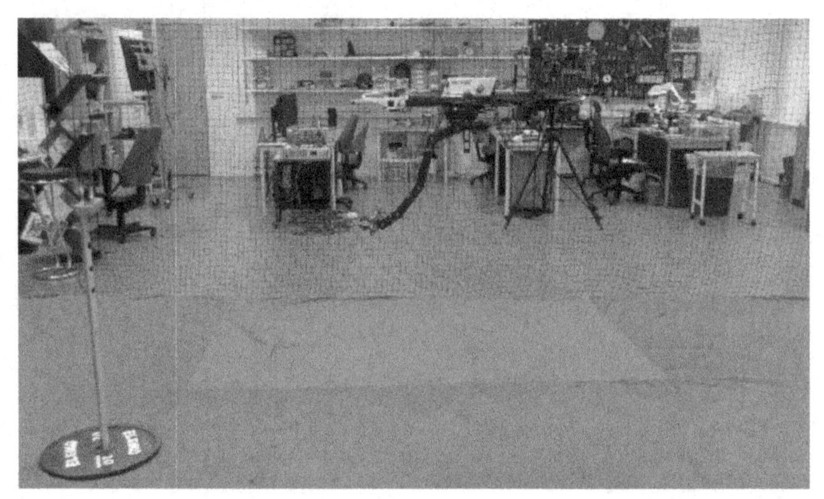

FIGURE 9.13 Snapshot from the experimental trials of the stereo-based guidance, depicting the MAV hovering in front of the second object.

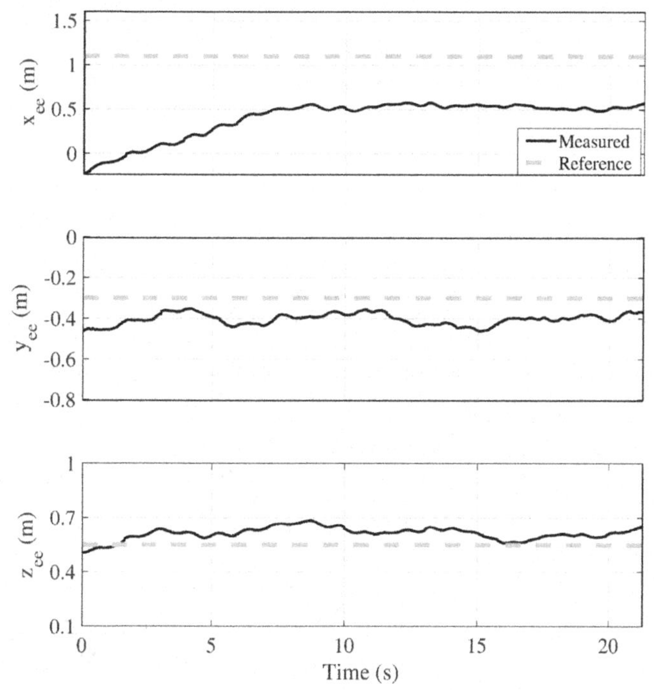

FIGURE 9.14 End-effector setpoints vs. the actual setpoints for the third experiment. The plots represent the waypoint following phase and the hovering part relative to the target.

ject has been placed in another part of the flying arena, and the main motion of the aerial vehicle was in the x axis. The plots depict the trajectory following and hovering parts of the manipulator guidance.

References

[1] S.S. Mansouri, C. Kanellakis, E. Fresk, D. Kominiak, G. Nikolakopoulos, Cooperative UAVs as a tool for aerial inspection of the aging infrastructure, 2018, pp. 177–189.

[2] A.C. Lee, M. Dahan, A.J. Weinert, S. Amin, Leveraging SUAS for infrastructure network exploration and failure isolation, Journal of Intelligent & Robotic Systems (2018) 1–29.

[3] C. Yuan, Z. Liu, Y. Zhang, Learning-based smoke detection for unmanned aerial vehicles applied to forest fire surveillance, Journal of Intelligent & Robotic Systems (Mar 2018), https://doi.org/10.1007/s10846-018-0803-y.

[4] C. Sampedro, A. Rodriguez-Ramos, H. Bavle, A. Carrio, P. de la Puente, P. Campoy, A fully-autonomous aerial robot for search and rescue applications in indoor environments using learning-based techniques, Journal of Intelligent & Robotic Systems (Jul 2018), https://doi.org/10.1007/s10846-018-0898-1.

[5] K. Kondak, A. Ollero, I. Maza, K. Krieger, A. Albu-Schaeffer, M. Schwarzbach, M. Laiacker, Unmanned aerial systems physically interacting with the environment: Load transportation, deployment, and aerial manipulation, in: Handbook of Unmanned Aerial Vehicles, Springer, 2015, pp. 2755–2785.

[6] Q. Lindsey, D. Mellinger, V. Kumar, Construction of cubic structures with quadrotor teams, in: Proc. Robotics: Science & Systems VII, 2011.
[7] F. Kondak, K. Huber, M. Schwarzbach, M. Laiacker, D. Sommer, M. Bejar, A. Ollero, Aerial manipulation robot composed of an autonomous helicopter and a 7 degrees of freedom industrial manipulator, 2014.
[8] F.R.V. Lippiello, Cartesian impedance control of a UAV with a robotic arm, in: 10th International IFAC Symposium on Robot Control, Dubrovnik, Croatia, 2012.
[9] D. Wuthier, D. Kominiak, E. Fresk, G. Nikolakopoulos, A geometric pulling force controller for aerial robotic workers, IFAC-PapersOnLine 50 (1) (2017) 10287–10292.
[10] M. Xu, A. Hu, H. Wang, Image-based visual impedance force control for contact aerial manipulation, IEEE Transactions on Automation Science and Engineering (2022).
[11] S. Kim, H. Seo, S. Choi, H.J. Kim, Vision-guided aerial manipulation using a multirotor with a robotic arm, IEEE/ASME Transactions on Mechatronics 21 (4) (2016) 1912–1923.
[12] H. Seo, S. Kim, H.J. Kim, Aerial grasping of cylindrical object using visual servoing based on stochastic model predictive control, in: 2017 IEEE International Conference on Robotics and Automation (ICRA), IEEE, 2017, pp. 6362–6368.
[13] A. Santamaria-Navarro, P. Grosch, V. Lippiello, J. Sola, J. Andrade-Cetto, Uncalibrated visual servo for unmanned aerial manipulation, IEEE/ASME Transactions on Mechatronics (2017).
[14] K. Steich, M. Kamel, P. Beardsley, M.K. Obrist, R. Siegwart, T. Lachat, Tree cavity inspection using aerial robots, in: 2016 IEEE/RSJ International Conference on Intelligent Robots and Systems (IROS), IEEE, 2016, pp. 4856–4862.
[15] P. Ramon Soria, B.C. Arrue, A. Ollero, Detection, location and grasping objects using a stereo sensor on UAV in outdoor environments, Sensors 17 (1) (2017) 103.
[16] A.W. Smeulders, D.M. Chu, R. Cucchiara, S. Calderara, A. Dehghan, M. Shah, Visual tracking: An experimental survey, IEEE Transactions on Pattern Analysis and Machine Intelligence 36 (7) (2014) 1442–1468.
[17] J.F. Henriques, R. Caseiro, P. Martins, J. Batista, High-speed tracking with kernelized correlation filters, IEEE Transactions on Pattern Analysis and Machine Intelligence (2015).
[18] H. Hirschmuller, Stereo processing by semiglobal matching and mutual information, IEEE Transactions on Pattern Analysis and Machine Intelligence 30 (2) (2008) 328–341.
[19] Q. Zhan, Y. Liang, Y. Xiao, Color-based segmentation of point clouds, 2009.
[20] R.B. Rusu, S. Cousins, 3D is here: Point cloud library (PCL), in: 2011 IEEE International Conference on Robotics and Automation (ICRA), IEEE, 2011, pp. 1–4.
[21] B. Oehler, J. Stueckler, J. Welle, D. Schulz, S. Behnke, Efficient multi-resolution plane segmentation of 3D point clouds, in: Intelligent Robotics and Applications, 2011, pp. 145–156.
[22] S. Lynen, M. Achtelik, S. Weiss, M. Chli, R. Siegwart, A robust and modular multi-sensor fusion approach applied to MAV navigation, in: Proc. of the IEEE/RSJ Conference on Intelligent Robots and Systems (IROS), 2013.
[23] K. Alexis, G. Nikolakopoulos, A. Tzes, Switching model predictive attitude control for a quadrotor helicopter subject to atmospheric disturbances, Control Engineering Practice 19 (10) (2011) 1195–1207.
[24] K. Alexis, G. Nikolakopoulos, A. Tzes, Model predictive quadrotor control: attitude, altitude and position experimental studies, IET Control Theory & Applications 6 (12) (2012) 1812–1827.
[25] M. Kamel, T. Stastny, K. Alexis, R. Siegwart, Model predictive control for trajectory tracking of unmanned aerial vehicles using robot operating system, in: A. Koubaa (Ed.), Robot Operating System (ROS): The Complete Reference, vol. 2, Springer, 2017.
[26] D. Wuthier, D. Kominiak, C. Kanellakis, G. Andrikopoulos, M. Fumagalli, G. Schipper, G. Nikolakopoulos, On the design, modeling and control of a novel compact aerial manipulator, in: 2016 24th Mediterranean Conference on Control and Automation (MED), IEEE, 2016, pp. 665–670.
[27] Compact AeRial MAnipulator (CARMA) assembly overview [Online]. Available: https://www.youtube.com/watch?v=rOV5K43cpho.

[28] A. Bochkovskiy, C.-Y. Wang, H.-Y.M. Liao, YOLOv4: Optimal speed and accuracy of object detection, arXiv preprint, arXiv:2004.10934, 2020.
[29] C.-Y. Wang, H.-Y.M. Liao, Y.-H. Wu, P.-Y. Chen, J.-W. Hsieh, I.-H. Yeh, CSPNet: A new backbone that can enhance learning capability of CNN, in: Proceedings of the IEEE/CVF Conference on Computer Vision and Pattern Recognition Workshops, 2020, pp. 390–391.
[30] S. Liu, L. Qi, H. Qin, J. Shi, J. Jia, Path aggregation network for instance segmentation, in: Proceedings of the IEEE Conference on Computer Vision and Pattern Recognition, 2018, pp. 8759–8768.
[31] J. Redmon, A. Farhadi, YOLOv3: An incremental improvement, arXiv preprint, arXiv:1804.02767, 2018.

Chapter 10

Machine learning for ARWs

Anton Koval, Sina Sharif Mansouri, and Christoforos Kanellakis
Department of Computer, Electrical and Space Engineering, Luleå University of Technology, Luleå, Sweden

10.1 Introduction

In recent years, several experimental demonstrations with Micro Aerial Vehicles (MAVs) have established increased robustness and technological performance in constrained and well-defined lab environments. Nevertheless, this technology is entering a new era with deployment in real-scale infrastructure environments and the ability to demonstrate levels of high autonomy [1,2]. A characteristic example of these novel developments is the integration of the Aerial Robotic Workers (ARWs) in various underground mine inspection operations. Recent technological advances allow human operators to use robots as a supportive tool in search and rescue [3], visual infrastructure inspection [4], environmental monitoring [5], etc., where a visible light camera is a common perception sensor. However, the quality of the visual information may significantly degrade, while working in night-time conditions. So, there are various lighting strategies to guaranty proper illumination of the environment [1]. Alternatively, robots might be equipped with thermal far-infrared cameras [6] that can perceive visual information in dark environments, making them useful for temperature inspections in agriculture, food-processing, and building industries, gas and fire detection, and military applications [6] as depicted in Fig. 10.1.

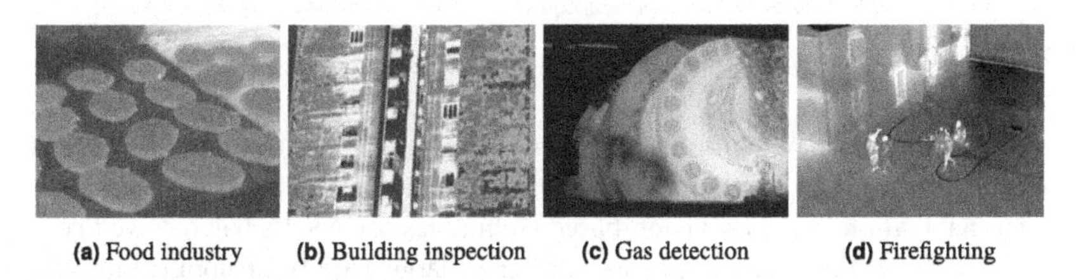

(a) Food industry (b) Building inspection (c) Gas detection (d) Firefighting

FIGURE 10.1 Thermal imagery application examples. Image courtesy of FLIR Systems, Inc.

Subterranean environments are harsh, posing obstacles for flying vehicles, including narrow and wide passages, reduced visibility due to rock falls, dust, wind gusts, and lack of proper illumination. Moreover, they can have complex

Aerial Robotic Workers. https://doi.org/10.1016/B978-0-12-814909-6.00016-0

geometries with many crossings among the tunnels. These crossings have a major impact on the ARWs navigation mission because, when not considered, they can lead to a crash on the tunnel surface or a wrong turn, thus decreasing the overall efficiency and performance of the ARWs to properly execute the mission. Therefore, it becomes a critical navigation capability for the aerial vehicle to not only execute a designated task but also identify junctions and to provide this information to the aerial planner for enabling a more safe and optimal overall mission execution.

Therefore, in this Chapter, we discuss an approach that allows ARW to identify junctions and human workers using onboard forward facing cameras for safe autonomous inspections of subterranean environments. The proposed approach is based on transfer learning [7] that is used as a training method. More specifically, the AlexNet [8] is selected for executing transfer learning, mainly due to the success of AlexNet pre-trained Convolutional Neural Network (CNN) features and the promising results that have been obtained from several image classification datasets with transfer learning on AlexNet [9]. Toward this approach, images are extracted from the datasets of: a) an underground mine in Chile [10], b) underground tunnels in Sweden [11], and c) FLIR ADAS dataset [12]. These datasets are further classified manually. Then the last three layers of AlexNet are replaced with a set of new layers for the classification based on selected categories of images, and the network is trained from the datasets. The obtained class for each image provides information from the local surroundings of the vehicle, which can later be utilized for the autonomous navigation and inspection in subterranean environments. Example of manual classification of junctions in four categories of junctions is depicted in Fig. 10.2.

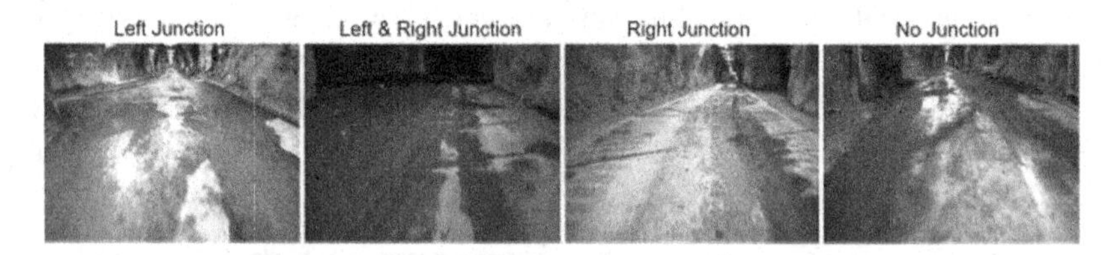

FIGURE 10.2 Example of junction types in the training datasets [10,11].

Autonomous navigation in unknown environments requires environmental awareness, such as recognition of obstacles, junctions, dead-ends, etc. Moreover, navigation based on vision-based techniques for ARWs has received considerable attention in recent years and has a large variety of application scenarios [13]. It should be noted that the navigation, based on a forward looking camera, has been based mainly either on computer vision algorithms or on machine learning methods.

Toward computer vision-based navigation, most of the works focus on obstacle detection methods. Thus in [14], a mathematical model to estimate the obstacle distance to the ARW was implemented for collision avoidance. However, the method provided poor results at high velocities and low illumination environments. In [15], the combination of multiple vision-based components for navigation, localization, mapping, and obstacle-free navigation of the MAV was described. The obstacle avoidance scheme consisted of 3 stereo cameras for a 360° coverage of the MAV's surroundings in the form of point clouds. However, the proposed method relied on sufficient illumination, landmark extraction, and high onboard processing power. In [16], random trees were generated to find the best branch; the method was evaluated in indoor environments, and the paths were calculated online, while the occupancy map of the perceived environment was conducted. This method required, in general, high computation power to process the images, calculate the best next point, and accurately localize and store the previous information of the map in order to avoid revisiting the area. Usually, the performance of the computer vision-based algorithms mainly relies on the surrounding environment having good distinctive features and good illumination and lighting conditions [14]. Furthermore, these methods require high computation power to process the images and extract landmarks, factors that could limit the usage of these methods in real-life underground mine applications.

There are few works using machine learning techniques for the problem of navigation in indoor and outdoor environments, mainly due to the fact that these methods require a large amount of data and a high computation power for training in most cases a CNN, which is an off-line procedure. However, after the training, the CNN can be used for enabling autonomous navigation with much lower computation power, especially, when compared to the training phase. The works using CNN for navigation, such as [17], [18], and [19], utilized the image frame of onboard camera to feed the CNN for providing heading commands to the platform. These works have been evaluated and tuned in outdoor environments and with a good illumination with the camera providing detailed data about the surrounding of the platforms, while none of the works consider the recognition of the junctions.

In [20], a CNN binary classifier was proposed for outdoor road junction detection. Besides, the authors also considered the possibility of using the proposed architecture for navigation and experimentally evaluated it on commercially available MAV Bebop 2 from Parrot. The problem of junction detection for outdoor environments was also addressed by [21], where machine learning approach was used. In [22], the authors offered an architecture combining CNN, Bidirectional LSTM [23] and Siamese [24] style distance function learning for junction recognition in videos. In [23], a road intersection detection module has been proposed. The developed method addressed the problem as a binary classification using Long-Term Recurrent Convolutional Network (LRCN) architecture to identify relative changes in outdoor features and, eventually, detect intersections. These methods, mainly use binary classifier, while in real-life scenarios more complex type of junctions exist and the junction recognition should recognize the different types of junctions.

The rest of the Chapter is structured as follows. In Section 10.2, we present the AlexNet architecture and the corresponding transfer learning. Then, in Section 10.3, the dataset collection, the training of the network, and the evaluation of the trained network are presented for the junction recognition and human detection scenarios. Finally, Section 10.3.3 concludes the chapter.

10.2 AlexNet architecture

AlexNet [8] is one of the most used and studied CNN methods [25] that has 60 million parameters and 650,000 neurons. Thus, a large dataset is required to train it. However, in this chapter, transfer learning is used due to the limited available datasets. In this approach, the input of the AlexNet is a Red, Green, and Blue (RGB) image with fixed size of $227 \times 227 \times 3$ pixels, and it follows with 2D convolutional layers of size 11×11 with an output size of $55 \times 55 \times 96$. Then it follows a 2D Max pooling layer of size 3×3 and an output of $27 \times 27 \times 96$, followed with 2D convolutional layers of 5×5 and an output of $27 \times 27 \times 256$. In the sequel, there is another max pooling of size 3×3 and with an output of $13 \times 13 \times 256$, which passes through 2D convolutional layers of 3×3 with the same output size. Next, another 2D convolutional layers of 3×3 with an output size of $13 \times 13 \times 256$, followed with a max pooling of size 3×3 with an output of $6 \times 6 \times 256$. The output passes through two fully connected layers, and the last layer results are fed into a softmax classifier with 1000 class labels. To summarize, AlexNet consists of eight layers, five of them are convolutional layers, and three of them are fully connected layers. Each one of the first two convolutional layers are followed by an Overlapping Max Pooling layer. The other three convolutional layers (third, fourth, and fifth) are connected directly. Finally, the last convolutional layer (fifth) is followed by an Overlapping Max Pooling layer. Fig. 10.3 depicts the overall utilized AlexNet structure.

Max pooling layers are usually used in CNNs in order to reduce the size of the matrices while keeping the same depth. On the other hand, overlapping max pooling uses adjacent windows that overlap each other to compute the max element from a window each time. This kind of max pooling has been proven to reduce top-1 and top-5 error rates [8].

One of the main aspects of AlexNet is also the use of the Rectified Linear Unit (ReLU) [26]. The authors [8] proved that AlexNet could be trained much faster using the ReLU nonlinearity than classical activation functions, like $sigmoid$ or $tanh$ [27]. Actually, they tested their hypothesis on the CIFAR-10 dataset [28], and the ReLU-AlexNet achieved the same performance (25% training error) with the Tanh-AlexNet in one sixth of the epochs.

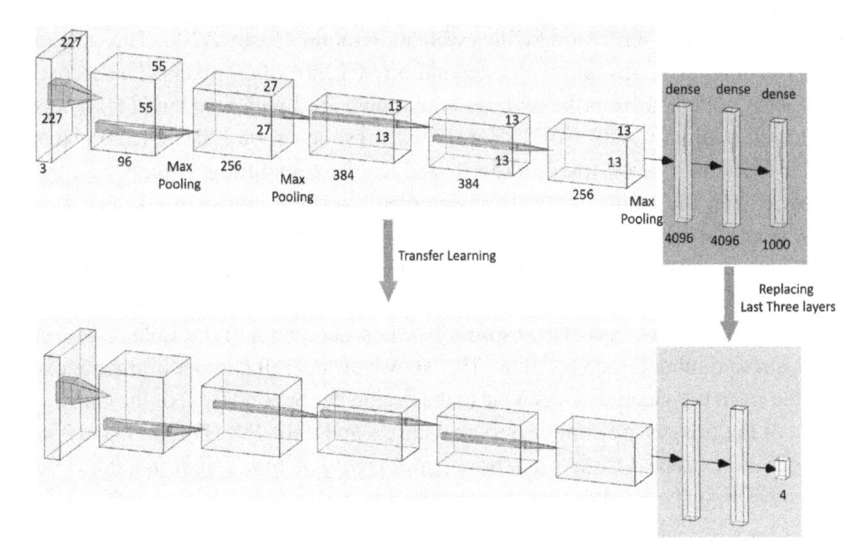

FIGURE 10.3 AlexNet architecture and the transfer learning method.

10.2.1 Transfer learning

Transfer learning [26,29] for CNNs is generally used to refer to the process of using an already trained CNN in another dataset, where the number of classes to be recognized are different from the initial dataset, because it has been used in different tasks and with various datasets. There are two main strategies for Transfer Learning, with both of them using the same weights from the trained AlexNet on the images from the ImageNet database [30].

The first one treats the CNN as a feature extractor by removing the last fully connected layer. Then, one can use the features extracted from the trained AlexNet in order to train a classifier like [31] for the new dataset. The second one replaces the last connected layer and retrains the whole CNN for the new dataset. This allows for a fine tuning of the trained weights.

10.3 Real-life scenarios

This section describes two real-life scenarios with utilization of CNN for subterranean missions.

10.3.1 Junction detection

10.3.1.1 Data-set

For transfer learning of the AlexNet, two datasets from underground mine and tunnels are selected. The first one is from the Chilean underground mine dataset [10], which is collected by a Point Grey XB3 multi-baseline stereo camera, mounted with a forward facing orientation on the Husky A200. The camera was operated at 16 fps and with a resolution of 1280 × 960 pixels. The second dataset is collected from the underground tunnels in Luleå Sweden [11]. It was collected manually with GoPro Hero 7 with resolution of 2704 × 1520 pixels and a frame rate of 60 fps. In order to reduce the over-fitting of the CNN, the images from the cameras are down-sampled, and few images are selected as the camera approaches and passes the junction. As an example, Fig. 10.4 depicts the multiple images collected for the branch on the left of the tunnel. It should be highlighted that the datasets from the Chilean underground mine contain more variety of branches, especially when compared to the Luleå Sweden underground tunnels dataset. Table 10.1 shows the overall number of images extracted from the video streams. Due to the Frame Per Second (fps) of the camera, most of the images are similar in both datasets and only 488, 339, 333, and 350 images are extracted for *left junction*, *left & right junctions*, *right junction*, and *no junction*, respectively.

FIGURE 10.4 Examples of extracted images from the visual camera in case of a left branch in the tunnel. The images are extracted while the camera approaches and passes the junction (continuous with a direct heading).

TABLE 10.1 The number of extracted images for each category from the two datasets, while the redundant images are excluded from the dataset.

	left	*left & right*	*right*	*no junction*
Chilean mine dataset	339	279	239	250
Sweden tunnels dataset	149	60	104	100

In the sequel, the dataset is manually classified into four categories of *left junction*, *right junction*, *left & right junctions*, and *no junction*, and the last three fully connected layers of the AlexNet are replaced with this set of layers that will classify 4 classes instead of 1000 classes as depicted in Fig. 10.3. Sample images of different areas of the Chilean underground mine dataset and the Luleå Sweden underground tunnels are depicted in Figs. 10.5 and 10.6, respectively.

10.3.1.2 Training and evaluations of the CNN

Both datasets are combined for training the AlexNet, while the junctions that are not included in the training dataset are used only for validation of the network.

FIGURE 10.5 Examples of acquired images from the Chilean underground mine dataset [10].

Moreover, the images are resized to 227 × 227 × 3 pixels. The network was trained on a workstation equipped with an Nvidia GTX 1070 GPU with mini-batch size of 10, maximum number epochs of 6, a selected initial learning rate of 10^{-4} and solved by the stochastic gradient descent [32] with momentum optimizer. The trained network provides an accuracy of 100% and 89.2% on training and validation datasets, respectively. Fig. 10.7 shows the accuracy and loss of the training and validation dataset, respectively, while the loss function for multi-class classification is defined as a cross entropy loss [26,33].

Moreover, Fig. 10.8 depicts the confusion matrix of the validation dataset, while the rows correspond to the predicted class from the validation datasets, and the columns correspond to the actual class of the dataset. The diagonal cells show the number and percentage of the correct classifications by the trained network. As an example, in the first diagonal, 45 images are correctly classified to the *left branch* category, which corresponds to 25.9% of the overall number of images. Similarly, 20 cases are correctly classified to *left & right branch* that

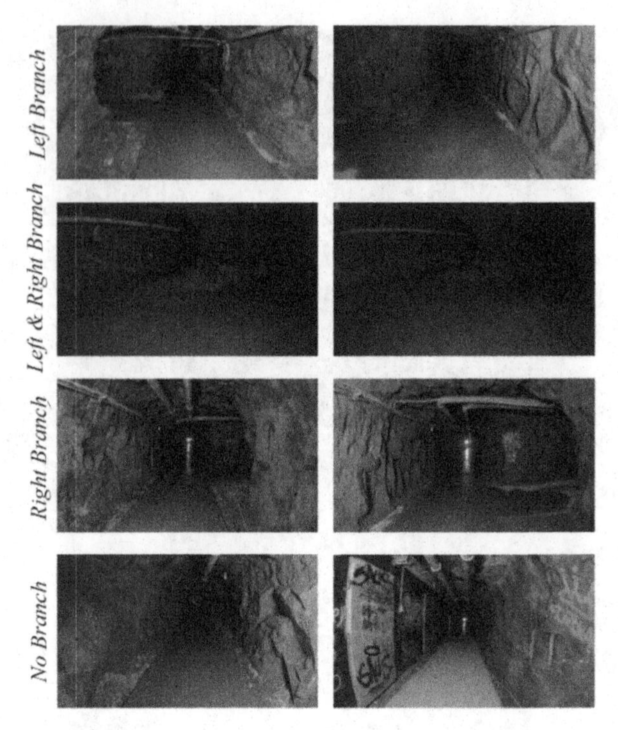

FIGURE 10.6 Examples of acquired images from the Luleå Sweden underground tunnels.

correspond to 19.5% of all the validation dataset. Moreover, the off-diagonal cells correspond to incorrectly classified observations, e.g., 2 and 3 images from the left branch are incorrectly classified to left & right branch and right branch, respectively, which corresponds to 10% of the left branch validation dataset or 1.1% and 1.7% of all the datasets, respectively. Similarly, for the right branch 12 images are incorrectly classified to left & right branches. Furthermore, the right gray column displays the percentages of all the images predicted to belong to each class that are correctly and incorrectly classified. On the other hand, the bottom row depicts the percentages of all the examples belonging to each class that are correctly and incorrectly classified. The cell in the bottom right of the plot shows the overall accuracy. Overall, 90.2% of the predictions are correct and 9.8% are wrong.

Moreover, Fig. 10.9 depicts 12 images from the validation dataset, while the correct label and the classification of the AlexNet are shown for each image. It should be highlighted that these images are excluded from the training datasets. It is observed that the network has incorrectly classified the *left junction* and *right junctions* images to *left & right junctions*, these images are from Chilean dataset.

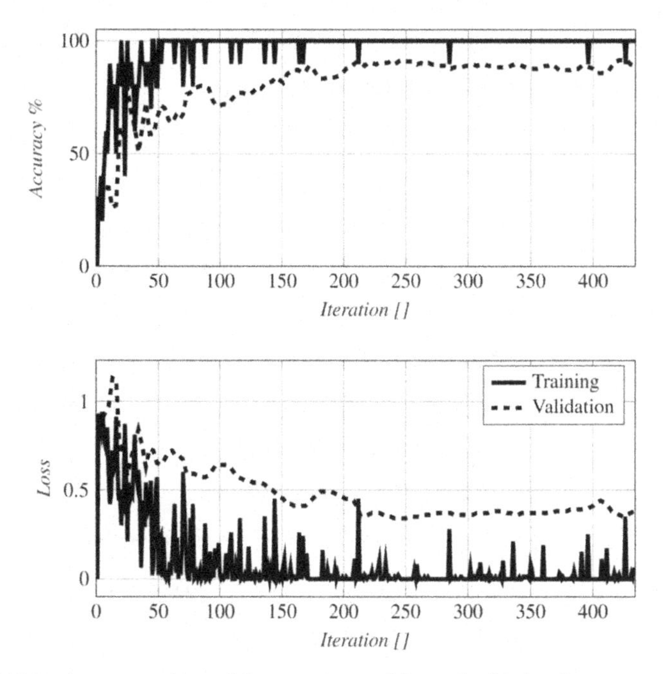

FIGURE 10.7 Accuracy and loss of the network on training and validation datasets.

Furthermore, Table 10.2 compares the training time, accuracy percentage, and loss [26] between three well-known pre-trained CNN architectures: AlexNet, GoogleNet [34] and Inceptionv3Net [35]. As one can see from this Table 10.2, the validation accuracy is smaller for the GoogleNet and significantly smaller for the Inception3Net. We think that this is due to the large number of Convolutional layers of these two Networks when compared to the AlexNet. Thus, more dataset is needed for these two networks, and this supports our choice for the AlexNet network.

TABLE 10.2 The comparison of transfer learning between AlexNet, GoogleNet, and Inceptionv3Net.

	AlexNet	GoogleNet	Inceptionv3Net
Training time [sec]	995	891	1438
Training accuracy	100%	100%	100%
Validation accuracy	89.2%	74.1%	63.79%
Training loss	0.01	0.01	0.17
Validation loss	0.29	0.82	0.93

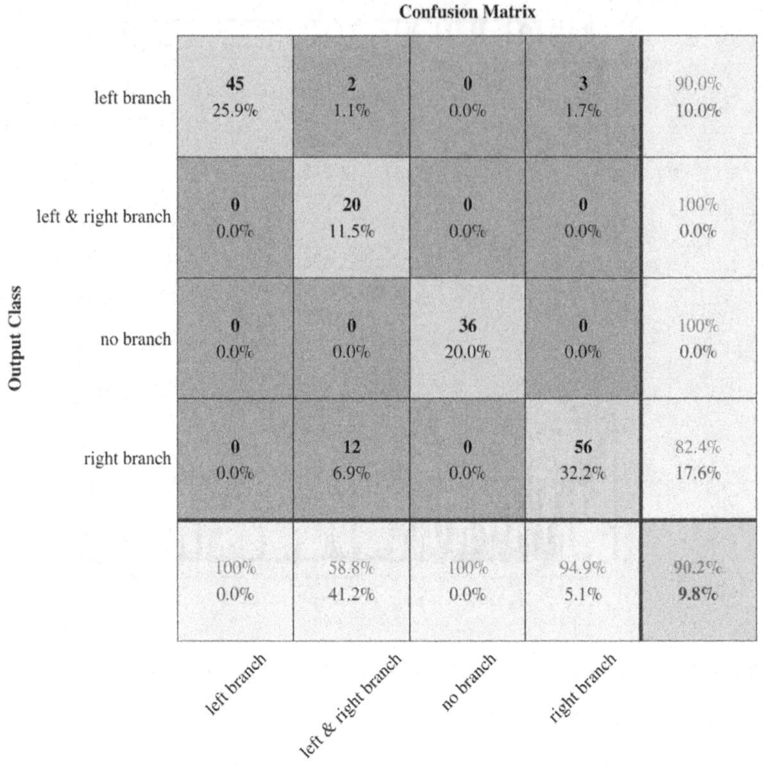

Confusion Matrix

Output Class					
left branch	45 25.9%	2 1.1%	0 0.0%	3 1.7%	90.0% 10.0%
left & right branch	0 0.0%	20 11.5%	0 0.0%	0 0.0%	100% 0.0%
no branch	0 0.0%	0 0.0%	36 20.0%	0 0.0%	100% 0.0%
right branch	0 0.0%	12 6.9%	0 0.0%	56 32.2%	82.4% 17.6%
	100% 0.0%	58.8% 41.2%	100% 0.0%	94.9% 5.1%	90.2% **9.8%**
	left branch	left & right branch	no branch	right branch	

FIGURE 10.8 The confusion matrix from the validation dataset.

10.3.2 Human detection

10.3.2.1 FLIR ADAS thermal dataset

In the transfer learning of the pre-trained AlexNet, a FLIR ADAS dataset was used that allows detecting and classifying walking and cycling persons, dogs, and vehicles under challenging conditions, including total darkness, fog, smoke, inclement weather, and glare [12]. The dataset was recorded with a FLIR Tau2 camera operated at 30 fps and with a resolution of 640×512 pixels. In our case, Fig. 10.10 depicts an example of the extracted images from the training dataset.

The second dataset is from Luleå Sweden underground tunnels. It was collected manually with a FLIR Boson 640 camera with the resolution of 640×512 pixels and a frame rate of 60 fps. Since the thermal camera resolution is significantly lower than that of a visible light camera, the acquired images were not down-sampled or cropped. Table 10.3 shows the total number of images that were extracted from the dataset. While Fig. 10.11 illustrates several images collected for the human detection in the underground tunnel.

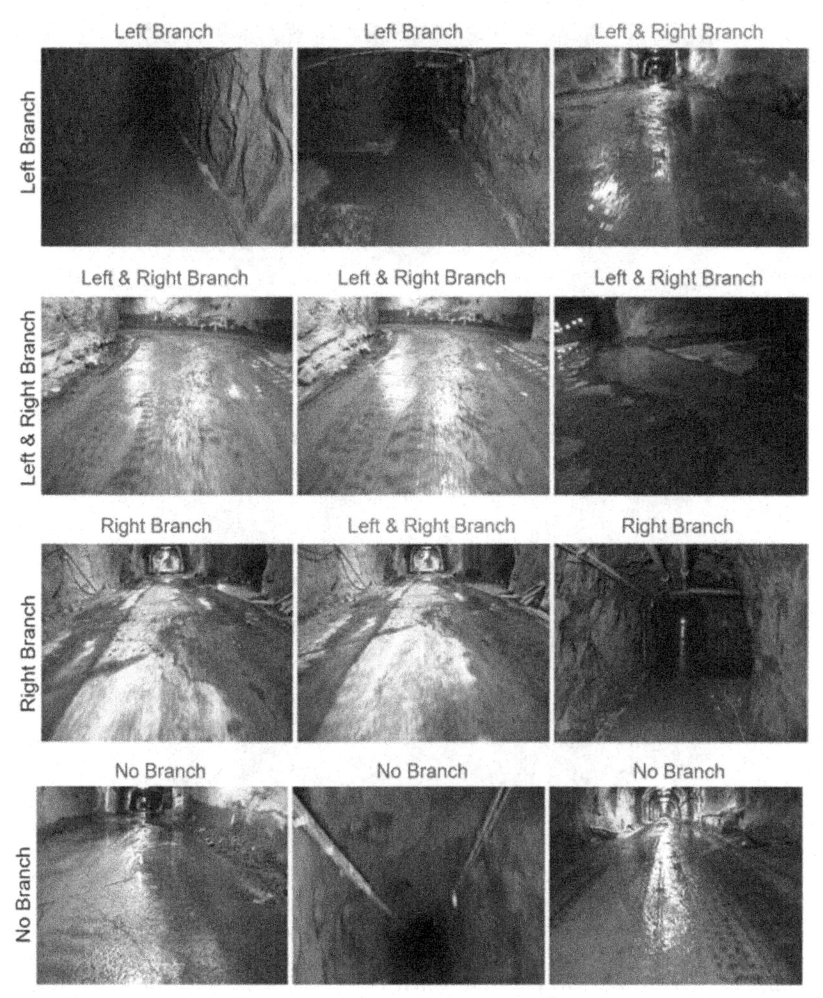

FIGURE 10.9 Examples of validation dataset, while the correct label is written on the left of the images and the estimated class form AlexNet is written on top of each image.

TABLE 10.3 The number of extracted images for each category from the training and the validation datasets, while the redundant images are excluded.

	human	*no humans*
FLIR ADAS dataset	79	174
Sweden tunnels dataset	169	68

In the continuation, the dataset is categorized manually into two classes of *human* and *no human*.

FIGURE 10.10 Examples of the extracted images from the FLIR ADAS dataset in the case of a human detection. The images are extracted while the camera approaches to the human.

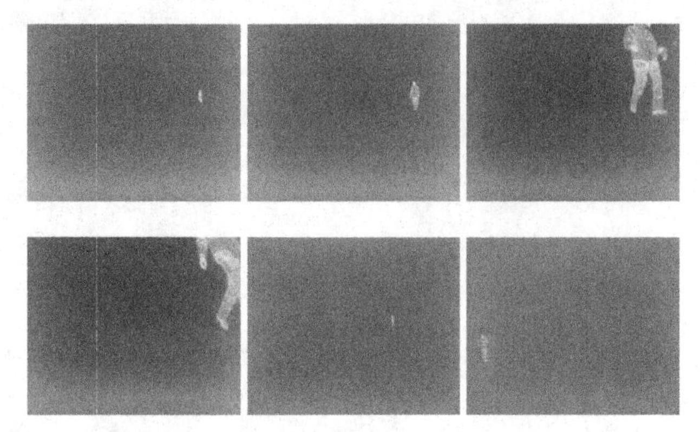

FIGURE 10.11 Examples of acquired images from the Luleå Sweden underground tunnels for the validation dataset.

10.3.2.2 Training and the evaluation

The FLIR ADAS dataset was used for training the AlexNet, while the Luleå Sweden underground tunnels dataset is used for the validation of the network. Additionally, as for the junction detection scenario, the images acquired from the FLIR ADAS dataset were resized to 227 × 227 × 3 pixels. The CNN was trained on a laptop equipped with an Nvidia MX 150 GPU while keeping the set of training parameters, initial learning rate, and solving method as in the first scenario. The trained network has a 100% accuracy on the training dataset, while the accuracy on the validation dataset was equal to 98.73%. The outcome of the CNN training is depicted in Fig. 10.12, which shows the accuracy and loss, while training and validation of the dataset. In this scenario, for classification, the losses were also defined as a cross-entropy loss function [26,33]. Furthermore, Fig. 10.13 shows the confusion matrix of the validation dataset. Where the rows from the validation dataset correspond to the predicted class,

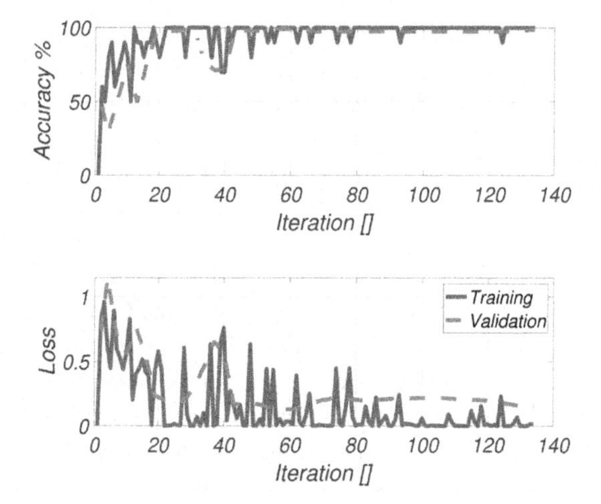

FIGURE 10.12 Accuracy and loss of the network on training and validation datasets.

Confusion Matrix

Output Class \ Target Class	human	no humans	
human	**71** 31.1%	**0** 0.0%	100% 0.0%
no humans	**0** 0.0%	**157** 68.9%	100% 0.0%
	100% 0.0%	100% 0.0%	100% 0.0%

FIGURE 10.13 The confusion matrix from the validation dataset.

and the columns refer to the actual class of the dataset. The diagonal cells show the number and percentage of the trained CNN's proper classifications. For example, 71 images are correctly categorized in the first diagonal into the category *human*, which corresponds to 31.1% of the total number of images. Likewise, 157 instances are correctly classified as *no humans* corresponding to 68.9% of the entire validation dataset. In addition, the off-diagonal cells correspond to wrongly labeled results, leading to 0.0% error. Additionally, the right gray column indicates the percentages of all the images that are expected to belong to each class that are correctly and faultily categorized. On the other hand, the bottom row displays the percentages of all the examples belonging to each class, which are listed correctly and faultily. The cell located on the plot's bottom right indicates the overall precision. Overall, 100.0% of predictions are right, and 0.0% are incorrect.

Similar to the junction detection scenario, the obtained results were compared with two other pre-trained networks, namely the GoogleNet [34] and the Inceptionv3Net [35], which are depicted in Table 10.4. As can be seen, the validation accuracy for GoogleNet is lower, and for Inception3Net, it is significantly lower. We suppose that such a result is caused by the larger amount of the convolutional layers in these two networks, and it allows in concluding that GoogleNet and Inception3Net, require bigger datasets. Such an outcome makes our choice of AlexNet more reasonable.

TABLE 10.4 The comparison of transfer learning performance between AlexNet, GoogleNet, and Inceptionv3Net.

	AlexNet	GoogleNet	Inceptionv3Net
Training time [sec]	186	265	2012
Training accuracy	100%	100%	100%
Validation accuracy	98.73%	77.64%	29.11%
Training loss	0.0036	0.008	0.17
Validation loss	0.041	0.51	1.04

10.3.3 Conclusions

In this Chapter, it was presented a framework based on CNN for detecting tunnel crossing/junctions and humans in underground mine areas, envisioning the application of ARWs autonomous deployment for inspection purposes. Within the emerging field of underground applications with ARWs, junction and human detection have been identified as fundamental capabilities for the ARW's risk-aware autonomous navigation. Moreover, inspired by the concept of lightweight aerial vehicles, the provided generic solution allows for keeping the hardware complexity low, relying only on a perception sensor. It feeds the image stream from the vehicle's onboard forward-facing camera in a CNN classification architecture expressed in four and two categories. The AlexNet model has been incorporated into a transfer learning scheme for the novel proposed classification categories, detecting humans and multiple branches in underground environments. The method has been validated by using datasets collected from real underground environments, demonstrating its performance and merits.

References

[1] C. Kanellakis, S.S. Mansouri, G. Georgoulas, G. Nikolakopoulos, Towards autonomous surveying of underground mine using MAVs, in: International Conference on Robotics in Alpe-Adria Danube Region, Springer, 2018, pp. 173–180.

[2] C. Kanellakis, E. Fresk, S.S. Mansouri, D. Kominiak, G. Nikolakopoulos, Autonomous visual inspection of large-scale infrastructures using aerial robots, UAV preprint, arXiv:1901.05510, 2019.

[3] T. Tomic, K. Schmid, P. Lutz, A. Domel, M. Kassecker, E. Mair, I.L. Grixa, F. Ruess, M. Suppa, D. Burschka, Toward a fully autonomous UAV: Research platform for indoor and outdoor urban search and rescue, IEEE Robotics & Automation Magazine 19 (3) (2012) 46–56.

[4] S.S. Mansouri, C. Kanellakis, E. Fresk, D. Kominiak, G. Nikolakopoulos, Cooperative coverage path planning for visual inspection, Control Engineering Practice 74 (2018) 118–131.

[5] B. Bayat, N. Crasta, A. Crespi, A.M. Pascoal, A. Ijspeert, Environmental monitoring using autonomous vehicles: a survey of recent searching techniques, Current Opinion in Biotechnology 45 (2017) 76–84.

[6] R. Gade, T.B. Moeslund, Thermal cameras and applications: a survey, Machine Vision and Applications 25 (1) (2014) 245–262.

[7] L. Torrey, J. Shavlik, Transfer learning, in: Handbook of Research on Machine Learning Applications and Trends: Algorithms, Methods, and Techniques, IGI Global, 2010, pp. 242–264.

[8] A. Krizhevsky, I. Sutskever, G.E. Hinton, ImageNet classification with deep convolutional neural networks, in: Advances in Neural Information Processing Systems, 2012, pp. 1097–1105.

[9] M. Huh, P. Agrawal, A.A. Efros, What makes ImageNet good for transfer learning?, UAV preprint, arXiv:1608.08614, 2016.

[10] K. Leung, D. Lühr, H. Houshiar, F. Inostroza, D. Borrmann, M. Adams, A. Nüchter, J. Ruiz del Solar, Chilean underground mine dataset, The International Journal of Robotics Research 36 (1) (2017) 16–23.

[11] S.S. Mansouri, C. Kanellakis, G. Georgoulas, G. Nikolakopoulos, Towards MAV navigation in underground mine using deep learning, in: IEEE International Conference on Robotics and Biomimetics (ROBIO), 2018.

[12] F.S. Inc, FLIR thermal dataset [Online]. Available: https://www.flir.com/oem/adas/adas-dataset-form/.

[13] C. Kanellakis, G. Nikolakopoulos, Survey on computer vision for UAVs: Current developments and trends, Journal of Intelligent & Robotic Systems (2017) 1–28.

[14] S. Saha, A. Natraj, S. Waharte, A real-time monocular vision-based frontal obstacle detection and avoidance for low cost UAVs in GPS denied environment, in: 2014 IEEE International Conference on Aerospace Electronics and Remote Sensing Technology, IEEE, 2014, pp. 189–195.

[15] F. Valenti, D. Giaquinto, L. Musto, A. Zinelli, M. Bertozzi, A. Broggi, Enabling computer vision-based autonomous navigation for unmanned aerial vehicles in cluttered GPS-denied environments, in: 2018 21st International Conference on Intelligent Transportation Systems (ITSC), 2018, pp. 3886–3891.

[16] A. Bircher, M. Kamel, K. Alexis, H. Oleynikova, R. Siegwart, Receding horizon next-best-view planner for 3D exploration, in: IEEE International Conference on Robotics and Automation (ICRA), 2016, pp. 1462–1468.

[17] S.P. Adhikari, C. Yang, K. Slot, H. Kim, Accurate natural trail detection using a combination of a deep neural network and dynamic programming, Sensors 18 (1) (2018) 178.

[18] L. Ran, Y. Zhang, Q. Zhang, T. Yang, Convolutional neural network-based robot navigation using uncalibrated spherical images, Sensors 17 (6) (2017) 1341.

[19] N. Smolyanskiy, A. Kamenev, J. Smith, S. Birchfield, Toward low-flying autonomous MAV trail navigation using deep neural networks for environmental awareness, UAV preprint, arXiv: 1705.02550, 2017.

[20] S. Kumaar, S. Mannar, S. Omkar, et al., JuncNet: A deep neural network for road junction disambiguation for autonomous vehicles, UAV preprint, arXiv:1809.01011, 2018.
[21] H. Haiwei, Q. Haizhong, X. Limin, D. Peixiang, Applying CNN classifier to road interchange classification, in: 2018 26th International Conference on Geoinformatics, IEEE, 2018, pp. 1–4.
[22] A. Kumar, G. Gupta, A. Sharma, K.M. Krishna, Towards view-invariant intersection recognition from videos using deep network ensembles, in: 2018 IEEE/RSJ International Conference on Intelligent Robots and Systems (IROS), IEEE, 2018, pp. 1053–1060.
[23] D. Bhatt, D. Sodhi, A. Pal, V. Balasubramanian, M. Krishna, Have I reached the intersection: A deep learning-based approach for intersection detection from monocular cameras, in: 2017 IEEE/RSJ International Conference on Intelligent Robots and Systems (IROS), IEEE, 2017, pp. 4495–4500.
[24] W.-T. Yih, K. Toutanova, J.C. Platt, C. Meek, Learning discriminative projections for text similarity measures, in: Proceedings of the Fifteenth Conference on Computational Natural Language Learning, Association for Computational Linguistics, 2011, pp. 247–256.
[25] H. Shin, H.R. Roth, M. Gao, L. Lu, Z. Xu, I. Nogues, J. Yao, D. Mollura, R.M. Summers, Deep convolutional neural networks for computer-aided detection: CNN architectures, dataset characteristics and transfer learning, IEEE Transactions on Medical Imaging 35 (5) (May 2016) 1285–1298.
[26] I. Goodfellow, Y. Bengio, A. Courville, Deep Learning, MIT Press, 2016, http://www.deeplearningbook.org.
[27] A. Saxena, Convolutional neural networks (CNNs): An illustrated explanation, https://xrds.acm.org/blog/2016/06/convolutional-neural-networks-cnns-illustrated-explanation/, 2016.
[28] A. Krizhevsky, V. Nair, G. Hinton, CIFAR-10 (Canadian Institute for Advanced Research) [Online]. Available: http://www.cs.toronto.edu/~kriz/cifar.html, 2010.
[29] S.J. Pan, Q. Yang, A survey on transfer learning, IEEE Transactions on Knowledge and Data Engineering 22 (10) (Oct 2010) 1345–1359.
[30] J. Deng, W. Dong, R. Socher, L.-J. Li, K. Li, L. Fei-Fei, ImageNet: A large-scale hierarchical image database, in: CVPR09, 2009.
[31] M.A. Hearst, Support vector machines, IEEE Intelligent Systems 13 (4) (Jul. 1998) 18–28, https://doi.org/10.1109/5254.708428.
[32] K.P. Murphy, Machine Learning: A Probabilistic Perspective, MIT Press, 2012.
[33] P. Kim, Matlab deep learning, in: With Machine Learning, Neural Networks and Artificial Intelligence, Springer, 2017.
[34] C. Szegedy, W. Liu, Y. Jia, P. Sermanet, S. Reed, D. Anguelov, D. Erhan, V. Vanhoucke, A. Rabinovich, Going deeper with convolutions, in: Proceedings of the IEEE Conference on Computer Vision and Pattern Recognition, 2015, pp. 1–9.
[35] C. Szegedy, V. Vanhoucke, S. Ioffe, J. Shlens, Z. Wojna, Rethinking the inception architecture for computer vision, in: Proceedings of the IEEE Conference on Computer Vision and Pattern Recognition, 2016, pp. 2818–2826.

Chapter 11

Aerial infrastructures inspection

Vignesh Kottayam Viswanathan, Sina Sharif Mansouri, and Christoforos Kanellakis

Department of Computer, Electrical and Space Engineering, Luleå University of Technology, Luleå, Sweden

11.1 Introduction

ARWs equipped with remote sensing instrumentation are emerging in the last years due to their mechanical simplicity, agility, stability, and outstanding autonomy in performing complex maneuvers [1]. A variety of remote sensors such as visual sensors, lasers, sonars, thermal cameras, etc. could be mounted onboard the ARW, while the acquired information from the ARW's mission can be analyzed and used to produce sparse or dense surface models, hazard maps, investigate access issues, and other area characteristics. However, the main problems in these approaches are: 1) Where am I (the basic question)? and 2) What should I do? To be more specific, the information on the location of the agent is critical for succeeding the missions and guaranteeing the full coverage of the area. These fundamental problems are directly related to the autonomous path planning of aerial vehicles.

This chapter demonstrates the application of an aerial tool for inspecting complex 3D structures with multiple agents. In this approach, the a priori coverage path is divided and assigned to each agent based on the infrastructure architectural characteristics to reduce the inspection time. Furthermore, to guarantee a full coverage and a 3D reconstruction, the introduced path planning, for each agent, creates an overlapping visual inspection area, that will enable the off-line cooperative reconstruction. This chapter presents the direct demonstration of the applicability and feasibility of the overall cooperative coverage and inspection scheme with the ARWs for outdoors scenarios without the utilization of any external reference system, e.g., motion capture systems. This demonstration has a significant novelty and impact as an enabler for a continuation of research efforts toward the real-life aerial cooperative inspection of the aging infrastructure. This concept that has never been presented before, to the authors' best knowledge, outdoor and with a real infrastructure as a test case. In the outdoors demonstrations, the ARWs have been autonomously operated on the basis of odometry information from Ultra-Wide Band and inertial sensor fusion and without any other support on localization, which adds more com-

Aerial Robotic Workers. https://doi.org/10.1016/B978-0-12-814909-6.00017-2

plexity and impact on the acquired results. The image and pose data on board the platform were post processed to build a 3D representation of the structure. The rest of the Chapter is structured as follows. The proposed C-CPP method is presented in Section 11.3, which is followed by a brief description of the 3D reconstruction part from multiple agents in Section 11.4. In Section 11.5, multiple simulation and experimental results are presented that prove the efficiency of the established scheme. Following with lessons learned in Section 11.6. Finally, the chapter is concluded in Section 11.7.

11.2 Problem statement

In the corresponding problem statement, multiple aerial robots will be employed and will address the problem of autonomous, complete, and efficient execution of infrastructure inspection and maintenance operations. To facilitate the necessary primitive functionalities, an inspection path-planner that can guide a team of ARWs to efficiently and completely inspect a structure will be implemented. The collaborative team of ARWs should be able to understand the area to be inspected, ensure complete coverage, and create an accurate 3D reconstruction to accomplish complex infrastructure inspection. Relying on the accurate state estimation, as well as the dense reconstruction capabilities of the collaborative aerial team, algorithms for the autonomous inspection planning should be designed to ensure full coverage.

Let $\Omega \subset \mathbb{R}^3$ be a given region that can have multi-connected components (complex structure), while we also consider the finite set

$$\Lambda = \{C_i : i \in I_n = \{1, 2, \dots, n\}\} \tag{11.1}$$

of cells

$$C_i = \left\{(x_i, y_i, z_i) \in \mathbb{R}^3 : (x, y, z)\right\} \tag{11.2}$$

denoting the camera specification and position. The placement of the cells C_i can be defined by the translation $u_i = (x_i, y_i, z_i)$ and the orientation vector $o = \{\phi, \theta, \psi_i\}$, $i \in I_n$, while the set of translated and orientated cells $C_i(u_i, o_i)$ is expressed by $\Lambda(u, o)$, where $u = \{u_1, u_2, \dots, u_n\} \in \mathbb{R}^{3n}$ and $o = \{o_1, o_2, \dots, o_n\} \in R^{3n}$ with $0 \leq o_i \leq 2\pi$.
The 3D polygonal

$$P(u_i, o_i, n) = \bigcup_{i=1}^{n} C_i(u_i, o_i) \tag{11.3}$$

represents the region covered by the union of the cells C_i, while Λ^* is a cover of Ω if there exists a solution such that

$$\Omega \subset P(u_i, o_i, n) = \bigcup_{i \in I_n} C_i(u_i, o_i) \tag{11.4}$$

Moreover, several cases arise in the interaction between the two cells $C_i(u_i, o_i)$ and $C_j(u_j, o_j)$ with $i \neq j$, $u_i = (x_i, y_i, z_i)$, $o_i = (\phi_i, \theta_i, \psi_i)$, $u_j = (x_j, y_j, z_j)$ and $o_j = (\phi_j, \theta_j, \psi_j)$, where mainly determined by:

$$\begin{aligned} C_i(u_i, o_i) \cap C_j(u_j, o_j) = \emptyset \\ C_i(u_i, o_i) \cap C_j(u_j, o_j) \neq \emptyset \end{aligned} \tag{11.5}$$

Additionally, the cases of $C_i(u_i, o_i) \subset C_j(u_j, o_j)$ and $C_j(u_j, o_j) \subset C_i(u_i, o_i)$ are not considered when dealing with the coverage problem because it is contrary to the optimality of the path to have a substantial overlapping for visual processing and cover the whole surface of the under inspection object.

11.3 Coverage path planning of 3D maps

Initially, for the establishment of the Collaborative Coverage Path Planning (C-CPP), the general case of a robot equipped with a limited Field of View (FoV) sensor is considered, determined by an aperture angle α and a maximum range r_{max}, as depicted in Fig. 11.1. Furthermore, $\Omega \in \mathbb{R}^+$ is the user-defined offset distance ($\Omega < r_{max}$), from the infrastructure's target surface, and $\Delta\lambda$ is the distance between each slice plane. $\Delta\lambda$ is equal to $\frac{\Omega}{2} \tan\alpha/2$ based on Fig. 11.1. In order to guarantee overlapping, the parameter $\beta \in [1, +\infty)$ was introduced and substituted in $\Delta\lambda$ formulation that resulted in $\Delta\lambda = \frac{\Omega}{\beta} \tan\alpha/2$, where β represents the ratio of overlapping. This means that in the case where $\beta = 1$, it results in a minimum overlapping, and when $\beta \to +\infty$, it reaches the maximum overlapping between each slice.

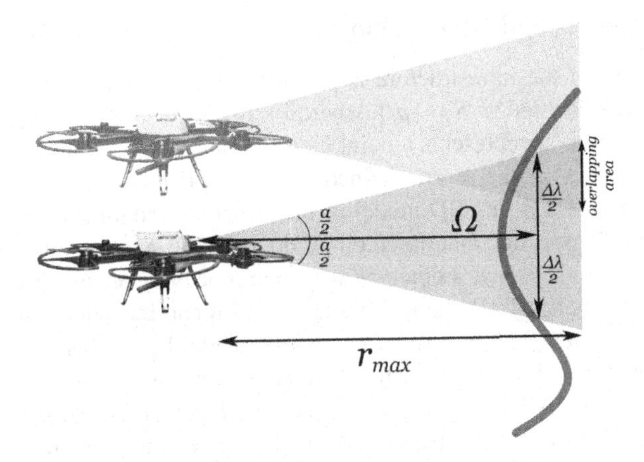

FIGURE 11.1 The Micro Aerial Vehicle (ARW) FoV (regenerated from [2]).

The proposed C-CPP method operates off-line, and it assumes that the 3D shape of the infrastructure is known a priori. The execution of the proposed cooperative inspection scheme is characterized by the following six steps: 1) slicing and intersection, 2) clustering, 3) adding offset from the infrastructure's surface, 4) path assignment for each ARW, 5) trajectory generation, and 6) online collision avoidance. This algorithmic approach is depicted in Fig. 11.2. It should be noted that this scheme has extended the approaches presented in [3] and [4] to multiple robots. Additionally, it should be highlighted that the main scope of the paper is not to find the optimal path but rather to solve the problem of cooperative coverage under geometric approaches. This problem could be formulated as a Mixed-Integer nonlinear programming one that has been already studied in the literature, and it is well known as the Multiple Traveling Salesman problem. This is a Non-deterministic Polynomial-time (NP) hard problem, where the complexity is exponentially increased by the number of cities/points. Thus, solving the complete problem is non-feasible and is not presented in this chapter.

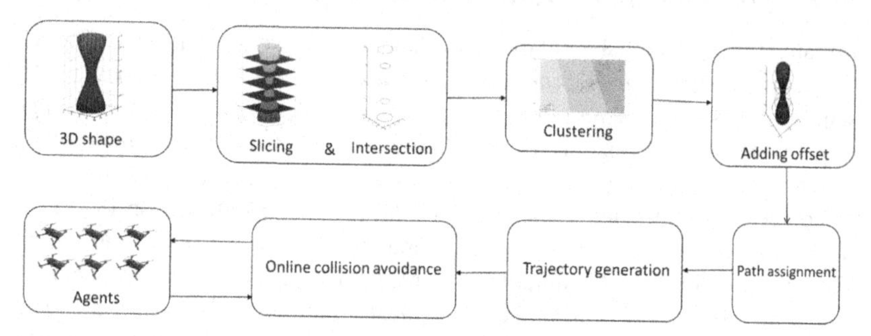

FIGURE 11.2 Block diagram of the overall proposed C-CPP scheme.

11.3.1 Slicing and intersection

The 3D map of the infrastructure is provided as a set $\boldsymbol{S}$ with a finite collection of points denoted as $\boldsymbol{S} = \{p_i\}$, where $i \in \mathbb{N}$ and $p_i = [x_i, y_i, z_i]^\top \in \mathbb{R}^3$. To be more specific, $\boldsymbol{S}$ represents a point cloud of the 3D object under inspection, where the points p_i, $i \in \mathbb{N}$ are defined by x, y, and z coordinates and can be extracted from CAD or a 3D model of the structure. Additionally, in the concept of aerial inspection and based on all the use case scenarios that have been examined, there is always a practical way in approaching an infrastructure with a simpler geometrical 3D shape. This assumption can be applied to almost all artificial structures without loss of generality, while the dimensions of this 3D shape simplistic approach can be realized based on in situ simple measurements or by blueprints. As it will be presented in the reconstruction results, the initial assumption of the 3D shapes is needed only for the determination of the path planning, while the final reconstructed results and point clouds have the required quality for creating a detailed and dense reconstruction. Thus, in the field of aerial inspection, even if a detailed 3D model of the structure is provided, the structure model can be further simplified in order to assist the generation of the flight paths for the ARWs.

As an illustrative example, in the complicated case of a wind turbine, the structure could be modeled by five different cylinder shapes for the tower, the hub, and the three blades. This geometrical assumption does not reduce the generality of the geometric approach; at the same time, it can be indirectly utilized as a means to create flying paths with an inherent level of safety for avoiding collisions with the infrastructure.

The map $\boldsymbol{S}$ is then sliced by multiple horizontal planes, defined as λ_i, with $i \in \mathbb{N}$. The value of λ_i started from $\min_z \boldsymbol{S}(x, y, z) + \Delta\lambda$ and ended with $max_z \boldsymbol{S}(x, y, z) - \Delta\lambda$. Thus, in this case, the horizontal plane translates vertically along the z-axis, while increasing the distance from the current slice, until reaching the maximum value. The intersection between the 3D shape and the slice can be calculated as follows:

$$\boldsymbol{\Sigma}_i = \{(x, y, z) \in \mathbb{R}^3 :< \vec{n}, \boldsymbol{S}(x, y, z) > -\lambda_i = 0\} \tag{11.6}$$

where $\boldsymbol{\Sigma}_i(x, y, z)$ are the points of the intersection of the plane, $\boldsymbol{S}$, $\vec{n} \in \mathbb{R}^3$ is the vector perpendicular to the plane, λ_i defines the location of the plane, and $< ., . >$ is the inner product operation. In the examined case, for simplicity the planes are chosen to be horizontal $\vec{n} = [0, 0, 1]^\top$. The slicing and intersection scheme is presented in Algorithm 1, while the overall concept is depicted in Fig. 11.3.

Algorithm 1 Slicing and intersection.

Require: Points of 3D shape $\boldsymbol{S}(x, y, z)$, $\Delta\lambda$
1: $\lambda = \min_z \boldsymbol{S}(x, y, z) + \Delta\lambda$
2: **while** $\lambda_i \leq max_z \boldsymbol{S}(x, y, z)$ **do**
3: $\quad \boldsymbol{\Sigma}_i = \{(x, y, z) \in \mathbb{R}^3 : < \vec{n}, \boldsymbol{S}(x, y, z) > -\lambda_i = 0\}$
4: $\quad$ Clustering() ; see Section 11.3.2.2
5: $\quad \lambda_{i+1} = \lambda_i + \Delta\lambda$
6: **end while**

11.3.2 Clustering

In the case of convex infrastructures, there is at most one cluster of points in any slice, while for complex structures, there will be more than one cluster in each slice. A cluster is defined by the number of branches of the structure, e.g., Fig. 11.4 shows the intersection points $\boldsymbol{\Sigma}_i(x, y, z)$ of the non-convex object with multiple branches in 2D, which results in a maximum of 3 clusters in λ_{i-k}. The set of points in $\boldsymbol{\Sigma}_i(x, y, z)$ (as obtained from Algorithm 1) is multi-modal. Thus, in order to complete the overall C-CPP task, it is critical to recognize the number of clusters and group the data elements (via clustering) for further task assignments of the agents.

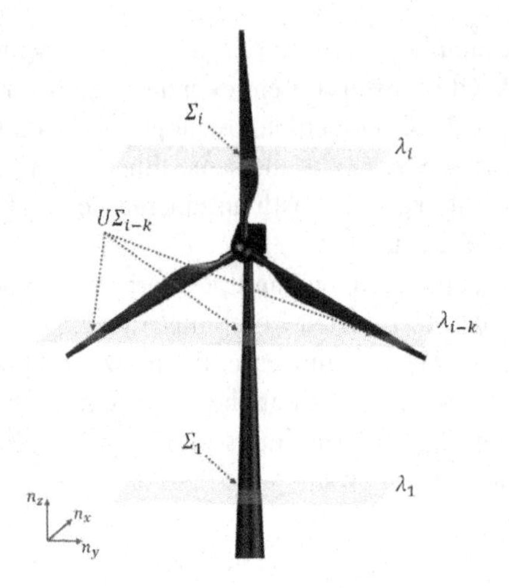

FIGURE 11.3 Concept of plane slicing and intersection points.

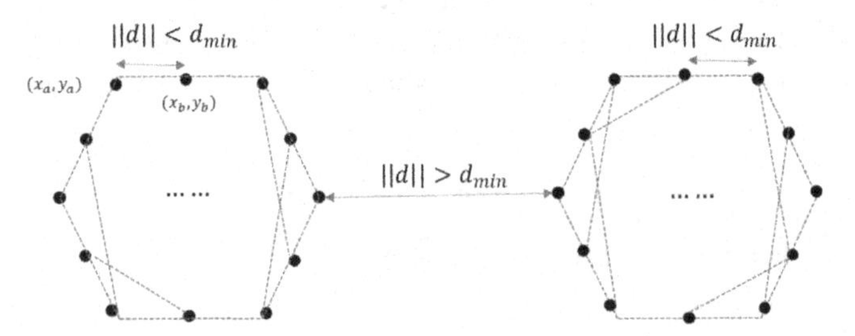

FIGURE 11.4 Intersection points of non-convex object with multiple branches.

11.3.2.1 Number of clusters

Clusters in $\mathbf{\Sigma}_i(x, y, z)$ are defined through connectivity by calculating the number of paths that exist between each pair of points. For some points to belong to the same cluster, they should be highly connected to each other. As shown in Fig. 11.4, they can be connected on the basis of their distances. In order to obtain the number of clusters, graph theory [5] will be utilized; thus, initially the adjacency matrix A is generated. The adjacency matrix A is formed on the basis of the Euclidean distance $||d||$ of the points. If the distance of two points is less than d_{min}, then the points are connected. Moreover, for selecting the value of d_{min}, the size of the agent is considered; if the agent cannot pass through two points, they are assumed to be connected. In the next step, the degree matrix $D(a, b)$ is generated, which is the diagonal matrix, and the elements of the diagonal $D(a, a)$ show the number of connected points to the ath point. Later on, the Laplacian matrix L is calculated from the subtraction of the adjacency matrix and the degree matrix $A - D$. Finally, the eigenvalues of L are calculated. The number of zero eigenvalue corresponds to the number of the connected graphs. The detailed process is presented in Algorithm 2.

Algorithm 2 Number of clusters.

Require: $\Sigma_i(x, y, z)$, d_{min}

1:
2: **for** a , $b \in \Sigma_i$ **do** ;a,b are two pairs of points
3: $||d|| = \sqrt{(x_a - x_b)^2 + (y_a - y_b)^2}$
4:
5: **if** $d < d_{min}$ **then**
6: $A(a, b) = 1$
7: **end if**
8: **end for**
9: $D(a,b) = \begin{cases} \text{\# of ones in } a^{th} \text{row of } A & a == b \\ 0 & a \neq b \end{cases}$
10: $L = A - D$; Laplacian matrix
11: $E = eig(L)$; eigen values of Laplacian matrix
12: $k = \#$ of $E == 0$; Number of clusters

11.3.2.2 Graph clustering

The existing global clustering approaches are capable of dealing with up to a few million points on sparse graphs [6]. More specifically, in global clustering, each point of the input graph is assigned to a cluster in the output of the method. Mixed Integer Programming (MIP) clustering and K-means clustering approaches can be used as a global method for graph clustering.

In general, the *K-means* method's objective function can be solved in different approaches, while in this chapter, the minimization of the point's distance to the center of the k clusters is considered. This formulation can be solved by different optimization algorithms, such as sequential quadratic programming [7] or MIP [8]. An alternative approach to this problem is to formulate the optimization problem as a minimization of the diameter of the clusters and solve it by MIP [9] or Branch-and-Bound algorithms [10,11]. Each method has its pros and cons that have been presented extensively in the relevant literature, e.g., in [6], where it has been presented that explicit optimization techniques such as MIP or Branch and Bound (BnB) algorithms can be used for reducing the necessary enumerations, but even with such savings, global searches remain impractical as that they are computationally expensive and complete enumeration of every possible partition is not possible [12]. On the other hand, the *K-means* algorithm can scan a large data set once and produces a clustering using small memory buffers, and thus it has been selected in the presented approach. In general, the clustering approach is a unique problem, and it will not be considered in detail in this chapter; however, *K-means* and MIP diameter clustering methods are studied and compared.

The challenge in using clustering algorithms is to determine the number of the clusters in a data set (Algorithm 2). The *K-means* clustering algorithm is used with a priori knowledge of the number of clusters k. The *K-means* clustering [13] is a combinatorial optimization problem that assigns m sets to exactly one of k ($k < m$) clusters, while minimizing the sum of the distances of each point in the cluster to its respective center. The utilization of the cluster algorithm can categorize the points, independently of the order of them or the direction of slicing to the object, while providing the center of each cluster, which is utilized for the calculation of the reference yaw angle for the agents. The algorithm is presented below.

Algorithm 3 K-means.

Require: $\boldsymbol{\Sigma}_i(x, y, z), k$

1: Choose k initial cluster centers C_i.

2: Calculate $\min\limits_{\boldsymbol{\Sigma}_i} \sum\limits_{i=1}^{k} \sum\limits_{\boldsymbol{\Sigma}_i} ||[x, y, z]^\top - C_i||$

3: Assign points to the closest cluster center.

4: Obtain k_{new} centers by computing the average of the points in each cluster

5: Repeat lines 2 to 4 until cluster assignments do not change

Moreover, in the MIP diameter clustering method, the objective function is to minimize the maximum diameter of the generated clusters with the goal of obtaining compact clusters. This criterion was previously studied in [9]. It is assumed that the number of desired k clusters is known from Section 11.3.2.1, while the mathematical formulation of the model MIP diameter, is provided in the following equation:

$$\min D_{max} \tag{11.7a}$$

$$s.t.\ D_c > d_{ij} x_{ic} x_{jc}\ \forall i, j = 1, \ldots, n,\ c = 1, \ldots, k \tag{11.7b}$$

$$\sum_{c=1}^{c=k} x_{ic} = 1 \tag{11.7c}$$

$$D_{max} \geq D_c\ \forall c = 1, ..., k \tag{11.7d}$$

$$x_{ic} \in \{0, 1\}\ \forall i = 1, \ldots, n,\ c = 1, \ldots, k \tag{11.7e}$$

$$D_c \geq 0\ \forall c = 1, \ldots, k \tag{11.7f}$$

where D_c is the diameter of the cluster c, D_{max} is the maximum diameter among the generated clusters, d_{ij} is the distance between two points, n is the total number of points, x_{ic} is the binary decision variable that represents the assignment to a particular cluster, and equals 1 if i is assigned to c and 0 otherwise. Eq. (11.7b) ensures that the diameter of cluster c is allowed to be at least the maximum distance between any two data points in cluster c, while Eq. (11.7c) guarantees that each point is assigned to only one cluster and Eq. (11.7d) sets the variable D_{max} equal to the value of the maximum diameter. The optimization has kn binary variables and $k+1$ continuous variables. As it is suggested in [9], a MIP CPLEX solver [14] and a heuristic approach are used to solve the problem. In Fig. 11.5, a dataset is presented where the global optimal solution for a clustering is known a priori. In each simulation a dataset with different number of points, n is studied, and the corresponding performance of the MIP clustering and *K-means* is compared. All the simulations have been performed in MATLAB® on a computer with an Intel Core i7-6600U CPU, 2.6 GHz and 8 GB RAM.

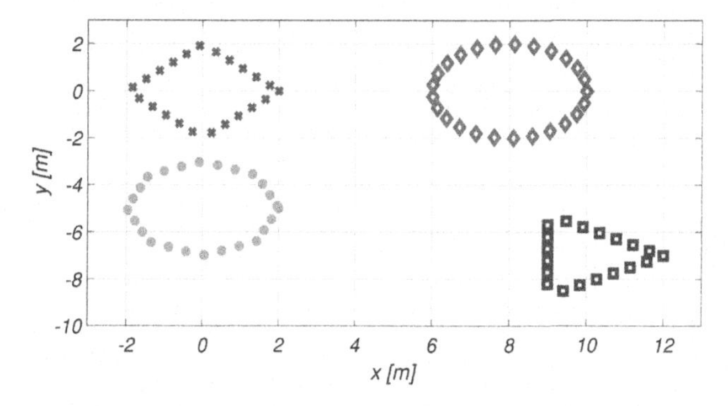

FIGURE 11.5 Dataset for 4 clusters.

Furthermore, Fig. 11.6 shows that the computational time of the MIP clustering method is increasing exponentially with the number of points when compared to the *K-means* algorithm. Thus, in this chapter, the *K-means* has been utilized without loss of generality with an overall aim for an online application of the proposed overall scheme, and, additionally, a case by case comparison is presented in Section 11.5 with a corresponding discussion.

11.3.2.3 Convex hull

It is important for the coverage algorithm that waypoints are not inside the object and guarantee full surface inspection. If the provided 3D model is not hollow, some waypoints are located inside the object. These points are not reachable during coverage mission and should not be considered. Therefore, a convex hull [15] algorithm is used to form the smallest convex set that contains the points in $\mathbb{R}^2$. However, using the convex hull, without clustering the point sets, results in a loss of some parts of the object in $\mathbb{R}^2$. Fig. 11.7 depicts the result of the convex hull before the generation of clustered point sets. In order to provide convex sets without loss of generality the $quickhull$ [15] algorithm is used (Algorithm 4).

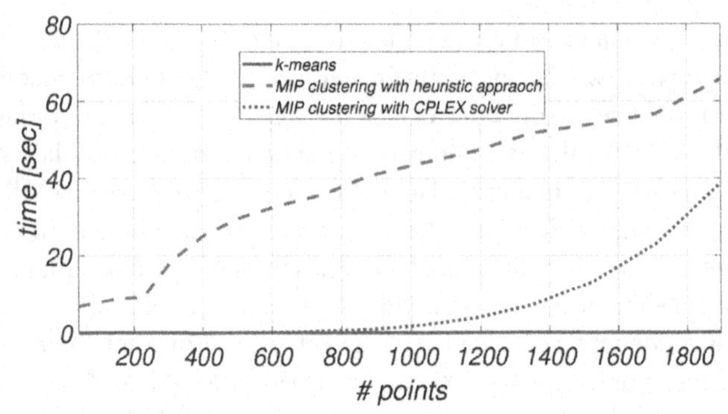

FIGURE 11.6 Computation time of the MIP clustering method and the *K-means* compared to the number of the points.

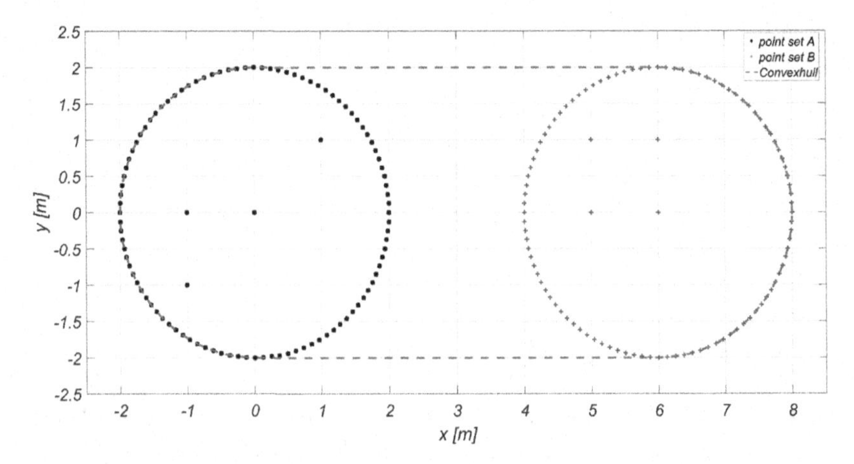

FIGURE 11.7 Convex hull example without clustering.

11.3.3 Adding offset

In the proposed C-CPP scheme, the ARW should cover the offset surface, which has a fixed distance Ω from the target surface, with the size of the agent also considered in Ω. A distance that can be considered as a safety distance Ω is kept constant, and it is continuously adapted to the infrastructure's characteristics. For a considered set of cluster $C(x, y, z)$, its offset path OP can be written as:

$$OP(x_j, y_j, z_j) = C(x_j, y_j, z_j) \pm \Omega \vec{n}_j, \tag{11.8}$$

where $\vec{n}_j$ is the normal vector at the jth point, and $\pm$ presents the direction of the normal vector; more details about the addition of the offset strategy can be found in [16]. Fig. 11.8 shows some examples in calculating OP for four different shapes of infrastructure, such as circular, square, hexagonal, and triangular. In each of these shapes, the desired offset path with $\Omega = 1$ m was calculated, while

Algorithm 4 *Quickhull* for the convex hull in $\mathbb{R}^2$.

Require: the set $Cluster_i$ of n points
  function QUICKHULL
  Find the points with minimum and maximum x coordinates $p_i = \{(x_{min}, y_i)\}$, $p_j = \{(x_{max}, y_j)\}$.
  Segment $p_i p_j$ divides the remaining $(n-2)$ points into 2 groups G_i and G_j
  convexhull=$\{p_i, p_j\}$
  FINDHULL(G_i, p_i, p_j)
  FINDHULL(G_j, p_j, p_i)
  end function
  function FINDHULL(G_k, p_i, p_j)
    if $G_k = \{\}$ **then**
  Return
    end if
  Find the farthest point p_f, from segment $p_i p_j$
  Add point p_f to the convex hull at the location between p_i and p_j
  $Cluster_i$ is divided into three subspaces by p_i, p_j, and p_f.
  convexhull=convexhull $\cup \{p_f\}$
  FINDHULL(G_i, p_i, p_f)
  FINDHULL(G_j, p_f, p_j)
  end function

in all the cases, the ARW flew parallel to the edges, with a constant distance of Ω until the next vertex was reached and while keeping the Ω distance while flying to the next edge.

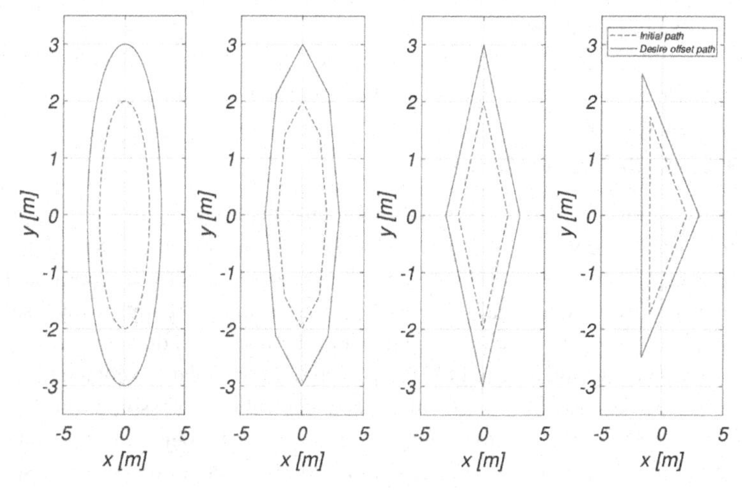

FIGURE 11.8 Example of adding offset to different shapes.

11.3.4 Path assignment

For sharing the aerial infrastructure inspection to multiple ARWs, two different cases have been considered; one where there is only one ($m = 1$) branch, and the other where there are multiple branches ($m > 1$). In the first case, each agent should cover a part of the branch, while in the case of two agents, θ_a and θ_b have a difference of 180° for equal space between the points. In the case of more than one branch, the ARWs should be equally distributed between the branches, where the policy of assigning each agent to each branch is shown in Algorithm 5.

Algorithm 5 C-CPP based assignment of ARWs to branches policy.

1: Assume to have n agents and m branches ($n \geq m$).
2: **if** $m == 1$ **then**
3: All n agents go in the same slice.
4: **end if**
5: **if** $m > 1$ AND $n > 1$ **then**
6: n agent equally distributed.
7: **end if**

11.3.5 Trajectory generation

The resulting waypoints are then converted into position-velocity-yaw trajectories, which can be directly provided to the utilized linear model predictive controller, cascaded [17] over an attitude-thrust controller. This is done by taking into account the position controller's sampling time T_s and the desired velocity along the path $\vec{V}_d$. These trajectory points are obtained by linear interpolation between the waypoints, in such a way that the distance between two consecutive trajectory points equals the step size $h = T_s||\vec{V}_d||$. The velocities are then set parallel to each waypoint segment, and the yaw angles are also linearly interpolated with respect to the position within the segment.

11.3.6 Collision avoidance

In order to avoid collisions, the following method is used to guarantee the maximum distance between the ARWs. Assuming n agents and a set of points $\mathcal{P}_i(x, y, z) \subset \mathbf{\Sigma}_i(x, y, z)$ assigned for i^{th} agent, the following optimization problem is solved sequentially for each agent to find the points with a safety distance with respect to the other agents and the minimum distance to the current position of the agent (Eq. (11.9)). The collision avoidance scheme that will be described later has been performed under the following assumptions: 1) the Ω is large enough to avoid the collision of agents to the object, 2) the distance between the branches of the object is more than the safety distance d_s, so the agents in separate branches cannot collide, and 3) entering and leaving of the agents, for each cluster, is a collision free path.

$$\begin{aligned}&\min_{p_i} ||p_i - p_i^*|| \\ &\quad ||\vec{d}_j|| > d_s \\ &\quad p_i(x_j, y_j, z_j) \in \mathcal{P}_i \\ &\quad j \geq 2\end{aligned} \tag{11.9}$$

where p_i is the current position, p_i^* is the future position of the ith agent, $||\vec{d}_i||$ contains the distances of the ith agent from the rest of the agents.
The above optimization is an *Integer Linear Programming* problem, as the optimization procedure should find the $(x_j, y_j) \in \mathcal{P}_i$ for each agent in order to minimizes the distance of the waypoints, guaranteeing collision free paths for each agent. The overall algorithm is also presented in Algorithm 6, where the operator "\" denotes the relative complement calculation and is defined as follows:

$$\mathcal{P}_i \setminus p_i = \{(x, y, z) \in \mathcal{P}_i | (x, y, z) \notin p_i\} \tag{11.10}$$

Algorithm 6 Collision avoidance between agent.

Require: $p_i, \mathcal{P}_i(x, y, z), n, d_s$
1: **for** $i = 2, \ldots n$ **do**
2: Compute p_i^* using (11.9)
3: $p_i \leftarrow p_i^*$
4: $\mathcal{P}_i \leftarrow \mathcal{P}_i \setminus p_i$
5: **end for**

Moreover, when the agents have to change a branch, they may collide with the object as the C-CPP only provides the initial point (x_s, y_s, z_s) and the destination point (x_d, y_d, z_d) for moving the agent from one branch to another branch. Thus, in general, it is possible that the intermediate points in this path are closer than the safety distance or inside the object. For avoiding these cases, the line is produced by connecting the initial point and destination. If the points of the line $\mathcal{L}(x, y, z)$ are closer than the safety distance d_{so} to the object, an offset value d_{of} is added until the distance is larger than the safety distance (Algorithm 7). The direction of adding the safety distance corresponds to the center of the cluster as calculated in (Section 11.3.3).

11.4 Multiple agent visual inspection

As stated, the C-CPP method is targeting the case of autonomous cooperative inspection by multiple aerial ARWs. Each aerial platform is equipped with a camera to record image streams and provide a 3D reconstruction of the infrastructure. More specifically, two main approaches have been considered to

Algorithm 7 Collision avoidance to the object.

Require: initial point=(x_s, y_s, z_s) and destination point=(x_d, y_d, z_d)

1: $\mathcal{L}(x, y, z) = \frac{x-x_s}{x_d-x_s} = \frac{y-y_s}{y_d-y_s} = \frac{z-z_s}{z_d-z_s}$
2: **for** $(x_a, y_a, z_a) \in \mathcal{L}(x, y, z)$ **do**
3: $||d|| = \sqrt{(x_a - x_b)^2 + (y_a - y_b)^2 + (z_a - z_b)^2}$
4: where $(x_b, y_b, z_b) \in \mathbf{S}(x, y, z)$
5: **if** $||d|| < d_{so}$ **then**
6: flag=1
7: **else**
8: Do NOTHING
9: **end if**
10: **end for**
11: **if** flag==1 **then**
12: Add offset value d_{of} to (x_s, y_s, z_s) and $(x_d, y_d, z_d) \rightarrow (x'_s, y'_s, z'_s)$ and (x'_d, y'_d, z'_d)
13: $\mathcal{L}(x, y, z) = \frac{x-x'_s}{x'_d-x'_s} = \frac{y-y'_s}{y'_d-y'_s} = \frac{z-z'_s}{z'_d-z'_s}$
14: go to line 2
15: **end if**

obtain the 3D model of the infrastructure using either stereo or monocular camera mapping. In both cases, the aim is to merge the processed data from multiple agents into a global representation. The selection between these two approaches depends mainly on the application's needs and the object's structure. The perception of depth using stereo cameras is bounded by the stereo baseline, essentially reducing the configuration to monocular at far ranges. For the monocular case, the employed Structure from Motion (SfM) approach, that provides a 3D reconstruction and camera poses using different camera viewpoints, induces the need to solve a large optimization problem. In this chapter, however, both methods were used as a proof of concept and will be presented in Section 11.5.

11.4.1 Stereo mapping

Generally, in the stereo visual mapping case, each robot provides an estimate of its pose and a reconstruction of the environment in its local coordinate frame. In the developed case, the agents localize themselves in the coordinate frame of the global localization system (e.g., motion capture system) and integrate this odometry information in the stereo SLAM [18] algorithm. All the generated pointclouds, from the individual agents, are expressed in the coordinate frame. In the case of n agents, each agent covers a specific area around the object of interest, reconstructing a unique map. Thereafter, each of the n created maps are merged through the Iterative Closest Point (ICP) [19] algorithm to form a global representation of the whole structure. This technique takes two point sets as inputs, Ξ and $\boldsymbol{Z}$ (Algorithm 8), and calculates the 3D transform τ that expresses the relative pose between them using a set of filtered corresponding points $f(\Xi)$ and $f(\boldsymbol{Z})$. The transform is estimated by minimizing the error $E_{er} = \sum_{j=1}^{N} (\|\tau(\xi_j) - \zeta_i\|^2)$. Then, τ is applied to align the point sets in a common frame. The output is a global point set $\boldsymbol{M}$ that represents the 3D reconstruction of the whole scene. Thus, it is important that the agents cover a common part of the scene in order to align the individual maps. The process is performed off-line.

Algorithm 8 Stereo 3D reconstruction.

1: **function** ICP FRAME ALIGNMENT
2: Map 1 point set $\Xi = \xi_1, \xi_2, \xi_3, \cdots, \xi_N$
3: Map 2 point set $\boldsymbol{Z} = \zeta_1, \zeta_2, \zeta_3, \cdots, \zeta_N$
4: Extract point features $(\Xi) \rightarrow f(\Xi)$
5: Extract point features $(\boldsymbol{Z}) \rightarrow f(\boldsymbol{Z})$
6: Match $f(\Xi)$ and $f(\boldsymbol{Z})$
7: Filter false matches and remove outliers
8: $\min_{\tau} E_{er} = \sum_{j=1}^{n} (\|\tau(\xi_j) - \zeta_i\|^2)$
9: Apply transform τ and align point sets
10: **end function**
11: **function** VISUAL SLAM
12: Frame location initialization (Feature Detection, Visual word vocabulary)
13: Loop closure detection
14: Frame location to pose graph
15: Pose graph optimization
16: Pointcloud with corresponding pose $\rightarrow \boldsymbol{M}$
17: **end function**
Assume to have n agents
18: **for** agent i **do**
19: Visual SLAM $\rightarrow \boldsymbol{M}_i$; 3D map generated for each agent
20: **end for**
21: Merge Maps
22: **for** $i : 1 : \#maps$ **do**
23: ICP FRAME ALIGNMENT($\boldsymbol{M}_i$) $\rightarrow$ global map $\boldsymbol{M}$; Agents with common scene coverage
24: **end for**

The basis for the 3D reconstruction is a state-of-the-art stereo mapping algorithm known as RTABMap [20] SLAM. This algorithm is suitable for large scale operations and fits well for complex structure mapping. More specifically, RTABMap is an appearance based Localization and Mapping algorithm that consists of three parts: the visual odometry, the loop closure detection, and the graph optimization part.

Initially, the algorithm identifies features in images and builds a visual word vocabulary [21] for loop closure processing. This process is used to determine whether a new image location has been previously visited. Then, the next step is to project the feature locations from the image to 3D using the depth measurements, through a filtering step to remove outliers. If features are identified in previous frames, the frame is added as a node in the pose graph, while a Bayesian filter keeps track of the loop closure cases. When loop closure is identified, the image location is added in the graph that holds the map $\boldsymbol{M}_i$. Finally, pose graph optimization and Bundle Adjustment are performed to refine the estimated location. The aforementioned process is described in Algorithm 8.

11.4.2 Monocular mapping

In the monocular mapping case, the incremental SfM [22,23] technique is used to build a reconstruction of the object under inspection. While the aerial agents follow their assigned path around the object of interest, the image streams from the monocular cameras of the agents are stored in a database. In the SfM process, different camera viewpoints are used off-line to reconstruct the 3D structure. The process starts with the correspondence search step, which identifies overlapping scene parts among input images. During this stage, feature extraction and algorithm matching between frames is performed to extract information about the image scene coverage. Following this, the geometric verification using the epipolar geometry [24] to remove false matches takes place. In this approach, it is crucial to select an initial image pair $\boldsymbol{I}_1$ and $\boldsymbol{I}_2$ with enough parallax to perform two-view reconstruction before incrementally registering new frames. Firstly, the algorithm recovers the sets of matched features f_1 and f_2 in both images. Next, it estimates the camera extrinsics for $\boldsymbol{I}_1$ and $\boldsymbol{I}_2$ using the 5-point algorithm [25], decomposing the resulting Essential matrix $\boldsymbol{E}_{es}$ with Singular Value Decomposition (SVD) and, finally, builds the projection matrices $\boldsymbol{P}_i = [\boldsymbol{R}_i|\boldsymbol{t}_i]$ that contain the estimated rotation and translation for each frame. Then, using the relative pose information, the identified features are triangulated to recover their 3D position $\boldsymbol{X}^{3D}$. Afterward, the two-frame Bundle Adjustment refines the initial set of 3D points, minimizing the reprojection error. After this initialization step, the remaining images are incrementally registered in the current camera and point sets. More specifically, the frames that capture the largest amount of the recovered 3D points are processed by the Perspective-n-Point (PnP) [26]. This algorithm uses 2D feature correspondences to 3D points to extract their pose. Furthermore, the newly registered images will extend the existing set of the 3D scene ($\boldsymbol{X}^{3D}$) using multi-view triangulation. Finally, a global Bundle Adjustment is performed on the entire model to correct drifts in the process. The aforementioned process is described in Algorithm 9. During the coverage tasks, the agents fly autonomously based on visual inertial odometry (Section 11.5). The absolute scale of the reconstructed object can be recovered by combining full-pose annotated images from the onboard localization of the camera.

Algorithm 9 Monocular 3D reconstruction.

1: **function** TWO-VIEW RECONSTRUCTION
2: Detect features (f_1, f_2) in frames $(\boldsymbol{I}_1, \boldsymbol{I}_2)$
3: Match f_1 and f_2 between $\boldsymbol{I}_1$ and $\boldsymbol{I}_2$
4: Remove false matches $\rightarrow$ inlier matches $(\hat{f}_1, \hat{f}_2)$; Geometric Verification
5: 5-point$(\boldsymbol{I}_1, \boldsymbol{I}_2, f_1, f_2) \rightarrow \boldsymbol{E}_s$; Essential Matrix
6: SVD$(\boldsymbol{E}_s) \rightarrow \boldsymbol{R}, \boldsymbol{t}$; Relative Camera Pose
7: Projection matrices $\boldsymbol{P}_1, \boldsymbol{P}_2 \rightarrow \boldsymbol{P}_1 = [\boldsymbol{I}|\boldsymbol{0}], \boldsymbol{P}_2 = [\boldsymbol{R}|\boldsymbol{t}]$ Triangulate$(\boldsymbol{I}_1, \boldsymbol{I}_2, f_1, f_2, \boldsymbol{P}_1, \boldsymbol{P}_2) \rightarrow \boldsymbol{X}^{3D}$; 3D points
8: Bundle Adjustment$(f_1, f_2, \boldsymbol{P}_1, \boldsymbol{P}_2, \boldsymbol{X}^{3D})$
9: **end function**
10: Import initial pair $\boldsymbol{I}_1, \boldsymbol{I}_2$
11: TWO-VIEW RECONSTRUCTION$(\boldsymbol{I}_1, \boldsymbol{I}_2, \boldsymbol{X}^{3D})$
12: **for** $i : 1 : \#frames$ **do**
13: Import new camera frame I_i
14: PnP$(\boldsymbol{I}_i, \boldsymbol{X}^{3D}) \rightarrow \boldsymbol{P}_i$; i_{th} projection matrix, new frame registered
15: Multi-view Triangulation $\rightarrow \boldsymbol{X}^{3D}_{new} = \boldsymbol{X}^{3D}, \boldsymbol{X}^{3D}_{i,k}$; Augment existing 3D structure.
16: Bundle adjustment$(f_i, \boldsymbol{P}_i, \boldsymbol{X}^{3D}_{new})$
17: **end for**

11.5 Results

In this section, simulation and experimental results are presented to prove the concept of the proposed method.

The following links provide the video summary of presented methods.

- ARW navigation in indoor and LTU fountain.
  Link: https://youtu.be/lUfCrDHL-Dc
- ARW navigation for wind turbine inspection.
  Link: https://youtu.be/z_Lu8HvJNoc

11.5.1 Simulation evaluation

Initially, in order to evaluate the performance of the method, a wind turbine, a 3D complex structure with multiple branches, is selected for simulation purposes. The tower diameter is 4 m at the base, 1 m at the top, and its height is 40 m. The blade length is 27 m with Cord length at the root of 2 m, cord length at the tip of 0.2 m, and the volume of the hub and nacelle is $4 \times 5 \times 4$ m^3. In the following simulation cases, the offset distance Ω from the inspection objects is 2 m. In Fig. 11.9, the paths, which are generated for one, two, and three ARWs for the cooperative aerial inspection, are depicted. This structure has up to three branches as depicted in Fig. 11.10, which can be recognized by the C-CPP. More specifically, the tower is considered to have one branch (green) until the height that two blades virtually intersect the horizontal plane of the tower. From that point, the structure is considered to have three branches (red) (dark grey in print), which are the two blades and the remaining part of the tower. Finally, the nacelle and the third blade of the structure are considered one branch (blue) (light grey in print). In Fig. 11.9 (left), the whole structure is covered by one agent, while in the case of more than one agents (middle and right), the branches are assigned to each agent, and in the case of a single branch, the area is shared between multiple agents. As a result, the inspection time is significantly reduced as presented in Table 11.1. The authors urge the reader to refer to the online version of this chapter to better understand the graphical depiction of the planned path for each aerial agent around the windturbine.

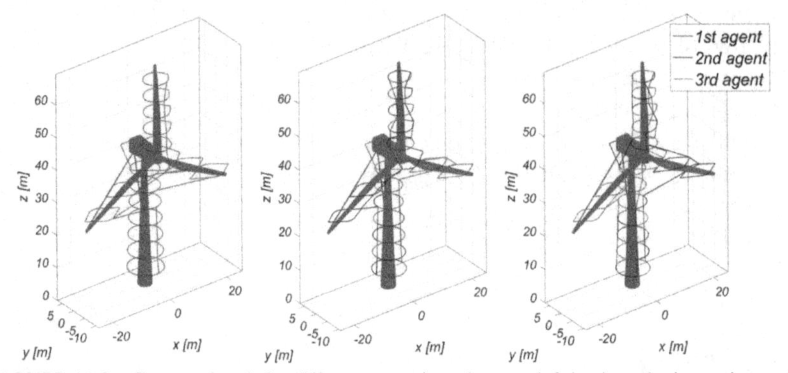

FIGURE 11.9 Generated path for different scenarios where on left is given the inspection path for a single agent, in the middle, inspection path for two aerial agents is shown and on the right, inspection path for three aerial agents is provided above.

FIGURE 11.10 Wind turbine with identified branches segmented by C-CPP.

TABLE 11.1 Inspection time for windmill inspection.

Number of agents	1	2	3
Inspection time [min]	24.86	17.63	11.36

Due to the lack of texture in the simulation, the quality of the reconstruction resulted from the recorded data is low and is therefore not included. More emphasis will be placed to the real-life experimental trials.

Fig. 11.11 presents the yaw references, which have been provided for each agent in different scenarios. As the number of agents increases, the coverage task is completed faster thus the yaw changes more frequently, as is indicated by the provided simulation results.

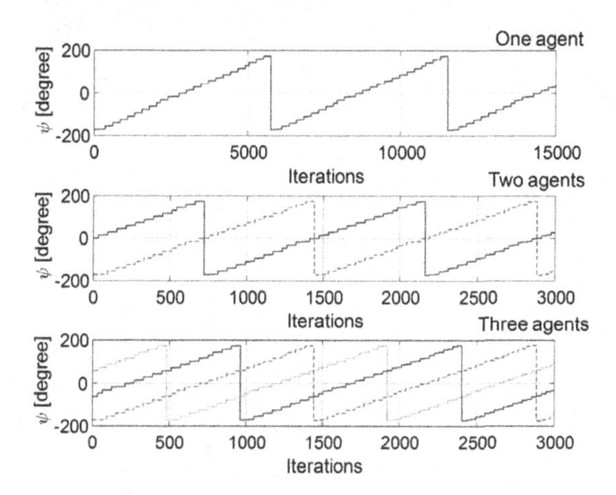

FIGURE 11.11 Yaw references for each scenario.

11.5.2 Experimental evaluation

11.5.2.1 Experimental setup

The proposed method has been evaluated utilizing the Ascending Technologies NEO hexacopter depicted in Fig. 11.12. The platform has a diameter of 0.59 m and height of 0.24 m. The length of each propeller is 0.28 m as depicted in Fig. 11.12. This platform is capable of providing a flight time of 26 min, which can reach a maximum airspeed of 15 m/s and a maximum climb rate of 8 m/s, with maximum payload capacity up to 2 kg. It has an onboard Intel NUC computer with a Core i7-5557U and 8 GB of RAM. The NUC runs Ubuntu Server 14.04 with Robot Operating System (ROS) was installed. ROS is a collection of software libraries and tools used for developing robotic applications [27]. Additionally, multiple external sensory systems (e.g., cameras, laser scanners, etc.) can be operated in this setup. Regarding the onboard sensory system, the Visual Inertia (VI) sensor (weight of 0.117 kg, Fig. 11.12) developed by Skybotix AG is attached below the hexacopter with a 45° tilt from the horizontal plane. The VI sensor is a monochrome global shutter stereo camera with 78° FoV, housing an Inertial Measurement Unit (IMU). Both the cameras and the IMU are tightly aligned and hardware synchronized. The camera was operated in 20 fps with a resolution of 752 × 480 pixels, while the depth range of the stereo camera lies between 0.4 and 6 m.

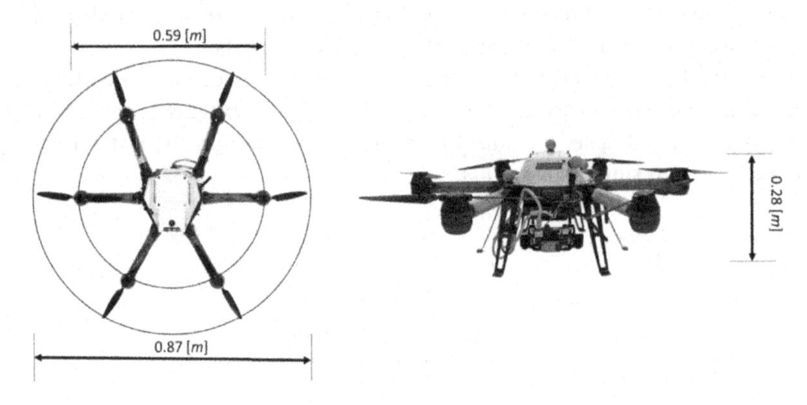

FIGURE 11.12 AscTec NEO platform with the VI sensor attached.

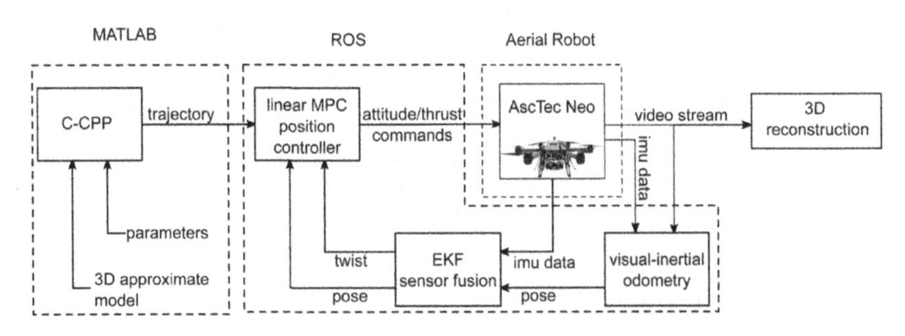

FIGURE 11.13 Software and hardware components used for conducting inspections.

The proposed C-CPP method, established in Section 11.3, has been entirely implemented in MATLAB. The inputs for the method are a 3D approximate model of the object of interest and specific parameters, which are the number of agents n, the offset distance from the object Ω, the FoV of the camera α, the desired velocity of the aerial robot V_d and the position controller sampling time T_s. The generated paths are sent to the NEO platforms through the ROS framework.

The platform contains three main components to provide autonomous flight: a visual-inertial odometry, a Multi-Sensor-Fusion Extended Kalman Filter (MSF-EKF) [28], and a linear Model Predictive Control (MPC) position controller [17,29,30]. The visual-inertial odometry is based on the Robust Visual

Inertial Odometry (ROVIO) [31] algorithm for the pose estimation. It consists of an Extended Kalman Filter (EKF) that uses inertial measurements from the VI IMU (accelerometer and gyroscope) during the state propagation, and the visual information is utilized during the filter correction step. The outcome of the visual inertial odometry is the position-orientation (pose) and the velocity (twist) of the aerial robot. Afterward, the MSF-EKF component fuses the obtained pose information and the NEO IMU measurements. This consists of an error state Kalman filter, based on inertial odometry, performing sensor fusion as a generic software package, while it has the unique feature of being able to handle delayed and multi-rate measurements while staying within computational bounds. The linear MPC position controller [30] generates attitude and thrust references for the NEO predefined low level attitude controller. In Fig. 11.14, the components of the controller are shown. Positions, velocity, and trajectory are sent to a Linear MPC position controller, which provides roll ϕ_d, pitch θ_d, and mass-normalized thrust T commands for the inner loop. From the initial control tuning experiments, it was proven that the low-level attitude controller is able to track the desired roll, pitch, and yaw trajectory and calculate the corresponding $n_1, n_2, \ldots, n_6$ rotor speeds for the vehicle. Details about the controller scheme and parameters' tuning can be found in [30].

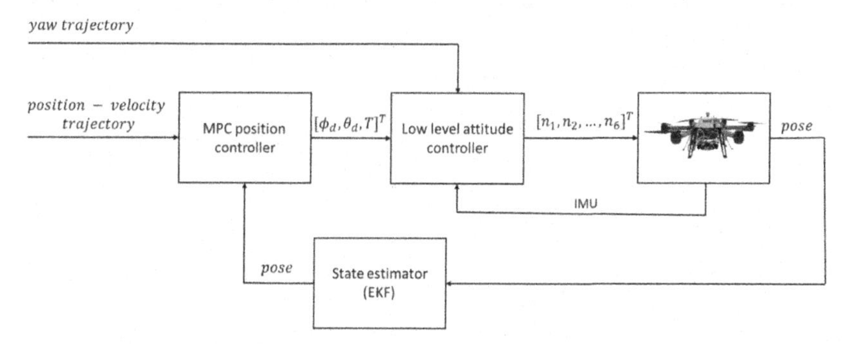

FIGURE 11.14 Controller scheme for NEO (regenerated from [30]).

The image stream from the overall experiment is processed using the method described in Section 11.4, while the overall schematic of the experimental setup is presented in Fig. 11.13.

11.5.2.2 Indoor artificial substructure inspection

An indoor artificial substructure was assembled as depicted in Fig. 11.15. The structure consisted of 6 boxes with dimensions of $0.57 \times 0.4 \times 0.3$ m with unique patterns and without any branches. The offset distance Ω from the inspection objects is 1 m in indoor experimental trials. In this case, two aerial agents were assigned to cover the structure. In these experiments, the Vicon Motion-capture (Mo-cap) system was used for the precise object localization. This information was utilized by the NEO for the autonomous flight. After the end of the experiment, the pose data from the Mo-cap system and the stereo stream were used by the mapping algorithm. The actual and the reconstructed structures are depicted in Fig. 11.15.

FIGURE 11.15 On the left is the simple indoor structure to be reconstructed and on the right the cooperative pointcloud of the structure.

Furthermore, Fig. 11.16 presents the actual and reference trajectories that the two agents followed. Moreover, the Mean Absolute Error (MAE) is summarized in Table 11.2.

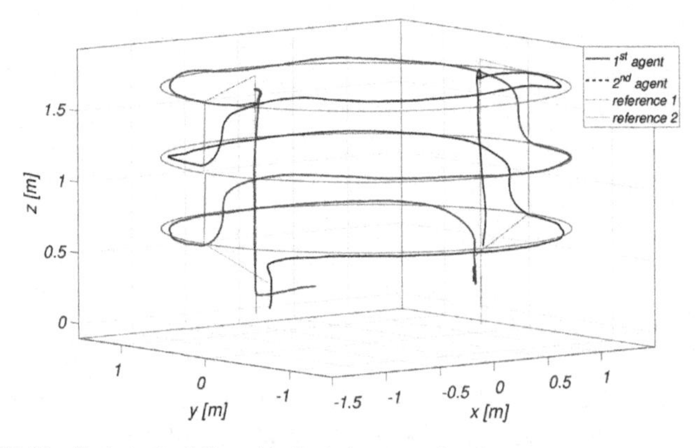

FIGURE 11.16 Trajectories followed in the indoor experiment.

TABLE 11.2 MAE of the trajectory followed by the first and second agents for the indoor experiment.

MAE	*x* [m]	*y* [m]	*z* [m]
1st agent	0.02	0.02	0.05
2nd agent	0.03	0.02	0.06

Additionally, the starting position of each has 180° difference and the agents completed the mission in 166 s instead of 327 s. The average velocity along the path was 0.2 m/s, and the points were fed to the agents so that to guarantee the maximum distance and avoid collision.

To retrieve the 3D mesh of the structure, the Autodesk ReCap 360 was used [32]. ReCap 360 is an online photogrammetry software suited for accurate 3D modeling. The reconstructed surface obtained from the image data is shown in Fig. 11.17.

FIGURE 11.17 3D mesh of the indoor structure.

11.5.2.3 Luleå University's fountain inspection

To evaluate the performance of the method for a real autonomous inspection task, an outdoor experiment was conducted. For this purpose the Luleå University's campus fountain was selected to represent the actual infrastructure for the cooperative aerial inspection. The fountain has a radius of 2.8 m and height of 10.1 m without branches. The offset distance Ω from the inspection objects is 3 m in experimental trials. Since in the outdoor experiments, motion capturing systems are rarely available, the localization of the ARW relied only on the onboard sensory system, in order to achieve a fully autonomous flight. Thus, the ARWs followed the assigned paths with complete onboard execution, based on visual inertial odometry localization. The actual and reference trajectories followed by both platforms are depicted in Fig. 11.18, while the MAE is provided in Table 11.3.

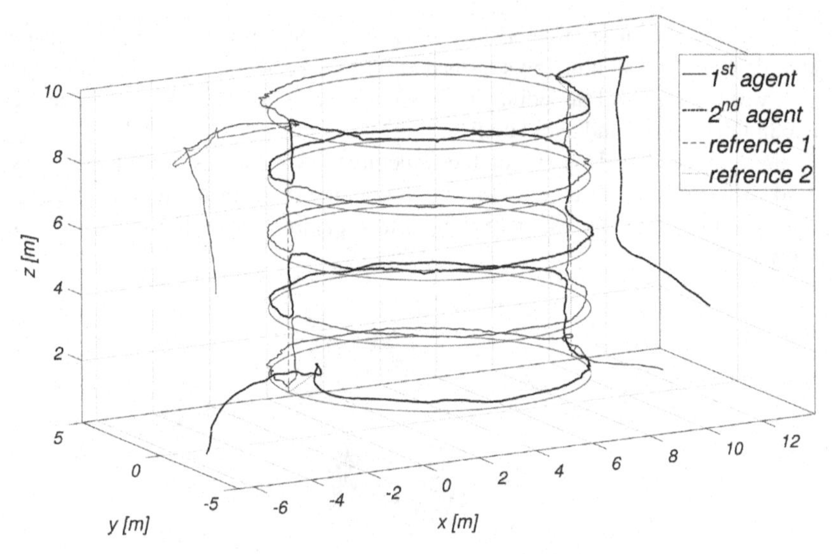

FIGURE 11.18 Trajectories followed in the outdoor experiment.

TABLE 11.3 MAE of the trajectory followed by the first and second agents for the outdoor experiment.

MAE	x [m]	y [m]	z [m]
1st agent	0.19	0.21	0.29
2nd agent	0.21	0.23	0.31

For the reconstruction, the image streams from both aerial agents were combined and processed by the SfM algorithm as described in Section 11.4. The fountain and its sparse 3D model are presented in Fig. 11.19.

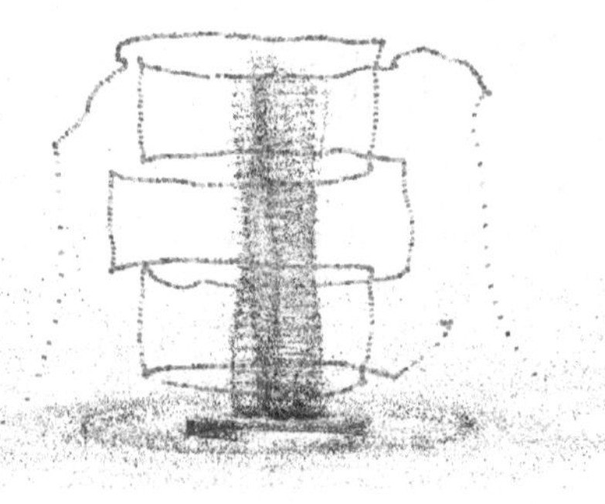

FIGURE 11.19 On the left is the Luleå University's outdoor fountain, and, on the right, the cooperative pointcloud of the structure with estimated flight trajectories.

In the proposed experiment, the same strategy as the indoor experiment is followed for two agents. The starting position of each of them has the maximum of distance with 180° difference. The overall flight time is reduced from 370 s to 189 s and the average velocity along the path was 0.5 m/s. The sparse reconstruction provided in Fig. 11.19 cannot be used for inspection tasks since it lacks texture information and contains noise. Similarly to the indoor experiment, the reconstructed surface obtained from image data is shown in Fig. 11.20. The results show that the collaborative scheme of the path planner could be successfully integrated for automating inspection tasks (https://youtu.be/IUfCrDHL-Dc).

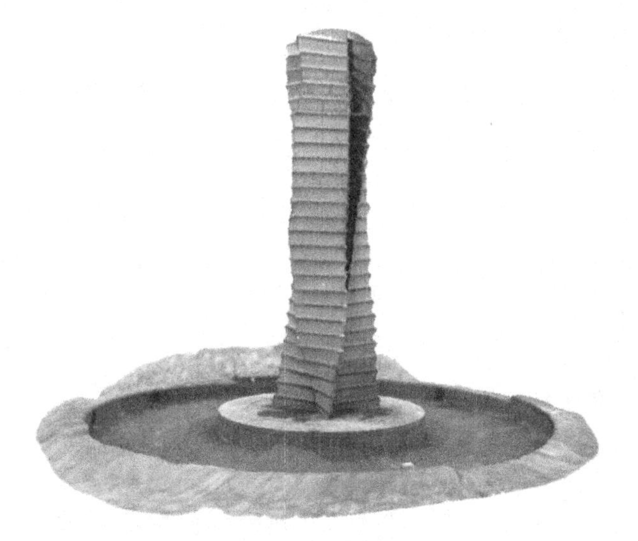

FIGURE 11.20 Cooperative 3D mesh of the outdoor structure.

11.5.2.4 Wind turbine inspection

The proposed method is evaluated for wind turbine visual inspection. For localization of the ARW the sensor fusion of Ultra WideBand (UWB) distance measurements and IMU are developed. In this approach, the aerial platforms navigate autonomously based on the UWB-Inertial fused state estimation, using a local UWB network, placed around the structure of inspection as depicted in Fig. 11.21. The initial step for the deployment of the system was to setup the ground station for monitoring the operations and fix 5 UWB anchors around the structure, with specific coordinates presented in Table 11.4, which constitute the infrastructure needed for the localization system of each aerial platform. The number of anchors as well as their position has been selected in a manner to guarantee UWB coverage around all parts of the wind turbine. From a theoretical point of view [33], only 3 anchors are needed; however, it is common that one anchor will be behind the wind turbine for the ARW's point of view, which gives rise to a minimum of 4 anchors to compensate, while the fifth anchor was added as redundancy. The resulting fixed anchor positions provide a local coordinate frame that guarantees repeatability of the system and with the significant ability to revisit the same point multiple times in case the data analysis shows issues that require further inspection. An important note for all the cases on the wind turbine and for the system in operation is that the blades are locked in a star position as shown in Fig. 11.21, which simplifies the 3D approximate modeling of the structure.

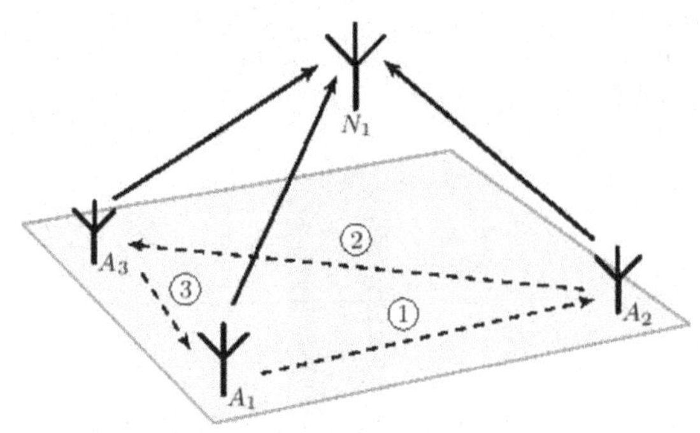

FIGURE 11.21 An overview of the UWB localization system, where A_1–A_5 are the stationary anchors, and N_1 is the tracked node mounted on the ARW, while the dashed lines highlight the measured distances.

TABLE 11.4 UWB anchor placement locations.

Coordinate	A_1	A_2	A_3	A_4	A_5
x	0 m	26.1 m	6.6 m	19.5 m	14.6 m
y	0 m	0 m	24.8 m	18.2 m	−21.7 m
z	0 m	0 m	0 m	0 m	0 m

In the proposed architecture, all the processing necessary for the navigation of the ARWs is performed onboard, while the overview of the mission and the commands from the mission operators (inspectors) is performed over a WiFi link, while the selection of WiFi is not a requirement and can be replaced with the communication link of choice, e.g., 4G cellular communication. The UWB based inertial state estimation runs at the rate of the IMU, which in this case was 100 Hz, and the generated coverage trajectory has been uploaded to the ARW before take-off, which is followed as soon as the mission started by the command of the operator. The paths have been followed autonomously, without any intervention from the operators on the site, and the collected data have been saved onboard, while after downloading the mission data post processing is performed in the ground station or in the cloud. The data provided by the system can be used for the position aware visual analysis, examining high-resolution frames, or they can be post-processed to generate 3D reconstructed models. The key feature to be highlighted from the task execution is that any detected fault can be fully linked with specific coordinates, which can be utilized by another round of inspections or for guiding the repair technician. The latter is a major contribution to the presented aerial system, since it needs the fundamental information that is needed to enable a safe and autonomous aerial inspection that has the potential to perform the human-based ones.

11.5.2.4.1 Wind turbine visual data acquisition

For the specific case of wind turbines, the C-CPP generated paths have been obtained with two autonomous agents in order to reduce the needed flight time and still be within the battery constrained flight time of the utilized ARW. However, due to the limited flight time of the ARWs in the field trials, the navigation problem has been split into the tower part and the blade part, where the specifics of each is presented in Table 11.5, while both can be performed at the same time with more ARWs to reduce the mission time even further. A common characteristic for both of the cases is that the generated path for each ARW keeps constant safety distance from the structure. At the same time, it is keeping it in view of the visual sensors, and maximizing the safety distance between agents, which gives rise to the agents being on opposite side of the wind turbine at all times. The area in which the field tests were performed is generally of high wind, and while the tower part is protected from wind, owing to the forest, the blade part is above the tree line. Thus, the aerial platform have been specifically tuned to compensate strong wind gusts that were measured up to 13 m/s, where the tuning was targeting the ARW's controller's weight on angular rate that has been increased to significantly reduce the excessive angular movement.

TABLE 11.5 Overview of the system configurations.

Mission configuration	Tower	Blade
Number of agents	2	1
Inspection time	144 sec	206 sec
Safety distance	7 m	9 m
Velocity	1 m/s	1.2 m/s
Starting height	8 m	30 m
Finishing height	24 m	45 m

11.5.2.4.2 Tower coverage

In the specific case of the wind turbine base and tower coverage, the generated paths are of a circular shape as depicted in Fig. 11.22, which results from the constant safety distance from the structure based on the C-CPP algorithm. As can be seen from the tracked trajectories, the controllers perform well with an Root Mean Square Error (RMSE) of 0.5464 m, while at the top of the trajectory, a more significant error can be seen that is induced from the specific ARW transitioning above the tree-line, where a wind gust caused the deviation from the desired trajectory where the ARW compensates and finishes its coverage trajectory.

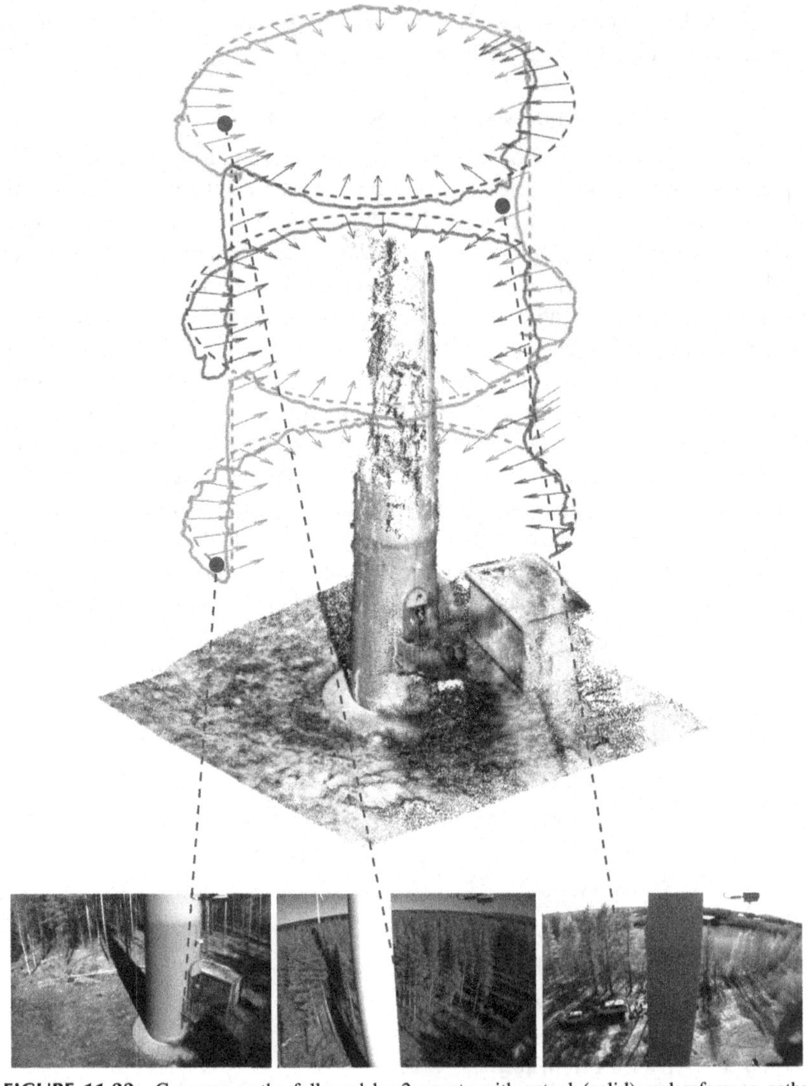

FIGURE 11.22 Coverage paths followed by 2 agents with actual (solid) and reference paths (dashed) together with desired direction, which resulted in the depicted 3D reconstruction and sample camera frames of the base and tower to be used by the inspector.

From the depicted reconstruction in Fig. 11.22, it is possible to understand that the base of the wind turbine, which is feature rich, provides a good reconstruction result, while as the ARW continues to higher altitudes, the turbine tower loses texture due to its flat white color, causing the reconstruction algorithms to not provide a successful reconstruction. However, the visual camera streams do have position and orientation for every frame as depicted in Fig. 11.22 for some instances, which allows for a trained inspector to review the footage and be able to determine if there are spots, which are needed extra inspection or repairs. For the reconstruction in Fig. 11.22, the [34] and [35] algorithms have been used, the former for pre-processing the images for enhance their contrast, while the latter was the SfM approach for providing the 3D model of the structure. The reconstruction took place on a PC with the configuration i7-7700 CPU and 32 GB of RAM, where the processing lasted approximately 4 hours.

11.5.2.4.3 Blade coverage

Compared to the base and tower coverage, for which the C-CPP algorithm generated circular trajectories, a similar approach was followed for the base case. This comes from the fact that this task is performed on the blade with a direction toward the ground and with the trailing edge of the blade toward the tower, which would cause the C-CPP algorithm to generate half-circle trajectories. However, in this case, the same agent can inspect the final part of the tower by merging both tower and blade trajectories as can be seen in Fig. 11.23, while minimizing the needed flight time and demonstrating at a full extend the concept of aerial cooperative autonomous inspection. With the available flight time of the ARW, it is possible to inspect the blade with only one operating ARW, allowing for the safety distance between agents to be adhered to, by the separation of the inspected parts. However, during the blade coverage task, the tracking performance of the ARW was reduced to an RMSE of 1.368 m, due to the constant exposure to wind gusts and the turbulence generated by the structure, and as these effects were not measurable, until the effects are observed on the ARW, it has reduced the overall observed tracking capabilities of the aerial platforms. The second effect of the turbulence was the excessive rolling and pitching of the ARW, which introduced a significant motion blur in the captured video streams, due to the fixed mounting of the camera sensor, introducing the need for adding a gimbal for stabilizing the camera and reducing the motion blur. Finally, as can be seen in the camera frames in Fig. 11.23, there are no areas of high texture on the wind turbine tower or blades, which caused 3D reconstruction to fail. However, the visual data captured is of high quality and suitable for review by an inspector.

11.5.3 Quantification of the design choices

In this Chapter, three different inspection scenarios were presented: Case 1) simulation results from a wind turbine inspection, Case 2) an inspection of artificial indoor structure, and Case 3) a real-world outdoor infrastructure inspection. In all the cases, there have been multiple limitations and challenges. The initial a priori information for enabling the aerial inspection of infrastructure needed is

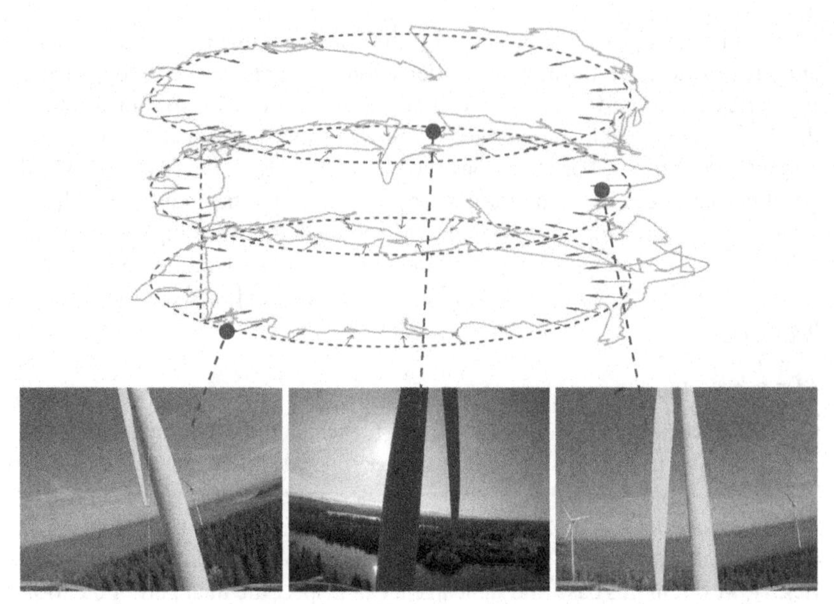

FIGURE 11.23 Coverage path followed by the agent with actual (solid) and reference path (dashed) together with desired direction, which resulted in the depicted 3D reconstruction and sample camera frames of the blade to be used by the inspector. Note the flat white color of the tower.

to have a 3D model of the structure. In cases that a non-detailed information can be obtained (e.g., blueprint or CAD model), as has been presented, the infrastructure can be approached by connected simple geometrical shapes without a loss of generality.

For all these experimental test cases and from a practical point of view, there are multiple limitations mainly due to: 1) the experimental setup and 2) the surrounding environment. The limitations in the experimental setup are mainly related to the utilized hardware and software that can directly affect the inspection mission. In the proposed experiment, the following components are critical for the successful mission: 1) localization, 2) control configuration, 3) visual processing, and 4) flight time. Additionally, from a practical perspective, important factors that influence the aerial coverage task in field trials are the heavy wind gusts (e.g., near wind turbines) that distort the followed path, direct sunlight that immediately degrades the performance of the vision algorithms, and increased humidity levels and low temperatures that cause malfunctions in the platform electronics. Finally, the need for a complete no rain environment and temperatures above +4°C should be highlighted. Except from these limitation factors, later, the available design choices and the corresponding effect on the results will be discussed.

Safety distance: The safety distance between the agents and the infrastructure is a critical factor for the experiment, as this selection directly affects the safety of the mission and the overall quality of the obtained images. For safety reasons and collision avoidance, it is preferable to have a proper safety margin from the structure. This margin can provide enough time for the safety pilot to avoid collisions in case of an accident. Additionally, the larger value of the Ω results in a larger camera footprint area. This has a direct effect on creating shorter paths for the ARWs that are now capable of covering bigger areas. However, the larger the value for Ω is, the more the quality of the images is reduced. This can be overcome by utilizing better cameras with a higher resolution. In the case of the indoor experiment, the safety distance was set to 1 m, as the flying area was limited, while in the outdoor experiment the Ω was set as 3 m, 7 m, and 9 m for the fountain, the windturbine tower, and the blade inspections, respectively, as there are extra disturbances to the system (such as wind gust), and the localization was accurate in a few centimeters.

Field of view: this parameter is dependent on the hardware that will be used in the experiment, and in most cases, it is considered and modeled as a constant. This value directly affects the camera footprint. To be more specific, the larger the FoV value is, the larger the camera footprint area becomes. However, the FoV and the camera resolution have a direct impact on the image quality. In this Chapter, a camera with a visual sensor with a 78° FoV was used.

Number of agents: the number of agents has a direct relation to the size of the structure and the flight time of the platform. For example, in the case of the wind turbine inspection, it is almost impossible to do the inspection with one agent as the flight time is 25 min. Thus, multiple agents are assigned for the inspection, and the flight time is reduced to 12 min for each agent in the case of three agents. From another point of view, the more the agents are, the longer time the optimization algorithm takes for creating the multiple ARW paths.

Controller: In this Chapter, MPC is used as the position controller. There are multiple parameters that can affect the performance of the control scheme, such as: 1) weights of the objective function terms, since the larger value of the weights are, the smaller the error in those terms is. To be more specific, in the case of the position error in the objective function, a higher weight results in a smaller positioning error, and 2) constraints: each platform has physical constraints such as maximum velocity, maximum acceleration, bound on propellers velocities, boundaries on angles, etc. During the experiments, it was observed that in the case of the outdoor trials, mainly due to the existence of the wind gusts, the MPC should have larger bounds on angles, and thus these were changed from 15° to 45°, which resulted in a better path tracking and wind gust compensation. It should also be highlighted that the error in the position and orientation may result in a different camera footprint area from the desired one, with a direct impact on the quality of the achieved inspection and reconstruction.

Velocity of the agent: this parameter is selected on the basis of the experiment and hardware setup. The velocity of the agent should not be larger than the camera frames rate, which will result in blurred images. Additionally, the velocity of the ARW should be higher than the wind gust in order to compensate for it, and due to safety reasons, the velocity should be also reasonable for the safety pilot to be able to take control of the ARW in unpleasant scenarios. Thus, in this Chapter, the velocity is chosen as 0.2 m/s for the indoors experiment and 0.5 m/s for the outdoors experiment.

Overlapping: Multi-view algorithms require the processing of the same surface, captured in multiple frames from different viewpoints, to provide a proper reconstruction. More specifically, a substantial overlapping between frames is considered. The absolute minimum overlapping required for each observed 3D point is only 2 frames. For increased robustness, practical implementations consider at least 3 images. In the presented thesis, the generated paths in the experimental trials considered at least 3 overlapping images for the reconstructed points. Additionally, 40% overlapping between slices was chosen.

Number of features: The basis of the SfM algorithms is mainly the correspondence search among frames. The correspondence step includes feature extraction (distinctive points in 2D frame), feature matching, and additional verification to filter out outliers. Generally, the feature extraction step identifies a high number of features, e.g., 500, but they are decreased after the matching and validation steps. Therefore, the number of inlier features in every frame is not fixed and varies depending on the extraction algorithm, the occlusions, and the object geometry. In the performed experimental trials, a mean value for every frame was around 200-300 features.

Camera frames: Multi-view algorithms can handle increased frame rates with proper view selection, when considering the data capturing. For the indoors and outdoors experimental trials the frame rate was fixed at 20 Hz. However, this parameter is highly connected with the image resolution, where with a higher resolution the algorithms can reconstruct in more details, since there exist fewer ambiguous matches in the captured frames. Generally, a higher resolution is preferred over higher frame rates.

Clustering: as it has been discussed in Section 11.3.2.2, in addition to the *K-means*, the MIP clustering was also investigated in the presented test cases. Thus, all the calculations, as stated before, have been performed on a computer with an Intel Core i7-6600U CPU, 2.6 GHz and 8 GB RAM and were implemented in MATLAB. In each scenario, both methods were evaluated and the computation time is provided in Table 11.6. In the first case, the number of points in each slice were between 500 and 8000, and the number of clusters was 1 or 3. The average computation time for *K-means* was 0.2 s, and the average computation time for the MIP clustering method with CPLEX solver and heuristic approaches were 75 s and 80 s, respectively. The computation time of the MIP clustering method can be reduced further by increasing the processing power, down sampling the structure, or utilizing different optimization solvers. In the second and third cases, there was only one branch and the number of points in each slice were between 60 and 150. Overall, the MIP clustering method provides a global solution, but the computation power is still the main restriction to real-life applications. However, if the utilization of MIP is not causing problems in the realization of the proposed approach, this method could be utilized for performing the necessary clustering.

TABLE 11.6 Computation time for clustering methods in different scenarios.

Cases	*K-means*		MIP CPLEX solver		MIP heuristic approach	
	avg.	max	avg.	max	avg.	max
1	0.2 s	0.45 s	75 s	100 s	80 s	170 s
2	0.10 s	0.24 s	2.4 s	3.4 s	5.3 s	16.2 s
3	0.14 s	0.32 s	3.5 s	4.2 s	6.6 s	18.1 s

11.6 Lessons learned

Throughout the experimental trials in wind turbine inspection, many different experiences were gained that assisted in the development and tuning of the algorithms utilized. Based on this experience, an overview of the lessons learned is provided in the sequel with connections to the different utilized field algorithms.

11.6.1 ARW control

When performing trajectory tracking and position control experiments indoors a dedicated laboratory many disturbances, which are significant in the field trials, can be neglected, and this is especially true for strong wind gusts and turbulence caused by the structure. In the case of indoor experimental trials, the ARW can be tuned aggressively to minimize the position tracking error, while in the full scale outdoor experiments, this kind of tuning would provide excessive rolling and pitching due to the controllers trying to fully compensate for the disturbances. However, this has the side effect of making the movements jerky and oscillatory and overall reducing the operator's trust in the system as it seems to be close to unstable. Furthermore, in the case that the controllers were tuned for a smooth trajectory following, larger tracking errors would have to be accepted in the trajectory following. During the field trials, some wind gust can even be above the operational limits of the ARW, causing excessive errors in the trajectory tracking. To reduce the effect in the outdoor experiments, the controller's weight on angular rate was increased to significantly reduce the excessive movement, while, in general, the tuning of the high level control scheme, for the trajectory tracking, is a tedious task, and it was found to be extremely sensitive to the existing weather conditions.

11.6.2 Planning

The path planner provides a path to guarantee for a full coverage of the structure; however, in the field trials, due to high wind gusts, there are variations between the performed trajectory and the reference. Thus, there is a need for an online path planner for considering these drifts and re-plan the path or to have a system that it is able to detect if a specific part of the structure has been neglected and provides extra trajectories to compensate. Additionally, due to the payload, the wind gusts and the low ambient temperature, the flight time was significantly less than the expected value from the ARW manufacturer. In certain worst cases, this time was down to 5 minutes, which is a severe limitation that should be considered in the path planning and task assignment to correctly select the correct number of agents for achieving a full coverage of the infrastructure.

11.6.3 System setup

One of the most challenging issues when performing large scale infrastructure inspection is to keep a communication link with the agents performing the inspection, which is commonly used for monitoring the overall performance of the system. In this specific case, WiFi was the communication link of choice, mainly due to its simplicity of directly performing as expected; however, it was quickly realized that the communication link was unstable due to height or occlusion of the ARW behind the wind turbine tower. To mitigate this issue, a different communication link should be used, e.g., the 4G cellular networks, and while WiFi can be used to upload mission trajectories it is not a reliable communication link at this scale.

Moreover, if it is desirable that the same mission can be executed again, the positions of the UWB anchors need to be kept. One possible way to achieve this is to consider the UWB anchors as supporting part of the infrastructure and have them permanently installed around the wind turbines, or to re-calibrate and consider the wind turbine as the origin, while only compensating for the rotation of the wind turbine depending on the mission setup.

11.6.4 3D reconstruction

Various visual sensors have been tested in the challenging case of wind turbine. The most beneficial sensor proved to be the monocular camera system. More specifically, the fixed baseline for stereo cameras can limit the depth perception and eventually degenerate the stereo to monocular perception. The reconstruction performance can also vary slightly, depending on the flying environment due to visual feature differences, therefore a robust and reliable, invariant to rotations feature tracker should be used. Another important factor for the reconstruction is the camera resolution, since it poses the trade-off between higher accuracy and higher computational costs. Additionally, the path followed around the structure affects the resulting 3D model, which in combination with the camera resolution can change the reconstruction results. Generally, the cameras should be calibrated, and it is preferred to have set manual focus and exposure to maintain the camera parameters for the whole dataset. For SfM techniques, it is required a large motion in rotation and depth among sequential frames to provide reliable motion estimation and reconstruction.

Moreover, a low cost LIDAR solution, that was tested during the field trials, failed to operate as sunlight interfering with the range measurements. This sensor technology, should be further examined with more tests since they could be useful in obstacle avoidance and cross-section analysis algorithms.

11.6.5 Localization

While UWB positioning was the main localization system in the presented approach, it should be noted that this should not operate stand-alone. In the case of infrastructure inspection, one reference system should not act as a single point of failure, and it should be the aim to fuse as many sensors as possible. In the case of a wind turbine, the Global Positioning System (GPS) does not provide a reliable position until the ARW is at significant height, and the UWB localization system works best at lower height. Hence, it should be the aim to fuse both and utilize the sensor that is performing optimally depending on the current height. Moreover, neither UWB localization nor GPS provides a robust heading estimate, and the wind turbine causes magnetic disturbances that causing the magnetometers to fail, and thus in this case, visual inertial odometry is a robust solution to provide heading corrections since the landscape can be used as a stable attitude reference.

11.7 Conclusions

This chapter addressed the C-CPP for inspecting complex infrastructures by utilizing multiple agents. A mathematical framework for solving the coverage problem by introducing branches and safety distances in the algorithm has been presented. The established theoretical framework provides a path for accomplishing a full coverage of the infrastructure, without simplification of the infrastructure (number of considered representation points), in contrast to many existing approaches that simplify the infrastructure to an area of interest and solve it by various optimization methods; methods that could not be applied otherwise due to the inherent NP-hard complexity of the problem. In addition, this Chapter has demonstrated the direct applicability and feasibility of coverage by multiple ARWs in field trials, while indicating the pros and cons when performing real-life tests. This effort integrated and adapted fundamental principles from control, image processing, and computer science, in a fully functional and efficient approach. Moreover, this Chapter has contributed to presenting a complete cooperative aerial coverage system that can be directly applied in any kind of infrastructure, with the reported limitations.

References

[1] C. Kanellakis, G. Nikolakopoulos, Survey on computer vision for UAVs: Current developments and trends, Journal of Intelligent & Robotic Systems (2017) 1–28.

[2] E. Galceran, R. Campos, P. Edifici IV, N. Palomeras, P. de Peguera, D. Ribas, M. Carreras, P. Ridao, Coverage path planning with realtime replanning and surface reconstruction for inspection of 3D underwater structures using autonomous underwater vehicles.

[3] P.N. Atkar, H. Choset, A.A. Rizzi, E.U. Acar, Exact cellular decomposition of closed orientable surfaces embedded in R^3, in: Proceedings 2001 ICRA, IEEE International Conference on Robotics and Automation, vol. 1, IEEE, 2001, pp. 699–704.

[4] E. Galceran, M. Carreras, Planning coverage paths on bathymetric maps for in-detail inspection of the ocean floor, in: 2013 IEEE International Conference on Robotics and Automation (ICRA), IEEE, 2013, pp. 4159–4164.

[5] D.B. West, et al., Introduction to Graph Theory, vol. 2, Prentice Hall, Upper Saddle River, 2001.

[6] S.E. Schaeffer, Graph clustering, Computer Science Review 1 (1) (2007) 27–64.

[7] J. Nocedal, S.J. Wright, Sequential Quadratic Programming, Springer, 2006.

[8] L.A. Wolsey, Mixed Integer Programming, Wiley Encyclopedia of Computer Science and Engineering, 2008.

[9] B. Sağlam, F.S. Salman, S. Sayın, M. Türkay, A mixed-integer programming approach to the clustering problem with an application in customer segmentation, European Journal of Operational Research 173 (3) (2006) 866–879.

[10] P. Hansen, B. Jaumard, Cluster analysis and mathematical programming, Mathematical Programming 79 (1–3) (1997) 191–215.

[11] W.L.G. Koontz, P.M. Narendra, K. Fukunaga, A branch and bound clustering algorithm, IEEE Transactions on Computers 100 (9) (1975) 908–915.

[12] B.S. Everitt, S. Landau, M. Leese, D. Stahl, Hierarchical clustering, in: Cluster Analysis, 5th ed., 2011, pp. 71–110.

[13] S. Lloyd, Least squares quantization in PCM, IEEE Transactions on Information Theory 28 (2) (1982) 129–137.

[14] I.I. CPLEX, V12. 1: User's manual for CPLEX, International Business Machines Corporation 46 (53) (2009) 157.

[15] C.B. Barber, D.P. Dobkin, H. Huhdanpaa, The quickhull algorithm for convex hulls, ACM Transactions on Mathematical Software 22 (4) (1996) 469–483.

[16] X.-Z. Liu, J.-H. Yong, G.-Q. Zheng, J.-G. Sun, An offset algorithm for polyline curves, Computers in Industry 58 (3) (2007) 240–254.

[17] K. Alexis, G. Nikolakopoulos, A. Tzes, Model predictive quadrotor control: attitude, altitude and position experimental studies, IET Control Theory & Applications 6 (12) (2012) 1812–1827.

[18] F. Endres, J. Hess, N. Engelhard, J. Sturm, D. Cremers, W. Burgard, An evaluation of the RGB-D slam system, in: 2012 IEEE International Conference on Robotics and Automation (ICRA), IEEE, 2012, pp. 1691–1696.

[19] P.J. Besl, N.D. McKay, Method for registration of 3-d shapes, in: Robotics-DL Tentative, International Society for Optics and Photonics, 1992, pp. 586–606.

[20] M. Labbé, F. Michaud, Memory management for real-time appearance-based loop closure detection, in: 2011 IEEE/RSJ International Conference on Intelligent Robots and Systems, IEEE, 2011, pp. 1271–1276.

[21] J. Sivic, A. Zisserman, Efficient visual search of videos cast as text retrieval, IEEE Transactions on Pattern Analysis and Machine Intelligence 31 (4) (2009) 591–606.

[22] J.L. Schönberger, J.-M. Frahm, Structure-from-motion revisited, in: CVPR, 2016.

[23] C. Wu, Towards linear-time incremental structure from motion, in: 2013 International Conference on 3D Vision – 3DV 2013, IEEE, 2013, pp. 127–134.

[24] R. Hartley, A. Zisserman, Multiple View Geometry in Computer Vision, Cambridge University Press, Cambridge, 2003.

[25] D. Nistér, An efficient solution to the five-point relative pose problem, IEEE Transactions on Pattern Analysis and Machine Intelligence 26 (6) (2004) 756–770.
[26] X.-S. Gao, X.-R. Hou, J. Tang, H.-F. Cheng, Complete solution classification for the perspective-three-point problem, IEEE Transactions on Pattern Analysis and Machine Intelligence 25 (8) (2003) 930–943.
[27] Robot Operating System (ROS) [Online]. Available: http://www.ros.org/.
[28] S. Lynen, M. Achtelik, S. Weiss, M. Chli, R. Siegwart, A robust and modular multi-sensor fusion approach applied to MAV navigation, in: Proc. of the IEEE/RSJ Conference on Intelligent Robots and Systems (IROS), 2013.
[29] K. Alexis, G. Nikolakopoulos, A. Tzes, Switching model predictive attitude control for a quadrotor helicopter subject to atmospheric disturbances, Control Engineering Practice 19 (10) (2011) 1195–1207.
[30] M. Kamel, T. Stastny, K. Alexis, R. Siegwart, Model predictive control for trajectory tracking of unmanned aerial vehicles using robot operating system, in: Robot Operating System (ROS), Springer, 2017, pp. 3–39.
[31] M. Bloesch, S. Omari, M. Hutter, R. Siegwart, Robust visual inertial odometry using a direct EKF-based approach, in: 2015 IEEE/RSJ International Conference on Intelligent Robots and Systems (IROS), IEEE, 2015, pp. 298–304.
[32] Autodesk RECAP 360 [Online]. Available: http://recap360.autodesk.com/.
[33] E. Fresk, K. Ödmark, G. Nikolakopoulos, Ultra wideband enabled inertial odometry for generic localization, IFAC-PapersOnLine 50 (1) (2017) 11465–11472.
[34] S.M. Pizer, E.P. Amburn, J.D. Austin, R. Cromartie, A. Geselowitz, T. Greer, B. ter Haar Romeny, J.B. Zimmerman, K. Zuiderveld, Adaptive histogram equalization and its variations, Computer Vision, Graphics, and Image Processing 39 (3) (1987) 355–368.
[35] J.L. Schonberger, J.-M. Frahm, Structure-from-motion revisited, in: Proceedings of the IEEE Conference on Computer Vision and Pattern Recognition, 2016, pp. 4104–4113.

Chapter 12

ARW deployment for subterranean environments

Björn Lindqvist, Sina Sharif Mansouri, Christoforos Kanellakis, and Vignesh Kottayam Viswanathan
Department of Computer, Electrical and Space Engineering, Luleå University of Technology, Luleå, Sweden

12.1 Introduction

Operating ARWs in subterranean environments is becoming more and more relevant in the field of aerial robotics. Despite the large spectrum of technological advances in the field, flying in such challenging environments is still an ongoing quest that requires the combination of multiple sensor modalities like visual/thermal cameras as well as 3D/2D lidars. Nevertheless, there exist cases in subterranean environments where the aim is to deploy fast and lightweight aerial robots for area reckoning purposes after an event (e.g., blasting). This chapter presents the baseline approach for the navigation of resource-constrained robots, introducing the aerial underground scout, with the main goal of rapidly exploring unknown areas and providing feedback to the operator. The main presented framework focuses on the navigation, control, and vision capabilities of the aerial platforms with low-cost sensor suites, contributing significantly to real-life applications. The merit of the proposed control architecture considers the flying platform as a floating object, composing a velocity controller on the x, y axes and an altitude controller for navigating along a mining tunnel. Moreover, this chapter presents a collection of different, experimentally verified, methods tackling the problem of ARWs heading regulation. In addition to the heading correction modules presented in Chapter 6, this chapter presents Convolutional Neural Network (CNN) classification and CNN regression heading correction modules while navigating in texture-less tunnel areas. The concept of the presented approaches is demonstrated in Fig. 12.1.

The rest of the chapter is structured as follows. Initially, Section 12.2 presents the problem formulation of the proposed method. Then Sections 12.3 and 12.4 present information on the developed state estimation and navigation. The discussed approaches are evaluated in an underground mine, and the results are presented in Section 12.5, followed by a corresponding comparison

Aerial Robotic Workers. https://doi.org/10.1016/B978-0-12-814909-6.00018-4

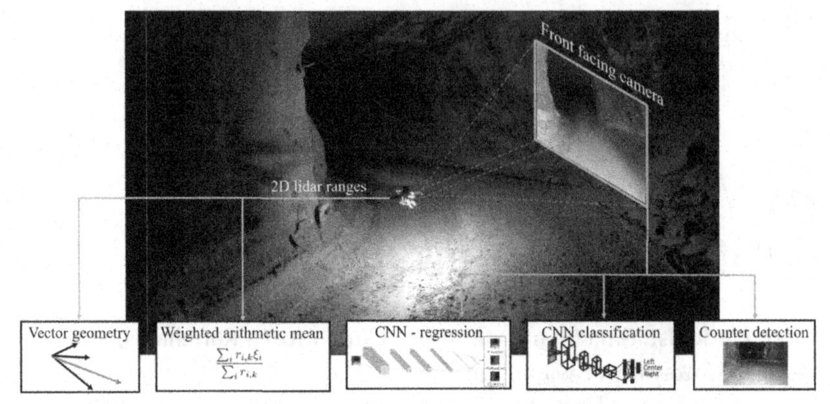

FIGURE 12.1 Photo of an underground mine in Sweden for indicating uneven surfaces and lighting condition in a mine tunnel, while the ARW is equipped with artificial light source. The concept of the five presented methods for defining heading angle is shown in white blocks.

and discussion in Section 12.6. Finally, Section 12.7 concludes the chapter by summarizing the findings and offering directions for future research.

12.2 Problem statement and open challenges

The mining industry targets the integration of the ARWs in application scenarios related to the inspection of production areas, like underground tunnels. In these cases, the inspection task includes deploying an aerial platform, endowed with a sensor suite, to autonomously navigate along the tunnel and collect mine-oriented valuable information (e.g., images, gas levels, dust levels, 3D models). In the sequel, the captured data will be used from the mine operators for further analysis to determine the status of the inspected area.

The deployment of ARWs in harsh underground mine environments poses multiple challenges [1]. In general, to increase the level of autonomy and provide stability and reliability in their operation, ARWs are equipped with high-end and expensive components. Nonetheless, the main aim for these platforms is the long-term underground operation, where their performance and integrity degrade over time [2], and they need to be maintained/replaced. This can be a bottleneck for expensive solutions since the maintenance/repairing costs accumulate to high levels. Therefore, the overall vision is to consider the aerial vehicles as consumable task-oriented scout platforms designed to accomplish specific needs in subterranean environments. These platforms could be deployed for various application scenarios, such as navigating quickly and providing a rough topological map or visual feedback of the area to the operator, or navigate and collect data from harsh areas.

The core part to enable autonomous navigation is considered to be the state estimation, which provides information of the local "positioning" of the ARW, e.g., position, orientation, velocities, etc. Due to limitations posed from both the mine environment and the platform, separate sensors have been assigned to specific states: 1) attitude $[\phi, \theta]^\top$ obtained from the onboard Inertial Measurement Unit (IMU), 2) linear velocities $[v_x, v_y]^\top$ from optical flow, and 3) height z estimation from single beam lidar. This architecture is characterized by low computational power, and thus, it results in the saving of computational resources for other components. Once state information is available, the corresponding navigation and control components could use it to generate and execute the navigation commands, respectively. The main challenge in the navigation commands is to extract proper heading to follow an obstacle free path along the tunnels. In this chapter, the heading commands are generated either from a lidar-based methods or from vision-based methods for finding open spaces. In the case of a 2D lidar, we consider two methods to find the open area: the first consists of summing vectors, and the second uses the weighted arithmetic mean of lidar ranges. In the case of a visual sensor, in the first method, we utilize the image stream to find the contour that captures the darkest area, corresponding to the continuation of the tunnel (mining drift). In the second method, we use a CNN approach that classifies images from a looking forward camera to three categories of *left*, *center*, and *right*. Finally, we introduce a regression CNN to provide the centroid position of open space from the onboard image. The controller has been designed to consider the ARW as a floating object on x and y axes, mainly selected to remove high dependencies on accurate localization schemes, while the platform is still able to perform the desired exploration task. An overview of the proposed concept for navigation is shown in Fig. 12.2.

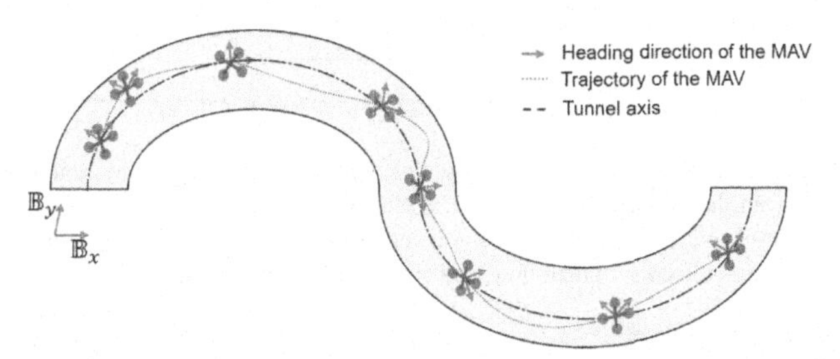

FIGURE 12.2 Top-view concept image of a mine tunnel with an ARW, which shows the necessity of heading correction in autonomous navigation. Body frame $\mathbb{B}_x$ and $\mathbb{B}_y$ are shown by red (dark gray in print) and green (lighter gray in print) arrows, respectively, while it is assumed that the heading of the ARW is toward the x-axis.

The dust is another challenge when flying in actual tunnels, but in our case, we have been visiting the production areas of mines where the ground was wet and only small dry parts were generating a negligible amount of dust. Therefore, in the presented field tests, although dust is a major issue, it was not considered since it heavily depends on the flight area.

Additionally, the potential fields method [1] is implemented to generate velocity references $[v_{r,x}, v_{r,y}]^\top$ to avoid collisions to the local surrounds using range measurements R of the 2D lidar placed on top of the ARW. Moreover, the same Nonlinear Model Predictive Control (NMPC) formulation is used for tracking the reference altitude and velocities without considering collision avoidance constraints. The NMPC objective is to generate attitude ϕ, θ and thrust commands T for the low-level controller $u = [T, \phi, \theta]^\top$, while the constant desired altitude z_d and $v_{z,d} = 0$ are defined from the operator. The low-level controller is integrated to the flight controller and Proportional Integral Derivative (PID) controllers are generating the motor commands $[n_1, \ldots, n_4]^\top$ for the ARW. The low-level attitude controller [3] is able to track the desired roll, pitch, and yaw rate and calculate the corresponding rotor speeds for the ARW. A block diagram representation of the proposed NMPC and the corresponding low-level controller is shown in Fig. 12.3, where I is the image stream from the looking forward camera, a_x, a_y, a_z, w_x, w_y, and w_z are linear and angular accelerations along each axis from IMU measurements, and v_x and v_y are linear velocities from a down-ward optical-flow sensor, and z is altitude from the single beam lidar.

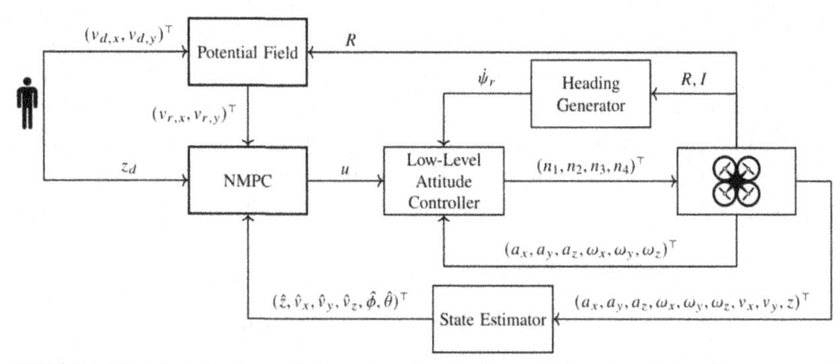

FIGURE 12.3 Control scheme of the proposed navigation module, where the heading commands are provided from the geometry approach. The NMPC generates thrust and attitude commands, while the low-level controller generates motor commands $[n_1, \ldots, n_4]^\top$. The velocity estimation is based on IMU measurements, optical flow, and downwards facing single beam lidar.

12.3 State estimation

Successful autonomous navigation of ARWs requires the combination of accuracy and low latency in the state estimation. Generally, for localization purposes, different sensor suites are used. The most common sensors are a combination of Global Positioning System (GPS) with IMU units; however, underground mines are GPS-denied environments. Moreover, motion capture systems provide localization with high precision and high sampling rate in GPS-denied areas, nonetheless these systems require multiple cameras in fixed positions to be installed beforehand, their performance is affected by dust and high humidity, and their cost is high, which makes them unsuitable for the case of the underground mine environments. Another group of localization systems are forward-looking monocular or stereo cameras for visual/visual-inertial pose estimation. These methods are widely used in indoor and outdoor environments; however, they rely on features from the environment, which can be limited in the case of a low-illumination environment. Additionally, due to payload and power constraints of ARWs, these algorithms should perform under computational limitations of embedded hardware [4].

In this work, the autonomous ARW is considered as a floating object that explores a 3D space. The measurement updates of the state vector $\boldsymbol{x} = [z, v_x, v_y, v_z, \phi, \theta]^\top$ is accomplished by assigning sensors to specific states. The altitude state z is provided from a downward pointing single beam lidar, while the linear velocity v_z is estimated from z using Kalman Filtering [5]. Moreover, ϕ and θ are provided from the flight controller through the filtering of IMU measurements from an Extended Kalman Filter (EKF). Finally, the states v_x and v_y are calculated from a down-ward optical-flow sensor. The optical flow sensor is equipped with an additional illumination source to avoid drifts in the measurements, while providing high update rates. To robustify the performance of the state estimation, the optical flow has been calibrated at a specific height, and the ARW is commanded to fly at a fixed altitude. More details about the software and hardware of the employed optical flow can be found in [6].

12.4 Navigation and collision avoidance

The potential field's objective is to generate linear x- and y-axis velocity commands in order to avoid collisions to the walls or any other obstacle standing in the way of the ARW. In this work, the potential field uses range measurements from 2D lidar placed on top of the vehicle to obtain repulsive velocity commands for both the x- and y-axis when flying close to obstacles, while the attractive velocity command is given a constant value on the x-axis. The potential field approach generates the desired velocities $v_{r,x}, v_{r,y}$ for the platform and is fed to the NMPC as depicted in Fig. 12.3.

In the classical potential field methods [1], the relative vector, between the agent position and destination, is required for obtaining the attractive force. However, in this article, the position estimation is not available; thus, it is assumed that the desired vector is constant and is toward the x-axis. The desired velocity vector is defined as:

$$\vec{\boldsymbol{v}} = [v_x, 0]^\top, \ v_x \in \mathbb{R}_{>0} \tag{12.1}$$

As mentioned above, the repulsive potential is structured around the 2D lidar information and is defined in the body frame of the aerial vehicle. Before describing the algorithm, it is essential to introduce the notation used in this case. Initially, the measurements are transformed from ranges into a 2D pointcloud Λ, removing invalid values. The ARW is considered always to lie at the origin of Λ, expressed always through $\vec{\boldsymbol{p}}_s = [0, 0]^\top$. For every time instance k the pointcloud is updated according to the new range measurements. For simplicity reasons, the time instance indexes are omitted in the following equations. Every point registered in Λ is considered an obstacle and is described by its 2D position $\vec{p}_o = [x, y]^\top$. Based on the $\vec{p}_s$ and $\vec{p}_o$, the repulsive potential is calculated, and the desired velocities are obtained similar to work [1].

12.4.1 Vision based methods

12.4.1.1 CNN classification

A CNN [7,8] is composed of a series of nonlinear processing layers stacked on top of each other. The typical layers presented in a CNN are: the convolutional, the pooling, the fully connected, and the nonlinear activation layers. The main component of the CNN is the convolutional layer, which operates on the local volumes of data through convolutional kernels, also called filters, that extract feature representations (feature maps). The pooling layer reduces the size of the original image and the subsequent feature maps and thus providing translation invariance. The nonlinear activation layers (usually consisted of Rectified Linear Units (ReLUs) that have almost completely substituted the traditional sigmoid activation functions), as in the case of the conventional Neural Networks (NNs) allows to the CNNs to learn nonlinear mappings between the input and the output. The final layers of a CNN are "flat" fully connected layers identical to the ones used in conventional MultiLayer Perceptrons (MLPs). A CNN is trained "end to end" to learn to map the inputs to their corresponding targets using gradient-descent-based learning rules.

In this chapter, the same CNN structure as presented in Fig. 12.4 [9] is used, while the layers and implementation details are discussed comprehensively. The CNN receives a fixed-size image as an input and provides three categories for each image. Similar to most neural networks, a CNN has an input layer, an output layer, and many hidden layers in between. These layers learn features specific to the task by altering the input data. The main difference between these novel architectures is that the features are learned during the training process instead of relying on tedious feature engineering processes. In the case of CNNs, this is basically achieved via the convolution filters. Each convolution filters learns to be activated by certain features of the image. One advantage of convolution connections especially compared to a fully connected architecture is the dramatic reduction in number of parameters due to weight sharing. ReLUs has been used as an additional operation after each convolution operation. ReLUs is an element-wise operation and allows for faster and more effective training by mapping negative values to zero and maintaining positive values. ReLUs introduces nonlinearity to the CNN, since most of the real-world data would be nonlinear and convolution is a linear operation. Another layer, which is not part of NNs, is the pooling layer. Pooling performs nonlinear down-sampling at the same time reducing the number of parameters that the network needs to learn while retaining the most important information. These operations are repeated over a large number of layers, with each layer learning to identify different features and extract the useful information from the input image, introduce nonlinearity, and reduce feature dimensions. The outcome of convolution and pooling layers represent high-level features of the input image; thus, the fully connected layer uses these information for classifying the images.

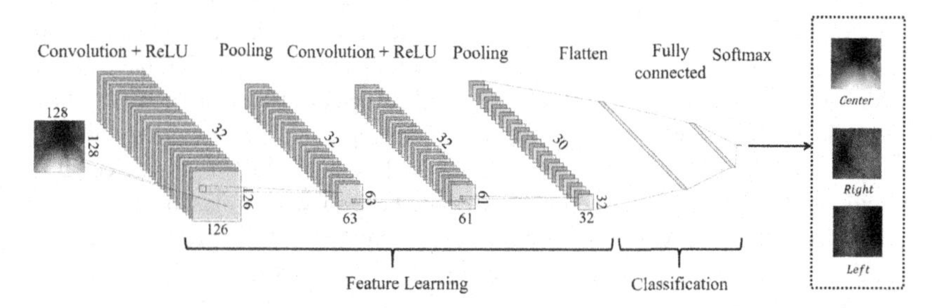

FIGURE 12.4 Architecture for the proposed CNN classification for correction of heading.

The input layer of the CNN is a matrix of 128 × 128 × 1, followed by 2D convolution layer with size of 2 × 2 and output of 126 × 126 × 32, followed by 2D max pooling with output of 63 × 63 × 32 and size of 2 × 2, then next 2D convolution layer with size of 3 × 3 to extract features with output of 61 × 61 × 32, after that the 2D pooling layer of size 3 × 3 with output of 30 × 30 × 32, followed by flatten layer and ended with fully connected layer with softmax activation for classifying images to three classes. Depending on onboard camera, the images can have different resolutions; however, to reduce computation time and size of data, the input image should be resized to 128 × 128 pixels. Moreover, object recognition based on gray-scale images can outperform RGB image-based recognition [10], and the RGB sensors do not provide any extra information about the dark mine environments. Thus, for reducing the computation time and noise, the images from the cameras are converted to gray-scale. Moreover, CNNs with large numbers of parameters and multiple nonlinear hidden layers are powerful techniques; however, the large networks require high computation power and suffer from the problem of overfitting. In the proposed method, the dropout [11] layer is added to reduce the overfitting during the training of a model and is not used when evaluating the trained model. The dropout layer removes units with all incoming and outgoing connections in the CNN architecture. Random dropout of neurons during training phase results in adaptation of other neurons to make prediction for the missing neurons. In this way, multiple independent internal representations are learned by the network. This results in a network with less size and less sensitivity to the specific weights of neurons. Moreover, during the training, the model cross-validation [12] method is used to reduce overfitting and provide models with less biased estimation. The training data set is split into subsets, and the model is trained in all subsets except one, which is used to evaluate the performance of the trained model. The process is continued until all subsets are evaluated as a validation set and the performance measure is averaged across all models.

For the proposed method, a loss function is categorical crossentropy, an optimization is Adam optimizer [13], and the network is trained on a workstation equipped with an Nvidia GTX 1070 GPU, which is located outside of the mine with 25 epochs and 200 steps per epoch. For each image, the CNNs provides the class of the image that can be *Left*, *Center*, *Right*. Each label represents the direction of the platform's heading, e.g., in the case of *Center*, the heading rate should be zero $\dot{\psi} = 0$, in case the of *Right*, the heading rate should have a positive value, in the case of *Left*, the heading rate should have a negative value to avoid the left wall. In the proposed CNN approach, the constant value of $\dot{\psi}_r = \{-0.2, 0, 0.2\}$ rad/sec is selected for each label; 0.2 rad/sec is selected to avoid sudden heading rate commands in the case of wrong classification from the CNN approach. Algorithm 1 provides an overview of the proposed method for generating heading rate commands based on the CNN classification method.

Algorithm 1 Calculate heading rate command based on the CNN approach.

Require: I
Ensure: $\dot{\psi}_d$
1: $I^{m \times n \times 3} \rightarrow I^{128 \times 128 \times 1}$ //converting the RGB image with $m \times n$ pixels to gray scale and resizing to 128×128 pixels
2: CNN($I^{128 \times 128 \times 1}$) $\rightarrow$ {*left*,*center*,*right*} //Output of the CNN
3: **if** CNN(I)==*left* **then** $\dot{\psi}_r = -0.2$ rad/sec
4: **else if** CNN(I)==*center* **then** $\dot{\psi}_r = 0.0$ rad/sec
5: **else** $\dot{\psi}_r = 0.2$ rad/sec
6: **end if**

12.4.1.2 *CNN regression*

A regression CNN method is proposed to enable autonomous navigation with a low-cost platform in unknown dark underground mines. Initially, the collected data sets are prepossessed in order to obtain the depth information by utilizing the work reported in [14]. The depth image is then segmented into regions, and the region with the highest depth is extracted, while in the sequel, the centroid of this region is calculated. This is a novel way to generate multiple training data sets for the CNN, when an absolute reference is not available, and the access to the field is limited. The trained regression CNN provides heading rate commands by extracting the centroid position in the horizontal axis of the image plane from a looking forward camera. Moreover, the trained regression CNN requires less computation power when compared to depth map estimation methods and provides online performance for enabling autonomous navigation of the ARW.

12.4.1.2.1 Centroid extraction

In this work, the CNN has been trained using information of the open space along the tunnel in sequential frames. This information is expressed through the extraction of the centroid of the identified free tunnel space. The overall concept is based on the depth map estimation of the scene using a single acquired image, while an image can be expressed using the atmospheric scattering model [15] as follows:

$$I_{observed}(x, y) = I_{initial}(x, y) \cdot T(x, y) + l[1 - T(x, y)] \tag{12.2}$$

where $I_{observed}$ is the observed image, $I_{initial}$ is the original image, l is the atmospheric light, $T(x, y)$ is the transmission term, (x, y) are the pixel coordinates where $x = 1, ..., M$ and $y = 1, ..., N$ with M the width and N the height of the image. The first term $I_{observed}(x, y) \cdot T(x, y)$ is called direct attenuation [16], and the second term $l[1 - T(x, y)]$ is airlight. The transmission term is usually defined as [17]:

$$T(x, y) = e^{-\theta D(x,y)} \tag{12.3}$$

where θ is the scattering coefficient of the atmosphere, and $D(x, y)$ is the depth of the scene for pixel coordinates (x, y). A method that is usually used to compute the transmission term is the Dark Channel Prior (DCP) method [18].

After the depth map has been computed, we perform a gray-scale morphological closing in order to smooth the image [19]. Then the *K-means* clustering algorithm [20] detects a number of clusters in the processed image. From the clusters, we choose the one with the minimum average intensity. For this cluster, we compute the centroid [21]. The centroid of this cluster will be at maximum distance from the camera as preliminary presented in [22].

A brief overview of the method decomposed in the number of steps can be seen below.

Algorithm 2 Calculate the heading rate correction based on the centroid extraction.

Input: RGB image acquired by the forward looking camera.

- Input: RGB image acquired by the forward looking camera.
  Start
- Step 1. Convert the RGB image to grayscale.
- Step 2. Extract the depth image.
- Step 3. Perform grayscale morphological close and cluster the result image with K-means.
- Step 4. For the cluster of minimum average intensity extract the centroid (s_x, s_y).
  End
- Output: Pixel coordinates of the heading point (s_x, s_y).

12.4.1.2.2 CNN architecture

The Deep Learning (DL) methods require large amount of data in order to train the model; however, proper data sets are not available in all the real-life application scenarios. Thus, the centroid information from the collected underground mine data sets is extracted off-line for training the CNN. In this way, data sets without depth information can be used, and it is not needed for collecting data sets with depth information, especially, when there is a limited access to the field. The overview of the proposed method is depicted in Fig. 12.5.

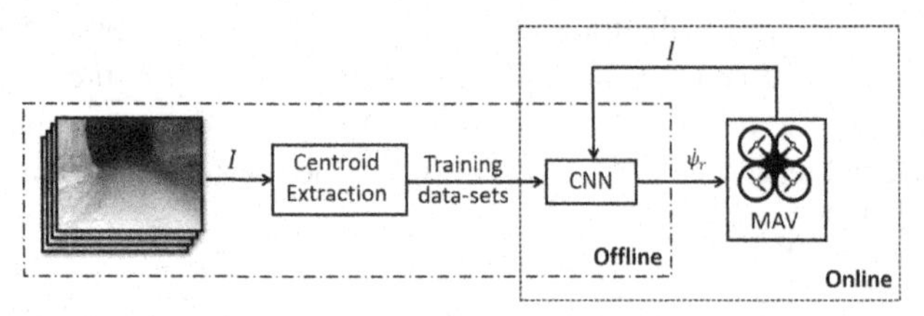

FIGURE 12.5 Overview of the proposed method, while the training data sets are generated off-line by centroid extraction method, and the CNN provides online heading rate commands based on images from the looking forward camera.

Fig. 12.6 depicts the architecture of the proposed CNN [7], while it receives a fixed-size image as an input and provides the centroid position of the tunnel open space. A CNN is composed of an input layer, a number of hidden layer and an output layer. The main advantage of these Neural Networks is the fact that they do not rely on feature extraction from the images, but the features are extracted automatically and learned during the training process. This is achieved via the convolution operation that is employed with different type of filters in the initial image and allows a number of features to be extracted from the initial image. Furthermore, the convolution operation reduces the number of parameters due to weight sharing. Finally, the extra layer of pooling simplifies the output by nonlinear down sampling using. e.g., the ReLUs function.

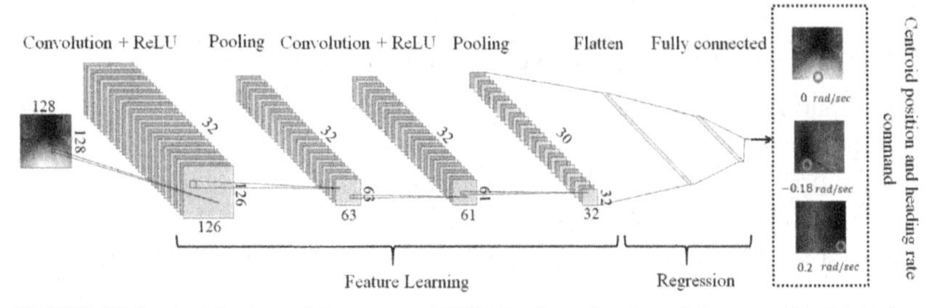

FIGURE 12.6 Architecture of the proposed CNN for the estimation of the centroid with highest depth.

The input layer of the CNN is a matrix of 128 × 128 × 1, followed by a sequence of two 2D convolutional and pooling layers as feature extractors, a fully connected layer to interpret the features, and with a dropout layer to reduce the over fitting [11], and, finally, an output layer with a *sigmoid* activation [23] to provide outputs between 0 and 1. Depending on the ARW equipped camera, the input image of the CNN can have different sizes; however, to reduce the computational power, the image stream of the camera is resized to 128×128 pixels, while for off-line centroid extraction, the data set is resized to 512×512 pixels for providing better information, then the centroid position s_x is mapped to [0,1] ($[0, 511] \rightarrow [0, 1]$) for training the CNN.

Thus, the output o_{cnn} of the CNN is a continuous value between $[0-1]$ for representing the centroid position s_x, e.g., 0, 0.5, and 1 are the location of the centroid in the left corner ($s_x = 0$), center ($s_x = 63$) and right corner ($s_x = 127$) of the image with 128×128 pixels, respectively. Then the output of the CNN is mapped to the heading rate command ($[0, 1] \rightarrow [-0.2, 0.2]$ rad/sec), where the heading rate of -0.2 rad/s, -0.0 rad/s, and 0.2 rad/s corresponds to the centroid in the left, center, and right corner in the image plane. Algorithm 3 provides the overview of the proposed method for generating heading rate commands. The CNN has been implemented in Python by Keras [24] as a high-level neural network Application Programming Interface (API). The loss function is Mean Absolute Percentage Error (MAPE), the optimization is based on an Adam optimizer [13], the learning rate is 0.001, and the learning rate decay is 5×10^{-6} over each update. Finally, a workstation has been utilized and equipped with an Nvidia GTX 1070 GPU for the training of the network with 200 epochs and 150 steps per epoch, while the trained network is evaluated online on the onboard ARW main processing unit. To train the CNN, the data sets collected from moving the camera by an operator in different directions and from flights in the underground mine are used. For training the CNN 5067 images, corresponding to 50 m tunnel length is selected, while the data set is shuffled, and 70% are used for training and 30% for the validation in an off-line procedure. The trained network provides MAPE of 10.4% and 12.9% on training and validation data sets.

Algorithm 3 Calculate heading rate command.

Require: I

Ensure: $\dot{\psi}_d$

1: $I^{m \times n \times 3} \rightarrow I^{128 \times 128 \times 1}$ //converting the RGB image with $m \times n$ pixels to gray scale and resizing to 128×128 pixels

2: $\text{CNN}(I^{128 \times 128 \times 1}) \rightarrow o_{cnn} \in [0, 1]$ //Output of the CNN

3: $\dot{\psi}_r = \frac{o_{cnn} - 0.5}{2.5}$ //Mapping $[0, 1] \rightarrow [-0.2, 0.2]$ rad/sec

12.5 Results

This section describes the experimental setup and the experimental trials performed in the underground environment. Each method is evaluated separately in different locations and for different periods of time. Thus, minor modification and updates can be noticed on the hardware and software components for every experimental trial. Initially, we present the overall experimental setup, while in the sequel each modification for the different methods is explained in detail before the presentation of the results.

The following links provide the video summary of proposed methods.

- ARW navigation with potential field and vector geometry method for heading correction.
  Link: https://youtu.be/sW35Q3wVpI0
- ARW navigation with NMPC with 2D collision avoidance constraints and weighted arithmetic mean method for heading correction.
  Link: https://youtu.be/-MP4Sn6Q1uo
- ARW navigation with potential field and CNN classification method for heading correction.
  Link: https://youtu.be/j3N8ij9MfSA
- ARW navigation with potential field and CNN regression method for heading correction.
  Link: https://youtu.be/WKHEvcovXqk

12.5.1 Experimental setup

In this work, a quadcopter developed at Luleå University of Technology based on the ROSflight [25] flight controller is used. The vehicle's weight is 1.5 kg and provides 8 min of flight time with 4-cell 1.5 hA LiPo battery. The flight controller is ROSflight, and the Aaeon UP-Board[1] is the main processing unit, incorporating an Intel Atom x5-Z8350 processor and 4 GB RAM. The operating system running on the board is Ubuntu Desktop to which Robot Operating System (ROS) has also been included. Regarding the sensor configuration, a 2D rotating lidar is placed on the top of the vehicle. The velocity estimation is based on the PX4Flow optical flow sensor at 20 Hz, while the height measurements are provided from the single beam Lidar-lite v3 at 100 Hz, both installed on the bottom of the vehicle pointing down. Furthermore, the aerial platform is equipped with two 10 W LED light bars in both front arms for providing additional illumination for the forward-looking camera and LEDs looking down to provide additional illumination for the optical flow sensor. Fig. 12.7 presents the platform at different viewpoints, highlighting its dimensions and the sensor configuration.

12.5.2 Lidar based methods evaluation

12.5.2.1 Vector geometry based approach

The location of the field trials was 790 m deep in an underground mine in Sweden without any natural illumination sources. Furthermore, the underground

[1] https://www.aaeon.com/en/p/up-board-computer-board-for-professional-makers.

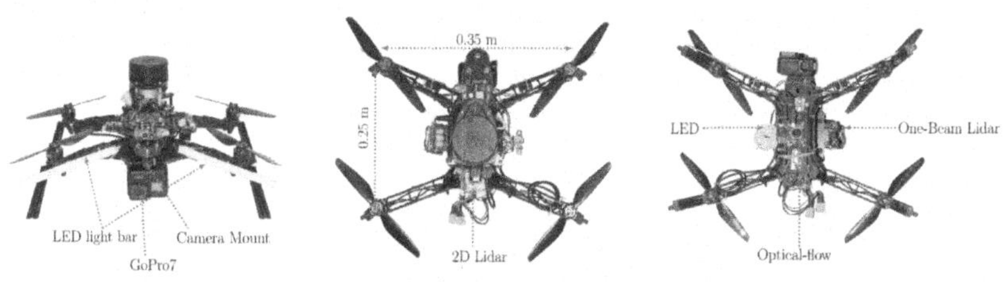

FIGURE 12.7 The developed quadcopter equipped with a forward looking camera, a LED lights, optical flow, 2D lidar, and single beam lidar.

tunnels did not have strong corrupting magnetic fields, while their morphology resembled an *S* shape environment with small inclination. Overall, the field trial area had width, height, and length dimensions of 6 m, 4 m, and 150 m, respectively. Fig. 12.8 depicts one part of the underground mine with uneven surfaces.

FIGURE 12.8 Photo of 790 m underground mine in Sweden.

The controller has been tuned to compensate light wind-gusts that were present in the tunnel; however, no wind measurement device was available to measure the speed of the wind.

The tuning parameters of the NMPC is presented in Table 12.1. The NMPC prediction horizon N is 40, and control sampling frequency T_s is 20 Hz. While the parameters of the ARW model is presented in Table 12.2.

In this case, the 2D lidar measurements are used for both potential fields and heading corrections, the optical flow sensor provides velocity state estimation, and the single beam lidar provides altitude estimations. In order to reduce 2D lidar measurement, noise, and uncertainties for calculating the heading rate, rather than relying on only one beam, an array of beams are selected and passed through a median filter. Moreover, the potential field module is relying on 360

TABLE 12.1 The tuning parameters of the NMPC.

Q_x	Q_u	$Q_{\Delta u}$	T	ϕ_{min}	ϕ_{max}	θ_{min}	θ_{max}
$[0, 0, 6, 2, 2, 3, 3, 3]^{\top}$	$[2, 10, 10]^{\top}$	$[0, 0, 0]^{\top}$	$[0, 1] \cap \mathbb{R}$	−0.4 rad/s	0.4 rad/s	−0.4 rad/s	0.4 rad/s

TABLE 12.2 The parameters of the ARW model.

g	A_x	A_y	A_z	K_ϕ	K_θ	τ_ϕ	τ_θ
9.8 m/s^2	0.1	0.1	0.2	1	1	0.5 s	0.5 s

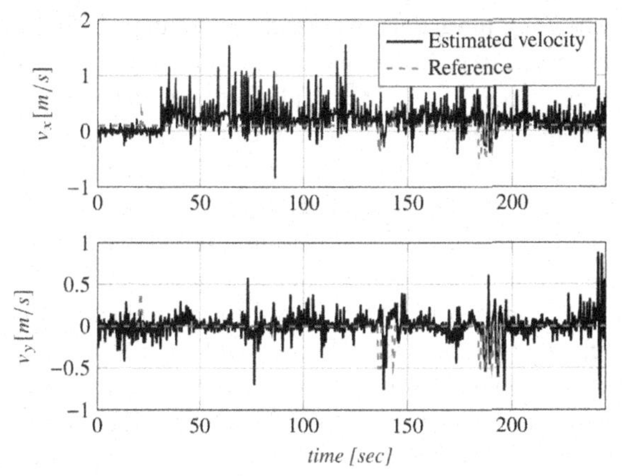

FIGURE 12.9 The velocity commands from the potential field, while the ROSflight platform navigates on the basis of potential field and vector geometry modules.

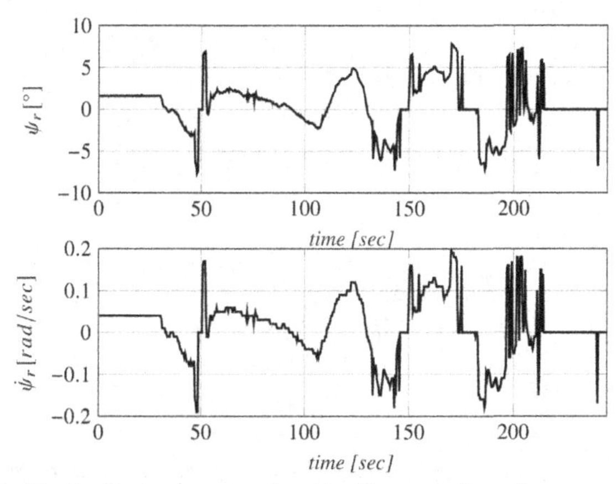

FIGURE 12.10 The heading rate command and heading angle from the vector-geometry-based approach, while the ROSflight platform navigates based on potential field and vector geometry modules.

beams of 2D lidar measurement. The maximum range of 2D lidar is 4 m, and the safety distance d_s for potential field is 1.5 m, which enables the repulsive forces, while the vector geometry module is always active. Moreover, the desired constant altitude of $z_r = 1$ m is considered for the ARW. Fig. 12.9 depicts the generated $v_{r,x}$, $v_{r,y}$ commands from the potential fields. The desired $v_{d,x}$ velocity is constant at 0.1 m/s, while the changes in $v_{r,x}$ and $v_{r,y}$ are due to repulsive forces from the tunnel walls.

Fig. 12.10 depicts the ψ_r and $\dot{\psi}_r$ from the vector geometry modules. It can be seen that when the $v_{r,y}$ has negative values, the $\dot{\psi}_r$ has positive values which means the platform is close to the left wall, trying to move away from the left wall with the heading toward the other side of the tunnel. Additionally, it can be seen that in some cases, the value of $v_{r,x}$ is changing as the ARW is looking at the walls, and the potential field provides different velocities. Moreover, due to $d_s = 1.5$ m in the potential field, the commands are mainly constant until the ARW is close to the obstacles, while the heading commands are always correcting the ARW to look at open areas. The Mean Absolute Error (MAE) between v_x, v_y and $v_{r,x}$, $v_{r,y}$ are 0.25 m/s and 0.13 m/s, respectively. The velocity measurements provided from the optical flow sensor are noisy, but it does not affect the overall navigation mission. Moreover, the sensor placement and parameter tuning significantly influence the performance of the measurements.

During the field trials, it was observed that the platform heading was corrected toward the open area. Nevertheless, due to the combination of the tunnel width with the limitations of the 2D lidar, the ARW path followed an $\mathbb{S}$ shape.

12.5.2.2 Weighted arithmetic mean approach

In this case, the ARW was evaluated in two tunnels with different dimensions. The β and k_p for the heading generator approach were 0.95 and 0.03, respectively. In the first tunnel environment, the ARW autonomously navigates with $z_r = 1.0$ m, $v_{x,r} = 0.5$ m/s, $v_{y,r} = 0.0$ m/s. Fig. 12.11 provides the heading rate generated, which guides the ARW toward open spaces.

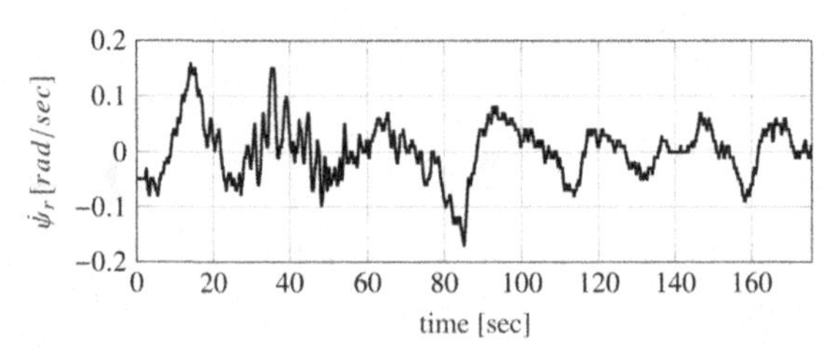

FIGURE 12.11 The heading rate command from the weighted arithmetic mean approach in the first tunnel environment, while the NMPC avoids obstacles.

In the second tunnel environment, the ARW autonomously navigates with $z_r = 1.0$ m, $v_{x,r} = 1.2$ m/s, $v_{y,r} = 0.0$ m/s. The generated heading rate command is depicted in Fig. 12.12. Moreover, the method is evaluated in blockage environment, while the constant references of $z_r = 1.0$ m, $v_{x,r} = 0.5$ m/s, $v_{y,r} = 0.0$ m/s are fed to the NMPC. When the ARW reaches the end of the tunnel, the return command is transmitted and the ARW returns to the starting point and passes the blockage again. This test shows the applicability of the method in applications regarding underground mines to navigate in blocked areas and return to the base for the reports. The generated heading rate command is presented in Fig. 12.13, while due to strong magnetic field and disturbances in the IMU, a high peak heading rate command is instantaneously calculated around 42 s as shown in Fig. 12.13 without affecting the navigation task.

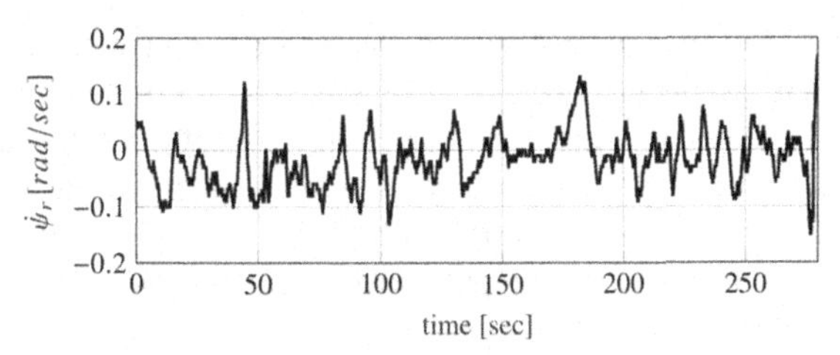

FIGURE 12.12 The heading rate command from a weighted arithmetic mean approach in the case of the second tunnel environment navigation, while the NMPC avoids obstacles.

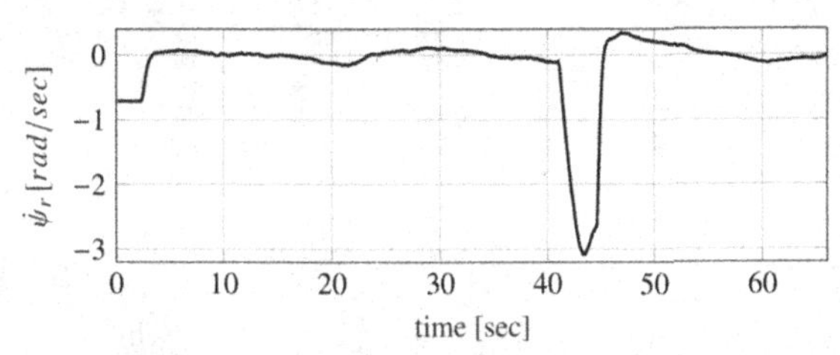

FIGURE 12.13 The heading rate command from the weighted arithmetic mean approach in the tunnel blockage environment navigation, while the NMPC avoids obstacles.

12.5.3 Vision based methods evaluation

12.5.3.1 Darkness contours detection

The performance of the proposed method is evaluated in an underground tunnel located at Luleå Sweden with lack of natural and external illumination in the tunnel. The tunnel did not have corrupting magnetic fields, while small particles were in the air. The tunnel morphology resembled an *S* shape and the dimensions of the area, where the ARW navigates autonomously, were 3.5(width) × 3(height) × 30(length) m^3. The platform is equipped with a PlayStation Eye camera with a resolution of 640 × 480 pixels and 10 fps. The front LED bars provide illumination of 460 lux from 1 m distance. The desired altitude and velocities for the ARW were set to 1 m, and $v_{d,x} = 0.5$ m/s, $v_{d,y} = 0.0$ m/s, respectively. The potential field was active for avoiding collision to the walls. Fig. 12.14 shows the heading rate command generated by the proposed method for the ARW.

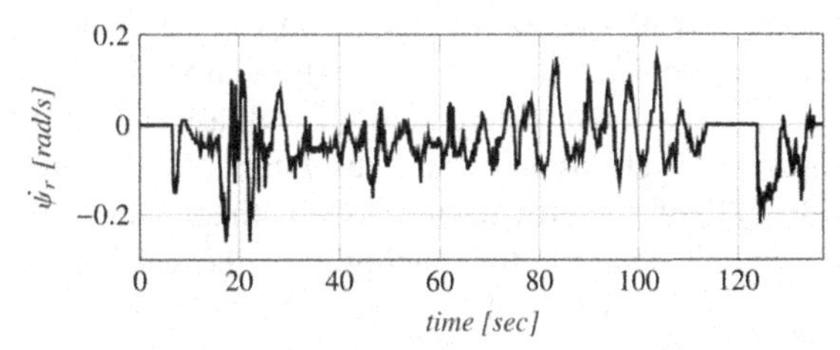

FIGURE 12.14 The heading rate commands generated from the darkness contours extraction method.

In Fig. 12.15, some examples from the onboard image stream during the autonomous navigation are depicted, while the centroids of the darkest contours are shown. Moreover, it is observed that in the case of branches in the tunnel, the proposed method cannot recognize them and select the darkest branch as darkest contour or combine both branches. As an example, Fig. 12.16 depicts tunnel with two branches and the proposed method consider both branches as one.

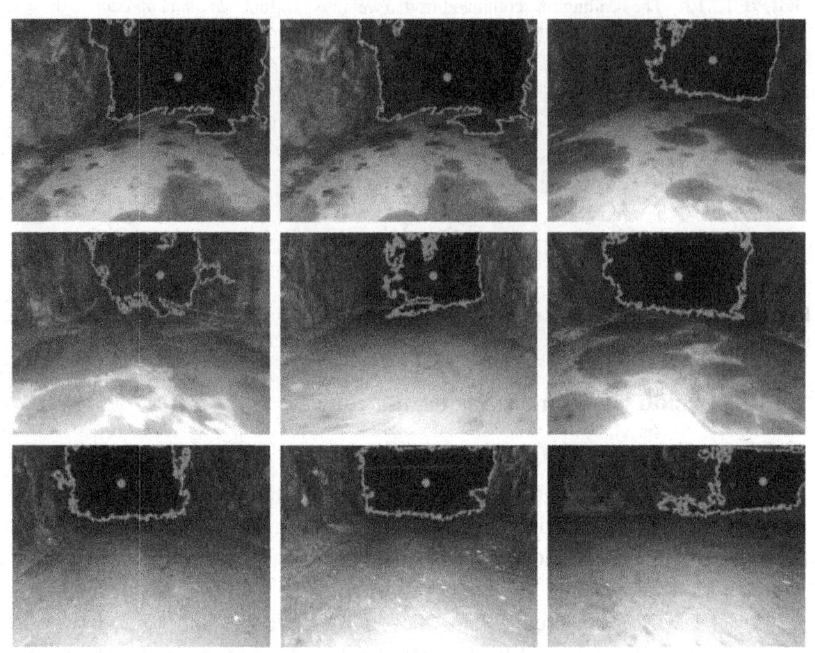

FIGURE 12.15 Sample images of the onboard forward looking camera, while the boundaries of the darkest contour is shown by red (dark gray in print) color.

12.5.3.2 CNN classification approach

The same platform and environment, as in Section 12.5.2.1, is used for evaluation of the potential field and the CNN classification approach. The ARW desired altitude is 1 m with constant desired $v_{d,x} = 0.1$ m/s. Additionally, the platform is equipped with PlayStation Eye Camera to provide image data for the CNN module as depicted in Fig. 12.17. The PlayStation Eye camera was attached to the camera mount, facing forward with a weight of 0.150 kg, this camera was operated at 20 fps and with a resolution of 640×480 pixels. Furthermore, the LED light bars provide 460 lux illumination in 1 m distance.

The CNN requires data sets for training, thus the setup in Fig. 12.18 has been used. The setup consists of three mounted cameras with separated LED light bars pointing toward the field of view of each camera. Different cameras, including Gopro Hero 7, Gopro Hero 4, and GoPro Hero 3, are used to reduce

FIGURE 12.16 Tunnel with two branches and the darkest area is extracted and is shown by red (dark gray in print) boundaries.

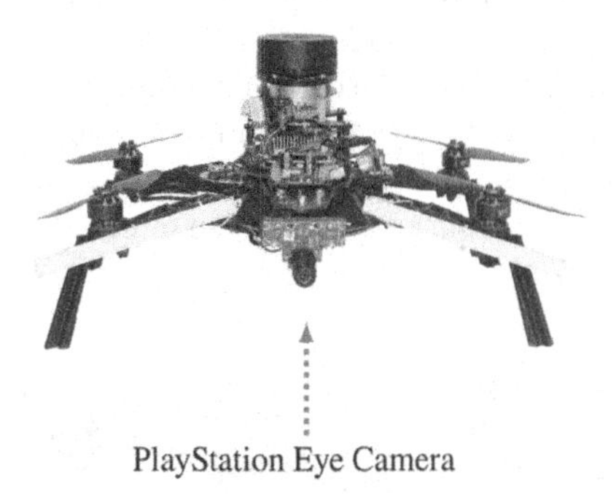

FIGURE 12.17 The ROSflight based platform equipped with PlayStation Eye camera.

the dependency of training data sets to specific cameras. Additionally, the light bars were calibrated to provide an equal illumination power, while data sets are collected with different illumination levels of 460 lux and 2200 lux from 1 m distance to consider uncertainties in illumination. During the data collection, each camera has a different resolution and Frame Per Second (fps) of 3840 × 2160 pixels with 60 fps, 3840 × 2160 pixels with 30 fps, and 1920 × 1080 pixels with 30 fps for Gopro Hero 7, Gopro Hero 4, and GoPro Hero 3, respectively. To reduce the redundancy of the images, the videos are down-sampled to 5 fps and converted to a sequel of images, however, without loss of generality faster fps could be selected. The images are re-sized to 128 × 128 pixels, converted to gray-scale mode and labeled based on the direction of the camera. During the data set collection, the triple camera setup is carried by a person, while guaranteeing that the middle camera is always looking toward the tunnel axis, and the camera is held 1 m above the ground.

FIGURE 12.18 The triple camera setup for obtaining training data sets for CNN.

Furthermore, the collected data sets are carefully selected to be not from the same tunnel area, which the ARW should be evaluated in for autonomous navigation. There has not been applied any camera/rig calibration process during the training process of the CNN. The captured images from the triple camera set have been resized and provided as input to the training phase of the learning workflow without any rectification thus with their fisheye distortions. Fig. 12.19 depicts the sample images from an underground mine collected by triple camera setup for training the CNN.

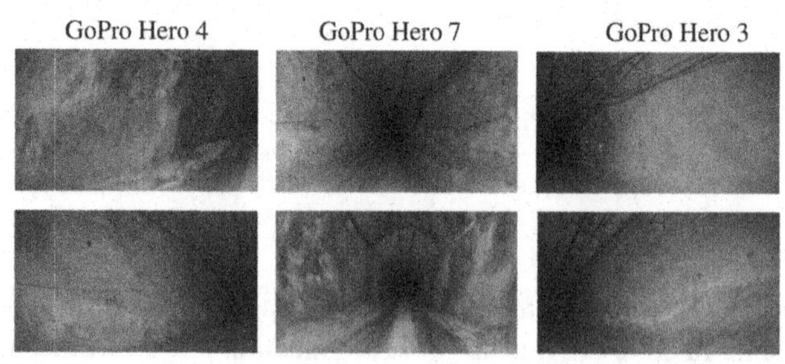

FIGURE 12.19 Examples of collected images from underground mine used for training the CNN. The left, center, and right images are from cameras looking toward left, center, and right, respectively.

After training, the CNN uses the obtained data sets on the workstation outside of the mine. The online video stream from the PlayStation Eye camera is re-sized, converted to gray-scale mode and fed to the CNN for generating the heading rate commands for the ARW to navigate autonomously in the mine. The output of the CNN is the class of the image, while for each label constant heading rate commands are generated for the ARW. The −0.2 rad/s, 0.0 rad/s, and 0.2 rad/s are the heading rate commands that correspond to the *left*, *center*, and *right* labels, respectively. Fig. 12.20 shows sample images from the PlayStation Eye camera, where the class is written above each image.

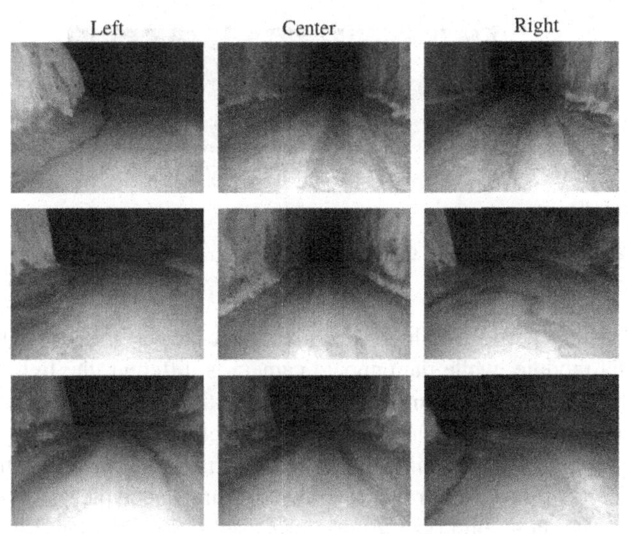

FIGURE 12.20 Examples of images from PlayStation Eye camera, while the ARW performs autonomous navigation in underground mine. The class of images from the CNN is written above the images.

A merit of the proposed CNN navigation method is the applicability of the network in unexplored areas of the similar structure. More specifically, the trained network can provide accurate classification of tunnel areas that have not been included in the training phase or have different surfaces with different level of illumination. Therefore, the CNN can be deployed in new tunnels for which it has not been trained for, reducing the complexity and time of the inspection task.

In this case, similar to Section 12.5.2.1, the potential field is active when d_s is less than 1.5 m. In the summary video, it can be seen that the platform heading is corrected toward the center of the tunnel, and the platform follows almost a straight line as the image streams provide sufficient information about the walls and center of tunnel. However, the heading of the ARW changes frequently with 2 [Hz] as depicted in Fig. 12.21.

This is due to the width of the tunnel and the corresponding camera field of view, small heading angle rotations results in different label of the images, e.g., from *right* label to *left* label.

Moreover, the Parrot Bebop 2 [26] is used for autonomous navigation in the mine, and only the CNN module for correcting the heading with velocity $v_{d,x} = 0.1$ m/s and constant altitude of $z_d = 1$ m is tested. The illumination of 310 lux from 1 m distance from the source is provided for this platform. The area differs from the collected training data sets and the same trained network as used

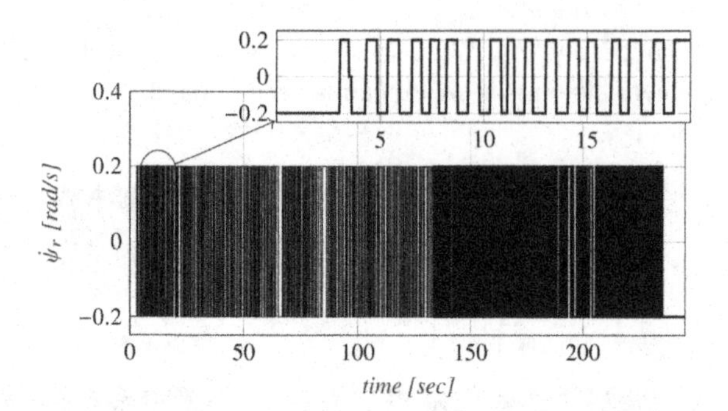

FIGURE 12.21 The $\dot{\psi}_r$ generated from CNN classification method.

before. The main purpose of this case is to evaluate the performance of the stand-alone CNN module, while the onboard camera is different, the light source is provided by a person following the platform, and the potential fields component is not available. Fig. 12.22 shows the examples of classification images, while flying autonomously in the mine. It can be seen that some classes of images are not identical, but the CNN provides correct heading and avoids collision to the mine walls. It can be observed that the proposed CNN obtained the necessary information for correct image classification without any brightness adjustment

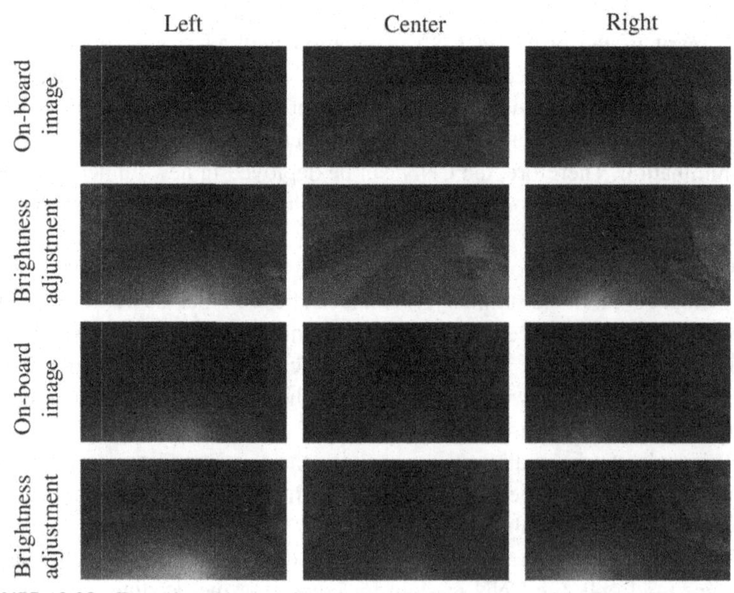

FIGURE 12.22 Examples of images from forward-looking camera of the Parrot Bebop 2, while the ARW performs autonomous navigation in the underground mine. The class of images from the CNN is written above the images. The brightness adjustment with Adobe Photoshop [27] is applied off-line for better visibility of images.

of the images. The frequency of heading rate commands generated from the CNN module is 4 Hz as depicted in Fig. 12.23. The heading rate command frequently changes for the same reasons mentioned before.

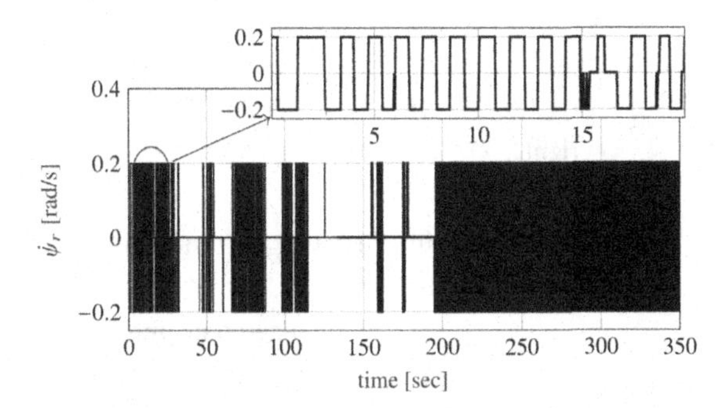

FIGURE 12.23 The $\dot{\psi}_r$ generated from CNN classification with Bebop platform.

In order to test the general applicability of the trained CNN, data sets from [28] are used to evaluate the performance of the method. The data sets are labeled to three categories, and the two underground tunnels are located in Luleå and Boden in Sweden with a different size dimension and structure when compared to the Boliden mine environment. Fig. 12.24 provides photos from two underground tunnels.

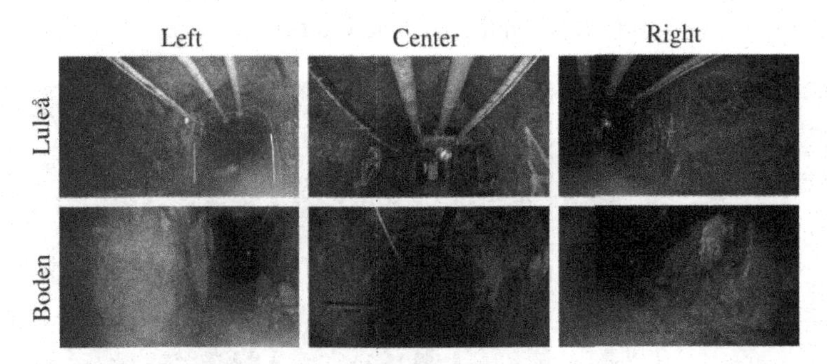

FIGURE 12.24 Photos from an underground tunnels in Sweden.

The trained CNN provides a 78.4% and 96.7% accuracy score for Luleå and Boden underground tunnels, respectively. The accuracy is calculated by the number of correct classified images over the total number of images. Moreover, Tables 12.3 and 12.4 provide the confusion matrix for each underground tunnel.

12.5.3.3 CNN regression approach

To train the CNN, two data sets are used, first the collected data set from moving the camera by an operator in different directions and secondly from looking forward camera of the ARW flights in the underground mine. Fig. 12.25 depicts

TABLE 12.3 The confusion matrix for the Luleå underground tunnel data set.

		Predicted outcome		
		left	center	right
Actual class	left	85.5%	3.1%	11.2%
	center	19.2%	66.8%	13.9%
	right	16.35%	0.7%	82.8%

TABLE 12.4 The confusion matrix for the Boden underground tunnel data set.

		Predicted outcome		
		left	center	right
Actual class	left	97.6%	2.4%	0%
	center	0.6%	98.1%	1.2%
	right	2.4%	2.4%	95.0%

few samples of the images, while the corresponding centroid location is also indicated. For training, the CNN 5067 images corresponding to 50 m tunnel length are selected, while the data set is shuffled, and 70% are used for training, and 30% for the validation in an off-line procedure. The trained network provides MAPE of 10.4% and 12.9% on training and validation data sets.

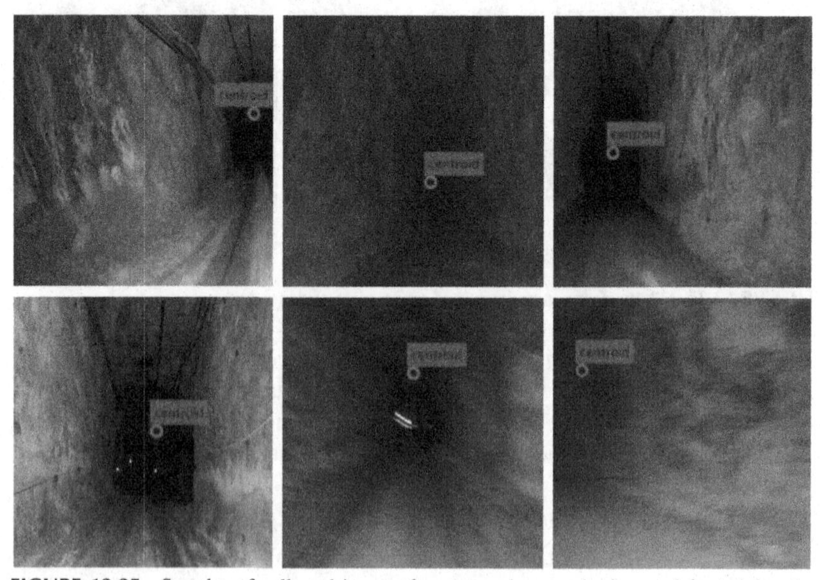

FIGURE 12.25 Samples of collected images from an underground mine used for training the CNN, while the centroid estimation is shown by a red (gray in print) circle.

The performance of the proposed CNN regression method is evaluated by the same hardware setup as in Section 12.5.3.2 and in the same tunnel environment as in Section 12.5.2.1. Fig. 12.26 shows the heading rate command generated by the CNN module, that due to the narrow width of the tunnel and the corresponding camera field of view, small heading angle rotations results in the replacement of the centroid in another direction, thus the heading rate commands are generated in different signs frequently.

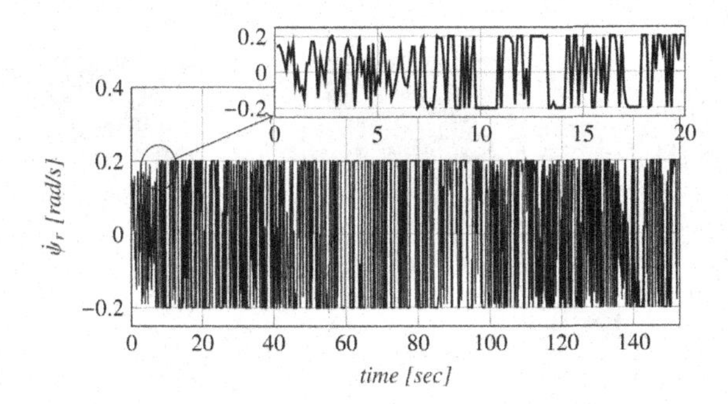

FIGURE 12.26 The heading rate commands generated from the utilized regression CNN.

In Fig. 12.27, some examples from the onboard image stream during the autonomous navigation are depicted, while the centroids obtained from the centroid extraction method and the CNN are compared. Moreover, it should be highlighted that the CNN only estimates the s_x centroid position, and the input image of the CNN is resized; however, for comparison, in the following figure, it is assumed that the s_y position of the centroid and resolution of the image are the same in both cases.

In few cases, the CNN failed to recognize the centroid position as well as branches in the tunnels as depicted in Fig. 12.28, which resembles a clear future area of research.

12.6 Lessons learned

Throughout the experimental trials for the underground tunnel inspection scenario, many different experiences were gained that assisted in the development and tuning of the algorithms utilized. Based on this experience, an overview of the lessons learned is provided below in relation to the different field algorithms used.

12.6.1 ARW control

Due to the nature of the floating object approach, identifying the explicit problem is challenging. However, there are multiple parameters that can affect the performance of the control scheme. The weights in the objective function have a direct impact on the controller's tracking performance, e.g., the higher weight

FIGURE 12.27 Comparison of s_x from CNN and centroid extraction method from ARW onboard camera. The CNN estimation is indicated by a blue (dark gray in print) circle, while the centroid extraction method is indicated by a red (lighter gray in print) circle.

FIGURE 12.28 Different centroid recognition in the case of mining tunnel branches. The CNN estimation is indicated by a blue (dark gray in print) circle, while the centroid extraction method is indicated by a red (dark gray in print) circle.

in altitude terms results in a smaller altitude error. Nonetheless, due to fluctuations in velocity estimation, eliminating the error in tracking cannot be obtained, while too large values on the weights results in an oscillation of the platform. Moreover, the physical constraints of the ARW, such as the maximum velocity, maximum acceleration, and bounds on angles, should be taken into consideration, based on the field trial conditions. As an example, due to the high wind gusts in the underground tunnels, the NMPC should have larger bounds on angles for better wind gust compensation. Nonetheless, it is observed that high wind gusts cannot be compensated even with maximum bounds on the angles, due to the larger wind velocity from the maximum velocity of the platform.

12.6.2 Localization

Overall, the performance was substantial, but alternative methods should be studied to improve the performance and stability of this particular component. More specifically, sensor fusion should be part of the localization module, combining the merits and redundancy of various sensors into a solid solution.

12.6.3 Navigation

Potential field: This component relies on 2D lidar and the update rate is limited to 5 Hz. Additionally, large angles for roll and pitch results in considering the ground as an obstacle.

Vector geometry based approach: This module, similarly to the potential field method, relies on a 2D lidar, thus the same limitations are observed. Moreover, due to the limited range of the 2D lidar and the low sampling rate, the generated heading rate commands result in an $\mathbb{S}$ shape path. Finally, it is observed that obtaining heading rate commands, based on only one beam measurement for each direction, is not reliable. Instead, an array of beams are selected for representing each direction, which results in smoother heading rate commands.

Weighted arithmetic mean approach: This module also relies on a 2D lidar, thus the same limitations are observed. However, the integrated z-axis of the gyro in the IMU results in better performance when compared to the Vector Geometry Based method.

Darkest contour extraction: This method relies on visual sensors and provides correct heading commands for collision avoidance. However, the main restriction of the method is relying on the dark environment. Thus, the present of the light sources in the environment will result in failure of the method.

CNN classification: The CNN relies on input images from a camera, thus the update rate is 30 Hz. However, the CNN requires training data sets, a LED light bar for the camera, and there is no guarantee that the proposed solution will work in completely different mine environments, thus there is a need for further investigation in this direction. Additionally, during the field trials, it has been observed that the stand alone CNN, in some cases, resulted in the ARW navigating close to the wall as depicted in Fig. 12.29, due to the fact that the CNN does not provide a y-axis reference velocity, and it only generates heading rate commands toward the tunnel axis. This is caused by the lack of a 2D lidar sensor on the platform, which results in unavailability of the potential field that is responsible for providing velocity commands on x- and y-axes toward the open area, while respecting the specified safety distance. However, this case is avoided utilizing a potential field or NMPC with collision avoidance constraints module on top of the CNN, while for such an operation, the CNN approach should be fused with a lidar, a fact that will increase the performance and the overall safety, but it will also decrease the payload. Finally, the method is providing discrete heading command rates compared to the other methods, which results in bang-bang heading movements.

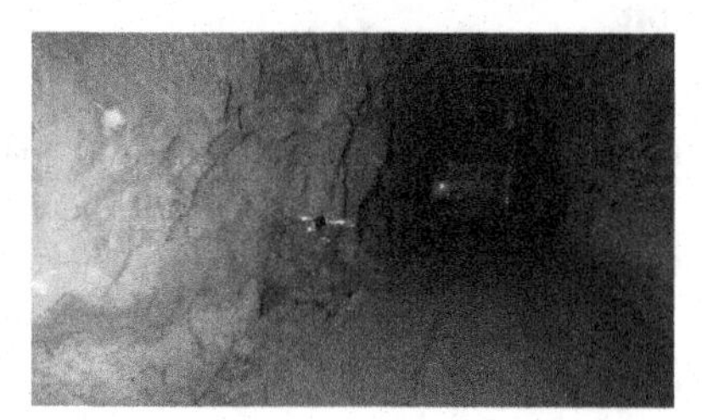

FIGURE 12.29 The ARW navigates close to the wall due to the stand-alone CNN module and the corresponding absence of a potential field module for obstacle avoidance in the Bebop case.

CNN regression: This method uses CNNs to extract the centroid of the open area. The data set is provided by the centroid extraction method to train the CNN. As a result, one can expect that the results of this method will be ideally as good as those of centroid extraction method standalone, and it cannot provide better results. Also, as in the CNN classification, we noticed a similar behavior in the heading rate commands due to the narrow width of the tunnel and the corresponding camera field of view; small heading angle rotations result in replacement of the centroid in another direction.

12.7 Conclusion

This chapter demonstrated the application of autonomous navigation for aerial vehicles in unknown underground tunnels with a resource-constrained low-cost platform. The presented navigation framework is evaluated in challenging real mining environments to evaluate the performance of the navigation modules. Moreover, the implemented control scheme ensures collision-free and optimal operations inside the mining environment. The framework is evaluated in mining tunnels without multiple branches and with the a width size in the range of a 2D lidar, thus, mainly, one open area can be recognized. Further work could focus on these kinds of environments as in Fig. 12.30 that depicts some schematics of these cases, where, in the first two cases, the ARW navigates in only one branch, and in the third case, the ARW is trapped in a void area. It should be highlighted that while the presented method provides collision-free paths, it cannot guarantee the full coverage of the area and there is a need for a higher-level mission planner.

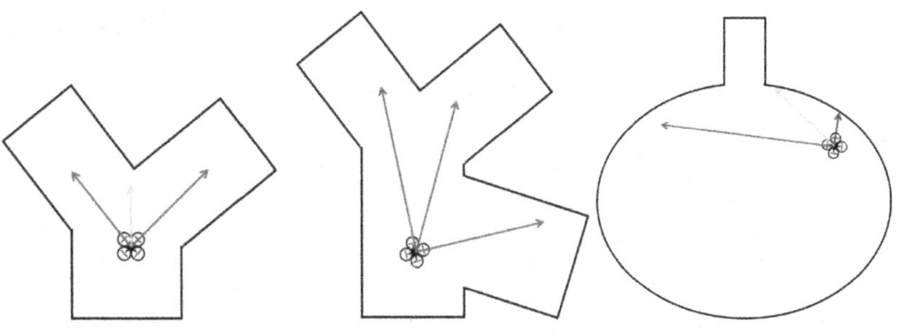

(a) Tunnel with multiple branches. (b) Tunnel with multiple branches. (c) Void area.

FIGURE 12.30 Schematic of the possible tunnel and void structures. The color of the vectors represent their distance to the obstacles, while green (gray in print), yellow (light gray in print), and red (black in print) are large, medium, and small distances to obstacles.

Furthermore, during the field trials, small particles presented in the air and small rocks on the ground moved due to the airflow generated from the ARW. Fig. 12.31 depicts some of the looking forward camera images in dusty areas visited by the drones. In all these cases, the aerial platform was able to provide reliable sensor measurements and avoid collisions. However, for a general conclusion of the overall applicability of the proposed scheme in other types of mines (except iron ore mines) requires further experimentation and tuning as well as improvement of the utilized visual perception in combination with novel vision based strategies, issues that are all crucial and important aspects of future work in toward the fully autonomous and robust aerial platforms operating into mines.

FIGURE 12.31 Sample examples of images from the ARW's forward looking camera, while small particles were floating in the air.

References

[1] C. Kanellakis, S.S. Mansouri, G. Georgoulas, G. Nikolakopoulos, Towards autonomous surveying of underground mine using MAVs, in: International Conference on Robotics in Alpe-Adria Danube Region, Springer, 2018, pp. 173–180.

[2] A. Manukyan, M.A. Olivares-Mendez, T.F. Bissyandé, H. Voos, Y. Le Traon, UAV degradation identification for pilot notification using machine learning techniques, in: 2016 IEEE 21st International Conference on Emerging Technologies and Factory Automation (ETFA), IEEE, 2016, pp. 1–8.

[3] J.-J. Xiong, E.-H. Zheng, Position and attitude tracking control for a quadrotor UAV, ISA Transactions 53 (3) (2014) 725–731.
[4] J. Delmerico, D. Scaramuzza, A benchmark comparison of monocular visual-inertial odometry algorithms for flying robots, Memory 10 (2018) 20.
[5] S. Haykin, Kalman Filtering and Neural Networks, vol. 47, John Wiley & Sons, 2004.
[6] D. Honegger, L. Meier, P. Tanskanen, M. Pollefeys, An open source and open hardware embedded metric optical flow CMOS camera for indoor and outdoor applications, in: 2013 IEEE International Conference on Robotics and Automation (ICRA), IEEE, 2013, pp. 1736–1741.
[7] A. Krizhevsky, I. Sutskever, G.E. Hinton, ImageNet classification with deep convolutional neural networks, in: Advances in Neural Information Processing Systems, 2012, pp. 1097–1105.
[8] D. Cireşan, U. Meier, J. Schmidhuber, Multi-column deep neural networks for image classification, arXiv preprint, arXiv:1202.2745, 2012.
[9] S.S. Mansouri, C. Kanellakis, G. Georgoulas, G. Nikolakopoulos, Towards MAV navigation in underground mine using deep learning, in: 2018 IEEE International Conference on Robotics and Biomimetics (ROBIO), IEEE, 2018, pp. 880–885.
[10] H.M. Bui, M. Lech, E. Cheng, K. Neville, I.S. Burnett, Using grayscale images for object recognition with convolutional-recursive neural network, in: IEEE Sixth International Conference on Communications and Electronics (ICCE), 2016, pp. 321–325.
[11] N. Srivastava, G. Hinton, A. Krizhevsky, I. Sutskever, R. Salakhutdinov, Dropout: a simple way to prevent neural networks from overfitting, Journal of Machine Learning Research 15 (1) (2014) 1929–1958.
[12] Y. Bengio, Y. Grandvalet, No unbiased estimator of the variance of k-fold cross-validation, Journal of Machine Learning Research 5 (Sep. 2004) 1089–1105.
[13] D.P. Kingma, J. Ba, Adam: A method for stochastic optimization, arXiv preprint, arXiv:1412.6980, 2014.
[14] S. Salazar-Colores, I. Cruz-Aceves, J.-M. Ramos-Arreguin, Single image dehazing using a multilayer perceptron, Journal of Electronic Imaging 27 (4) (2018) 043022.
[15] F. Cozman, E. Krotkov, Depth from scattering, in: Proceedings of IEEE Computer Society Conference on Computer Vision and Pattern Recognition, June 1997, pp. 801–806.
[16] R.T. Tan, Visibility in bad weather from a single image, in: 2008 IEEE Conference on Computer Vision and Pattern Recognition, June 2008, pp. 1–8.
[17] D. Berman, T. Treibitz, S. Avidan, Air-light estimation using haze-lines, May 2017, pp. 1–9.
[18] K. He, J. Sun, X. Tang, Single image haze removal using dark channel prior, IEEE Transactions on Pattern Analysis and Machine Intelligence 33 (12) (Dec. 2011) 2341–2353, https://doi.org/10.1109/TPAMI.2010.168.
[19] P. Soille, Morphological Image Analysis: Principles and Applications, 2nd ed., Springer-Verlag, Berlin, Heidelberg, 2003.
[20] S. Theodoridis, K. Koutroumbas, Pattern Recognition, 4th ed., Academic Press, Inc., Orlando, FL, USA, 2008.
[21] R.C. Gonzalez, R.E. Woods, Digital Image Processing, 3rd ed., Prentice-Hall, Inc., Upper Saddle River, NJ, USA, 2006.
[22] C. Kanellakis, P. Karvelis, G. Nikolakopoulos, Open space attraction based navigation in dark tunnels for MAVs, in: 12th International Conference on Computer Vision Systems (ICVS), 2019.
[23] A. Saxena, Convolutional neural networks (CNNs): An illustrated explanation, https://xrds.acm.org/blog/2016/06/convolutional-neural-networks-cnns-illustrated-explanation/, 2016.
[24] F. Chollet, et al., Keras, https://github.com/fchollet/keras, 2015.
[25] J. Jackson, G. Ellingson, T. McLain, ROSflight: A lightweight, inexpensive MAV research and development tool, in: 2016 International Conference on Unmanned Aircraft Systems (ICUAS), June 2016, pp. 758–762.
[26] S. Parrot, Parrot bebop 2, Retrieved from Parrot.com: http://www.parrot.com/products/bebop2, 2016.

[27] Adobe Photoshop [Online]. Available: https://www.adobe.com/products/photoshop.html.
[28] S.S. Mansouri, C. Kanellakis, G. Georgoulas, G. Nikolakopoulos, Towards MAV navigation in underground mine using deep learning, in: IEEE International Conference on Robotics and Biomimetics (ROBIO), 2018.

Chapter 13

Edge connected ARWs

Achilleas Santi Seisa and Anton Koval
Department of Computer, Electrical and Space Engineering, Luleå University of Technology, Luleå, Sweden

13.1 Introduction

Even though Aerial Robotic Workers (ARWs) have constructional and operational constrains, that can lead to computational limitations, they have to complete complex tasks under the increased requirements for autonomous robots' capabilities. At the same time, these procedures must be completed in real-time. For these cases the edge computing is mostly needed, to provide the desired high computation power, while retaining an overall low latencies. The development and utilization of edge computing technologies is drastically boosting the levels of robotic autonomy and result in overcoming existing onboard hardware limitations. In comparison to cloud computing, edge computing has the capacity to enable local computing thus dealing more efficiently with big volumes and velocities of generated local data [1]. Moreover, it will provide lower latencies due to the significantly smaller distance between the edge connected robots and the edge data centers.

13.2 Edge computing

The operations of ARWs are usually based on complex methodologies and algorithms, which are computationally demanding. ARWs' platforms can not handle the such processes since they do not have the required computational power onboard. Thus, it is crucial to increase the computational capability of ARWs. This is one of the main reasons, why utilization of external resources for robotic applications has been studied so extensively. Similar to onboard computers, edge platforms can handle huge amount of data at a very high processing speed, providing strong computation capacity and real-time data transmission. Additionally, since the hardware of edge computing is located in a close distance to the robots, the time delays for the data to be transfer from the robots to the edge and vice versa, are relatively small.

Aerial Robotic Workers. https://doi.org/10.1016/B978-0-12-814909-6.00019-6

13.2.1 Cloud – Fog – Edge

The need for computational power and storage resources has led to the development of various computing architectures [2]. Researchers from several fields of technology like deep learning [3], video analytics [4], and industry 4.0 in general [5] are trying to utilize the available architectures or propose new ones, in order to solve various challenges. All these architectures are based on four different computing and storage layers as depicted in Fig. 13.1, where the bottom layer is formed from the connected devices, up to the top layer that forms the cloud. These layers are the basis of any architecture for distributed resources, and edge connected ARWs are no exception.

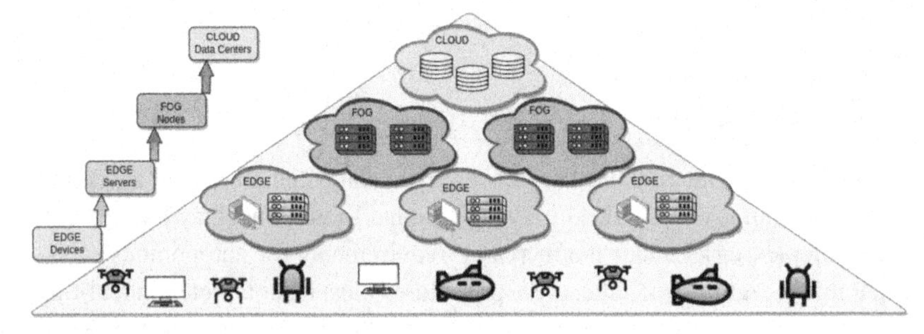

FIGURE 13.1 Data processing and storage layers.

Giving a closer look at each architectural layer, from the robotics perspective will allow a better understanding of the opportunities and challenges that can be addressed using one or another architecture.

At its base, the device layer is represented by individual robots, like ARWs, that are equipped with application driven sensors like cameras and lidars for perception, navigation, and inspection tasks. These sensors can produce gigabytes of data, which has to be processed onboard in order to enable the autonomous capabilities of the robotic platform. However, due to the limited computational power, only light data processing, which is required, for example, for a single robot Simultaneous Localization and Mapping (SLAM) and low-level control, can be performed onboard. While in more demanding applications, which involve multiple ARWs, there is a need to extend this layer with edge or fog layers. These layers are located close to device layer and allow real-time data processing of large data batches that are streamed from the network connected robots to the edge server, where heavy computations that require low latencies are carried out. One of such examples is the multi-session SLAM in which multiple SLAM missions are performed collaboratively by multiple ARWs that explore the same environment. Besides, the fog layer allows extending the exploration mission to multiple environments thus creating a network of heterogeneous interconnected nodes. The cloud layer, which is located far away from the device layer, excels over the edge and fog layer in terms of storage and computational capabilities. Additionally, cloud layer is commonly characterized by long response latencies, which can be explained by the long distance between the device and the cloud layer, and thus the robotic challenges for the next generation robotic systems cannot be fully addressed by solely this layer. Thus, one can observe various architectures that are based on the collection of these multiple layers that create a computing ecosystem.

13.2.2 Virtual machines – containers – kubernetes

With the advent of cloud computing and virtualization, many applications apart from robotics are exploring the optimal possible utilization of such technologies. Virtualization options are a known discussion in the academia and industry. In this subsection, three most popular virtualization technologies, as well as their characteristics, are presented and compared in terms of benefits and drawbacks for cloud robotics applications.

13.2.2.1 Virtual machines

A virtual machine (VM) is a computer resource that utilizes software instead of a physical computer to deploy applications and run programs and processes [6]. VMs also provide a complete emulation of low-level hardware devices like CPU, Disk, RAM, networking devices, and others. Moreover, VMs can be configured accordingly, to support any application topology and provide a stable platform regarding dependency issues (operating system compatibility, specific software packages, etc.). Specifying computational resources, resource scaling, redeploying VM applications on different host computers, isolating sensitive applications, as well as running VMs in the edge cloud, are some of the benefits that VMs offer in cloud robotics-related applications. Furthermore, VMs provide a more dynamic and interactive development experience when compared to other virtualization technologies. The latter derives from the fact that VM encompass a full-stack system, as well as utilizing an entire operating system and what one may provide, e.g., a Graphical User Interface.

13.2.2.2 Containers

Containers are software packages that include all the software dependencies required to execute the contained software application. The main difference between container and VM is that VM emulates an entire machine down to the hardware layers, while containers only emulate the software components. Such software components might be system libraries, external software packages, and other operating system level applications [7]. Because containers are only emulating software components, they are more lightweight and easy to iterate. Additionally, most container runtime systems provide robust pre-made container (image) repositories.

A popular example in robotic applications could be the Robot Operating System (ROS) image. Another key benefit of container technologies is that software packages, contained in the constructed images, can be stacked in levels and produce a novel and more complete applications. These flexible characteristics of containers, i.e., lightweight and easy to iterate, are responsible for the birth of another technology, container orchestration.

The architecture of a containerized applications for robotic systems is shown in Fig. 13.2. As it is indicated, data are sent from the robots to the edge and then commands generated from the containers are sent back to robots. Data processing, high level controllers, and advanced algorithms can be deployed to the application containers for various robotic missions.

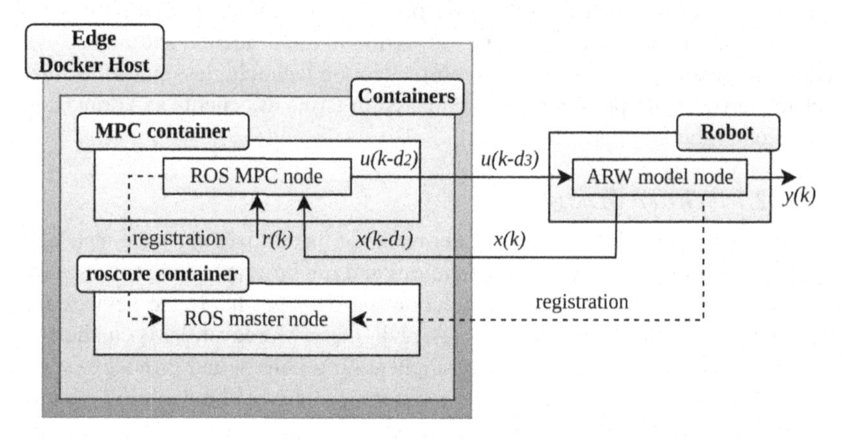

FIGURE 13.2 Docker-based [8] edge architecture for offloading the model predictive controller.

In the case depicted in Fig. 13.2, the Model Predictive Control (MPC) has been offloaded to the edge in the form of a docker [8] container. For this docker-based edge architecture, two docker images were deployed with all the necessary libraries and dependencies in order to run the controller. The first docker container is running the ROS master where all the ROS nodes have to register in order to communicate with each other, and the second docker container contains the MPC application and is running the controller. The MPC is responsible for controlling the trajectory of the ARW. The states $x(k)$, which are the linear position, linear velocity, and quaternions, are generated by the robot and are published to the odometry ROS topic. The MPC ROS node subscribes to that topic as well to the reference ROS topic which publishes the referenced trajectory signal of the ARW, $r(k)$. The state signal arrives to the MPC ROS node with time delay due to the travel time between the robot and the edge; thus, the state signal on the MPC is $x(k-d_1)$. The generated by the MPC control signal is denoted as $u(k-d_2)$ and is published to the control topic to which the robot ROS node subscribes in order to receive that control signal. The control signal arrives to the robot with some delay, so it is stated as $u(k-d_3)$. Finally, the output of the system is represented as $y(k)$.

13.2.2.3 Kubernetes

Kubernetes (k8s) is an open-source platform that orchestrates container runtime platforms across network-connected hardware infrastructure [9]. The k8s was initially developed by Google that needed an efficient way of managing billions of container applications weekly. In such systems, appeared the additional needs like reliability, scalability, robustness, security, and others, which are the essential requirements in real-world applications. The k8s bundles a set of containers into a group (called pod) and manages their life cycle.

An example of such a set could be an application server, an SQL database, a MPC running apart from the robot as in Fig. 13.3, a set processing SLAM, and others. The k8s will manage these pods performing multiple essential tasks, such as reducing network overhead, increasing resource usage efficiency (load balancing between copies of the same application, either to the same machine or across the cloud cluster), and hardware resources designated for your specific configuration, monitoring, and much more, since k8s is a rapidly progressing state-of-the-art technology.

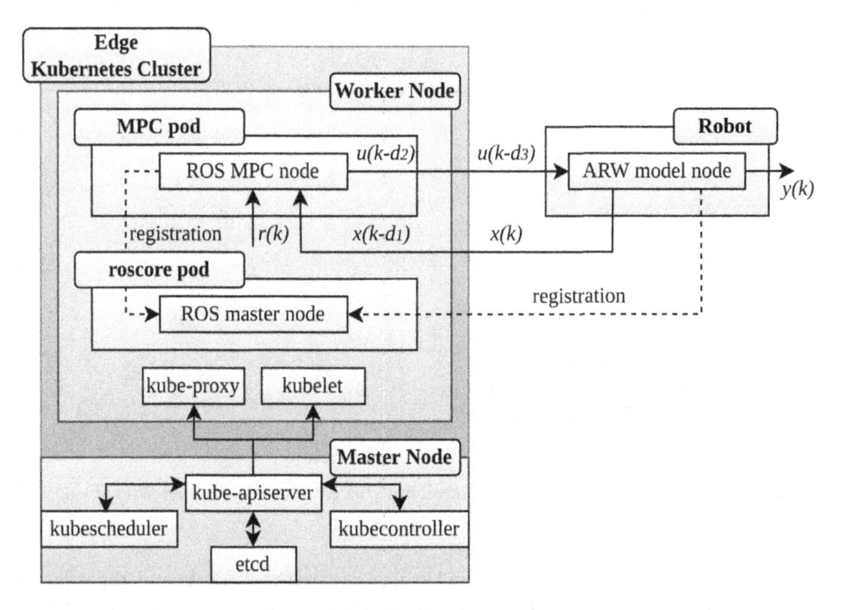

FIGURE 13.3 Kubernetes-based edge architecture for offloading the model predictive controller.

In Fig. 13.3, a small topology architecture example is illustrated. In this case, like in Fig. 13.2, the ARW is connected to the edge, but this time, the applications needed for the operation of the MPC are orchestrated by the k8s. In this example, the k8s is responsible for scheduling resource-demanding computational jobs (on different worker nodes), managing network traffic and can also be utilized for redundancy requirements. Kubernetes cluster consists of two nodes, the master node, which controls and manages the worker node, and the worker node where the application pods are deployed. The worker node consists of two pods. The first pod consists of the ROS master, which is necessary for the ROS operation, and the second pod, which consists of the controller and all the necessary libraries and dependencies.

13.3 Communication layer

The communication between the computing layers is an important factor of the mentioned architectures. Many different protocols, networks, infrastructures, and configurations have been used for enabling cloud, fog, and edge computing. There are many studies where networks, such as WiFi and LTE [10], have been used for robotic platforms, while the integration of 5G networks is gaining a lot of focus.

The communication between the device layer and the edge layer can be achieved both with WiFi or cellular networks [11]. WiFi is an inexpensive option and easy to use. The utilization of WiFi for data transferring from a ARW to the edge and vice versa has been studied, and the indicative values are shown in Figs. 13.4, 13.5, 13.6, and 13.7.

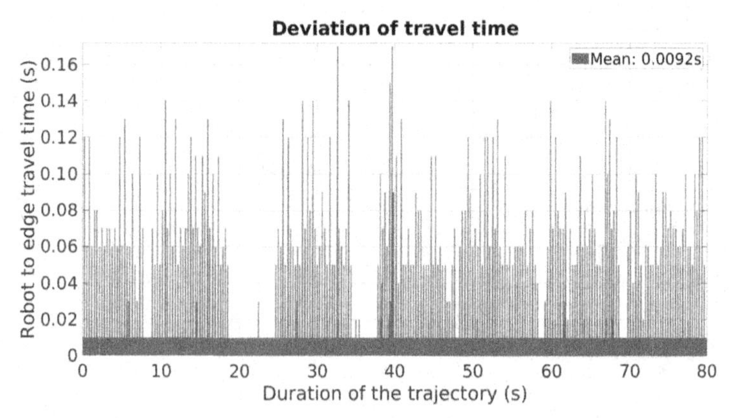

FIGURE 13.4 Travel time for a signal to travel from the ARW to the edge over WiFi of a docker-based edge architecture.

In Figs. 13.4 and 13.5, the travel time between the ARW and the edge is depicted, when controlling the trajectory of a ARW through the edge over WiFi, and the edge architecture is docker-based. The measured mean travel time for a signal, containing a message about the states of the robot, to travel from the ARW to the edge is 9.2 ms, while the measured mean travel time for a signal, containing a message with the commands for the robot, to travel from the edge to the ARW is 14.5 ms.

A similar setup, as in Figs. 13.6 and 13.7, was used for capturing the travel time between the ARW and the edge. Again WiFi was utilized when controlling the trajectory of a ARW though the edge, but in this case, a kubernetes-based

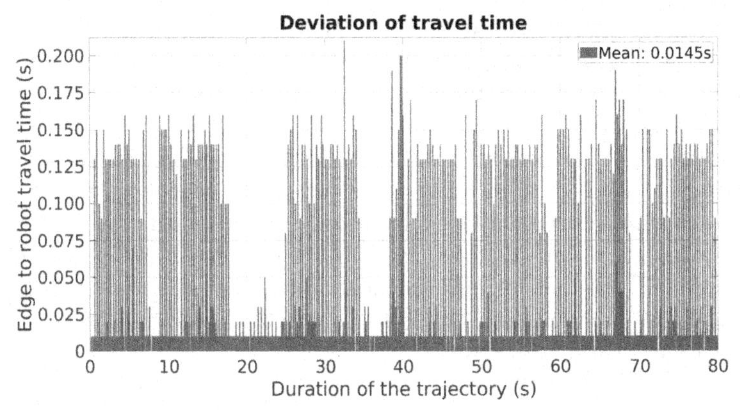

FIGURE 13.5 Travel time for a signal to travel from the edge to the ARW over WiFi of a docker-based edge architecture.

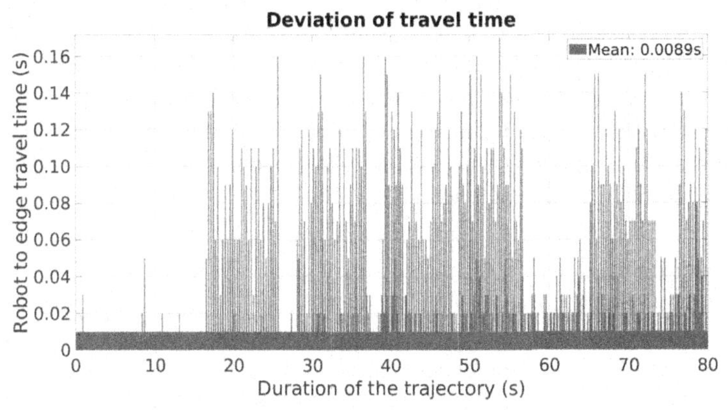

FIGURE 13.6 Travel time for a signal to travel from the ARW to the edge over WiFi of a kubernetes-based edge architecture.

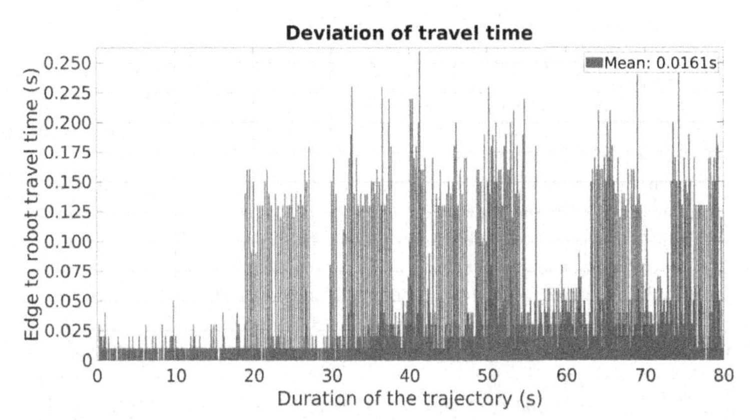

FIGURE 13.7 Travel time for a signal to travel from the edge to the ARW over WiFi of a kubernetes-based edge architecture.

architecture was implemented. In this example, the measured mean travel time for the state signal to travel from the ARW to the edge is 8.9 ms, while the measured mean travel time for the command signal to travel from the edge to the ARW is 16.1 ms.

Although WiFi offers, in many cases, efficient performance, it has some drawbacks, such as limited transmission range. 5G technology, on the other hand, can have a complimentary role to edge computing and cover broader areas providing real-time solutions due to its low latencies. Additionally, 5G can maximize the system throughput and bandwidth and provide a novel infrastructure with significant features, such as task scheduling, prioritizing, routing data, network slicing, etc. [12,13].

13.4 Conclusion

In this chapter, the benefits and the challenges of edge computing for aerial robots were presented, as well as the main computing layers, the communication between them and some key components used in many edge-based architectures. The chapter concludes with presenting some measured data of edge connected ARWs over WiFi. These technologies can contribute significantly to enhancing the performance of ARWs and enabling new capabilities that will foster moving toward safer aerial robot autonomy.

References

[1] Y. Jararweh, A. Doulat, O. AlQudah, E. Ahmed, M. Al-Ayyoub, E. Benkhelifa, The future of mobile cloud computing: Integrating cloudlets and mobile edge computing, in: 2016 23rd International Conference on Telecommunications (ICT), IEEE, 2016, pp. 1–5.

[2] M. De Donno, K. Tange, N. Dragoni, Foundations and evolution of modern computing paradigms: Cloud, IoT, edge, and fog, IEEE Access 7 (2019) 150936–150948.

[3] F. Wang, M. Zhang, X. Wang, X. Ma, J. Liu, Deep learning for edge computing applications: A state-of-the-art survey, IEEE Access 8 (2020) 58322–58336.

[4] Y. Wang, W. Wang, D. Liu, X. Jin, J. Jiang, K. Chen, Enabling edge-cloud video analytics for robotics applications, IEEE Transactions on Cloud Computing (2022).

[5] P. Pace, G. Aloi, R. Gravina, G. Caliciuri, G. Fortino, A. Liotta, An edge-based architecture to support efficient applications for healthcare industry 4.0, IEEE Transactions on Industrial Informatics 15 (1) (2018) 481–489.

[6] T.V. Doan, G.T. Nguyen, H. Salah, S. Pandi, M. Jarschel, R. Pries, F.H. Fitzek, Containers vs virtual machines: Choosing the right virtualization technology for mobile edge cloud, in: 2019 IEEE 2nd 5G World Forum (5GWF), IEEE, 2019, pp. 46–52.

[7] C. Pahl, B. Lee, Containers and clusters for edge cloud architectures – a technology review, in: 2015 3rd International Conference on Future Internet of Things and Cloud, IEEE, 2015, pp. 379–386.

[8] R. White, H. Christensen, Ros and docker, in: Robot Operating System (ROS), Springer, 2017, pp. 285–307.

[9] J. Shah, D. Dubaria, Building modern clouds: using docker, kubernetes & Google cloud platform, in: 2019 IEEE 9th Annual Computing and Communication Workshop and Conference (CCWC), IEEE, 2019, pp. 0184–0189.

[10] M.A. Ali, Y. Zeng, A. Jamalipour, Software-defined coexisting UAV and WiFi: Delay-oriented traffic offloading and UAV placement, IEEE Journal on Selected Areas in Communications 38 (6) (2020) 988–998.
[11] Y.-H. Wu, C.-Y. Li, Y.-B. Lin, K. Wang, M.-S. Wu, Modeling control delays for edge-enabled UAVs in cellular networks, IEEE Internet of Things Journal (2022) 1.
[12] Z. Ning, P. Dong, M. Wen, X. Wang, L. Guo, R.Y.K. Kwok, H.V. Poor, 5g-enabled UAV-to-community offloading: Joint trajectory design and task scheduling, IEEE Journal on Selected Areas in Communications 39 (11) (2021) 3306–3320.
[13] S. Barbarossa, S. Sardellitti, E. Ceci, M. Merluzzi, The edge cloud: A holistic view of communication, computation, and caching, in: Cooperative and Graph Signal Processing, Elsevier, 2018, pp. 419–444.

Index

D

G

H

I

K

Q

R

Printed in the United States
by Baker & Taylor Publisher Services